To the student: Two Workbooks accompany the textbook and are available through your college bookstore under the title

Workbooks I and II to accompany
The Graphic Languages of Engineering
by Robert F. Steidel Jr. and Jerald M. Henderson

These Workbooks can help you with course material by acting as a review and study aid. If the Workbooks are not in stock, ask the bookstore manager to order a copy for you.

THE GRAPHIC LANGUAGES OF ENGINEERING

THE GRAPHIC LANGUAGES OF ENGINEERING

ROBERT F. STEIDEL, Jr.
University of California, Berkeley

JERALD M. HENDERSON
University of California, Davis

JOHN WILEY & SONS
New York / Chichester / Brisbane / Toronto / Singapore

cover photo by Paul Silverman

Library of Congress Cataloging in Publication Data

Steidel, Robert F., 1926-
    The graphic languages of engineering.

    Includes index.
    1. Engineering graphics.   I. Henderson, Jerald M.
II. Title.
T353.S815 1983   604.2'43   82-13704
ISBN 0-471-86759-4

Printed in the United States of America

10 9 8 7 6 5 4 3 2 1

PREFACE

More people should read prefaces because prefaces can be very informative, sometimes surprising, and even entertaining. Prefaces tell you how a book came to be. They tell you why an author or, in our case, authors, has written this book at this time. One of the authors of this book regularly assigns the prefaces of textbooks as reading.

The best time to read a preface to a book is before you select it. We have written this preface as if you have not looked at any part of the book that follows. With that premise, please read on to see why we have constructed our book as we have.

This book is designed to fill a void that we perceive in graphic communication in undergraduate engineering education between engineering drawing and engineering. To us, this void is obvious. Engineering students used to receive extensive training and education in engineering graphics, mostly engineering drawing. Now they do not. Engineering graphics has been all but eliminated from engineering curricula. What remains is a shortened version that has little lasting benefit. The engineering graphics instruction that remains is viewed by engineering faculties as archaic, tedious, ritualistic, and boring. These criticisms have some foundation; engineering graphics has been taught at times almost as the language of a secret society. On the other hand, the instructors who teach engineering graphics resent the growing isolation of their subject from the rest of engineering. They recognize the importance of engineering graphics, and they know they cannot instruct properly within the limited time allowed them. No one is satisfied.

We therefore propose to introduce a different kind of graphics text, a text for engineering students at the university level, designed to emphasize not the doing but the reading and understanding of engineering drawings and engineering graphics. Of course, learning is doing. As instructors, we know this to be true. However, the practice of engineering drawing and engineering graphics is a professional field by itself; each is a living subject that is continuously changing. Engineering drafters and illustrators devote a lifetime to learning their profession. As professional engineers, aren't we a bit arrogant to think that we can learn someone else's profession and ours, too? It cannot be done, at least not in one or two undergraduate courses. We repeat what we propose for this book so that there can be no misinterpretation. *We propose to emphasize the reading and understanding of graphics in engineering.* We do not propose a simple survey of engineering graphics.

To be learned fully, engineering graphics must permeate the entire engineering curriculum, and that cannot happen if beginning graphics instruction is isolated from the rest of the curriculum. It is this isolation that first must be corrected. If instructors of graphics courses do not participate in later engineering courses, graphics will not be a part of these courses. If graphics instructors do participate, graphics will permeate the curriculum. It is that simple. This book is to be used by instructors who are engineering professionals, not graphics professionals. We want this book, and the courses for which it is used, to be a regular part of the engineering curriculum, taught by regular engineering faculty. Engineering colleges and schools can no longer afford specialized faculty in engineering graphics. Even if they could afford specialized faculty, they can not find them. Engineering graphics courses should be taught by the same faculty that teaches fluid flow, thermodynamics, mechanics, and strength of materials. With this book and the outline we propose, they can be.

Our purpose is to present the graphic languages of engineering in a clear but technical way so that students and instructors can understand why and when we use the forms of engineering graphics that we do. We want to be simple and direct, scholarly but not pedantic.

All of this book is meant to be used, but it can be separated into two halves. The first half, Chapters 1 to 6, covers the geometry of engineering graphics—that is, sketching, orthographic projection, and descriptive geometry—essentially the format of graphics. The second half, Chapters 7 to 12, contains engineering drawing, modeling, and graphic mathematics—essentially the manipulation and presentation of engineering information. Note that engineering drawing is in the middle of the book, mixed in with all the other important graphic communication tools. In keeping with our philosophy, the text ends with the engineering report. In this book the engineering report is awarded the climactic and terminal place because writing reports is what most engineers will do. They may or may not read drawings, they may or may not sketch ideas, but they *will* write reports, they will write them endlessly, and they will use their engineering graphic skills in them.

Most of our subject matter has been reduced to what we consider an irreducible minimum. Descriptive geometry is a beautiful subject, important to all engineering and all engineers, but it can be overdone. There are just three basic problems in descriptive geometry: the true length of a line, the point view of a line, and the true shape of a plane. All other descriptive geometry problems can be reduced to these three. Our intent is to present each subject and to present descriptive geometry and other graphic subjects in their barest but essential forms.

Even though we have tried to limit the text, problems are not neglected. Studying about the various graphic reading and writing skills in this book does not take the place of experiencing the material through exercises and problems. The doing of problems must accompany the reading. This book contains over 500 problems, including 70 thoroughly worked out and explained sample problems.

Chapter 6 is a brief exposition of computer graphics. It is not our original work, but a shortened version of a much more complete chapter written by R. Golden and R. Salomon, professors of engineering at the *New Jersey Institute of Technology*.

Some have criticized us for including introductory material on complex numbers, matrix algebra, dimensional analysis, and selected topics in graphic mathematics because these subjects are beyond the background of the students using the book. We are convinced that lower-division engineering students are ready (and eager) to absorb this subject matter. Of course, all these subjects will be covered in more depth later in most engineering cirricula, but we consider it important to identify early the graphical nature of these important communicative and analytical tools.

In Chapter 7 we have presented the practices and conventions of engineering drawing, as they exist today. We have tried to make this chapter language perfect. To do this, we have drawn heavily on the expert background and knowledge of George Tokunaga, Chief Drafter of the Lawrence Livermore National Laboratory, and we deeply appreciate his collaboration.

A chapter on modeling, new to a text such as this, has been included after the chapter on engineering drawing. In the design of large systems, the design model, one of the modeling types discussed, has all but eliminated the assembly drawing. In large engineering firms drafting tables are being replaced by modeling benches. Drafters work side by side with model makers. Student engineers must learn early that today's engineers are model builders. Engineering instruction in model building has never been tried, but we must move with out profession. Chapter 8 instructs students in model building.

Writing a book is hardly a solitary effort, and this book is no exception. Behind this productivity are the efforts and encouragement of a number of people. We would like to express our gratitude to four people who willingly or otherwise influenced our ideas in this book. The first is the former engineering editor of John Wiley & Sons, Thurman Poston, whose clever mind saw the need and conceived this book as a solution to that need. The second person is emeritus professor and former dean of engineering at Duke University, J. Lathrop Meriam, who was instructor and mentor to us both. He taught us how to teach from basic concepts and set a prime example to follow with his own use of graphic language. Many ideas and illustrations in this book are his, and we thank him for allowing us to use them. The third person to whom a debt is owed is Professor Alexander Levens, of the University of California, Berkeley. He affected us both, but in different ways. We both worked with him or for him in our early careers, and he passed on to us his enthusiasm for seeing graphics in every facet of engineering. We have moved away from a traditional approach to engineering graphics, but our work, in part, is a tribute to his teaching. The fourth is George Tokunaga.

We would like to acknowledge the cross fertilization of our ideas with Mary Kummer, of the Pennsylvania State University. Steve Slaby of Princeton, who is probably the foremost descriptive geometer of our time, also influenced us profoundly. We have not used his ideas in this book because we ethically cannot, but we do respect and appreciate them. We cannot take things from his head, but the things he drove into ours we will remember. We also thank Lesa Havert for her careful typing of the manuscript and Bruce Shawver for creating many of the text problems. One author expresses his thanks to his father, S. Milton Henderson now a retired engineering professor, who demonstrated the need for proper graphic languages in engineering and engineering education very early to his son.

Last, we would like to thank our wives, Jean Steidel and Fran Henderson, who persistently and patiently encouraged us through the many months of writing and rewriting, rewriting and writing.

Robert F. Steidel, Jr.

January 1983   *Jerald M. Henderson*

CONTENTS

1
THE GRAPHIC LANGUAGES OF ENGINEERING

1.1 WHAT IS ENGINEERING?

However engineering is defined, the definition of engineering is constantly changing. If you had lived 50, 100, 200, 500, or 2000 years ago, you would have had different answers to this question at each time in history. Indeed, a specific answer today will be slightly different from an answer given tomorrow. The reason for this is seated in the meaning of engineering. Engineering is related to the needs of human society, when human society changes, the needs of society change, and then the definition of engineering changes.

Many people confuse science and engineering, and others simply dismiss engineering as applied science, but engineering is more than applied science. There is an art to engineering, and judgment and experience are a part of that art.

The function of the scientist is to seek new knowledge. Science is a body of knowledge, and engineering is an activity related to science. The engineer applies knowledge, but engineering and science are quite distinct. Of course, there is overlap. Both engineers and scientists engage in research, but the type of research and the objectives are different. Both engineers and scientists are interested in the same body of knowledge, but they use it for different purposes.

Engineers do not pursue engineering for its own sake, as scientists pursue science and mathematicians pursue mathematics. It is true that engineers are often attracted to advanced thought, but there is always an application in mind. It is this application of knowledge that an engineer pursues. Engineers try things out, test, adapt, change, and change again. Engineering is an experience; conversely, to be a complete engineer, you need this experience.

The English word *engineering* is derived from the Latin word *ingenerare* ("to create"). Other words derived from the same root are *engine* and *ingenious.* Ingenuity and engineering are still very much tied together.

The Accreditation Board for Engineering and Technology (formerly the Engineering Council for Professional Development) defined engineering in a short paragraph.

> Engineering is the profession in which a knowledge of mathematical and natural sciences gained by study, experience and practice is applied with judgment to develop ways to utilize, economically, the materials and forces of nature for the benefit of mankind.

The Accreditation Board for Engineering and Technology is a 42-member council of engineers named from 21 separate professional engineering societies; it accredits engineering curricula proposed by the various engineering schools and colleges.

Simply stated, engineering is the application of science to fulfill the requirements of a socioeconomic system, with the application tempered by judgment and experience.

1.2 A HISTORY OF ENGINEERING

In ancient times, most engineering was associated with the military. The military engineer built catapults, battering rams, siege platforms, bridges, and fortifications. The military engineer also built highways, bridges, harbors, canals, aqueducts, and buildings, mostly because the military and civilian conduct of our early civilizations was indistinct.

The first person who could really be identified as an engineer was Imhotep, a great physician, architect, and statesman of ancient Egypt. He built the first known pyramid at Saqqara about 2650 B.C. as a tomb for King Zoser. Built in a series of steps, it is known as the Step Pyramid and marked the first recorded use of hewn stone for any structure. Over the next 150 years, many

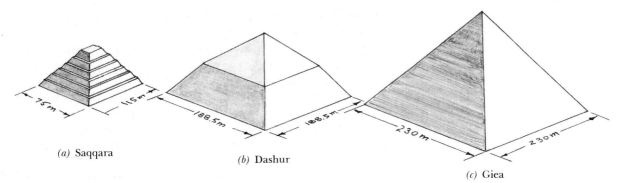

(a) Saqqara   *(b)* Dashur

(c) Giea

Figure 1.1 The Pyramids of Egypt.

other pyramids were built—undoubtedly our first civil works, built with the manual labor of hundreds of thousands of people. These feats were truly remarkable, but for engineering they also marked the first application of arithmetic, geometry, and a smattering of empirical data to solve problems. As an example of the latter, one of the first changes made was to "face" the pyramids, making them smooth. At Dashur, about 10 miles south of Saqqara, there are two pyramids. The first was built with the lateral sides inclined at 52 degrees. Evidently, after it was more than two-thirds built, the builders changed the angle to $43\frac{1}{2}$ degrees. This pyramid is known as the Bent Pyramid. The entire second pyramid was built at $43\frac{1}{2}$ degrees, but all later pyramids have the original face inclination of 52 degrees (see Figure 1.1).

All the largest pyramids were built within a 150-year period, 2650–2500 B.C. For some still-elusive reason, Egyptian society required the development of the pyramid during those 150 years—never before and never afterward. Recall that engineering adapts to the needs of human society, whatever they are.

The Egyptians were also skilled surveyors. Annually, the Nile River overflowed its banks, obliterating all land markers. Using very primitive measuring equipment, the Egyptians used geometry to resurvey land boundaries each year. The Nile flood was a life-giving event, and resurveying land was part of this event.

The Greeks and Romans were also great engineers. They built ships and roads that served both military and civil needs. Commerce flourished and cities grew. The architecture of these times was and still is magnificent. The Parthenon in Athens, the lighthouse at Alexandria, and the Coliseum in Rome were all monuments to the imagination, skill, and knowledge of ancient engineers. One treatise survives, *De Architectura*, written by the engineer-architect Vitrivius in the first century A.D. It contains many descriptions of building materials and construction meth-

Figure 1.2 The Roman Arch.

ods, hydraulics, surveying, and planning. It is our only written source of knowledge about engineering methods in Roman times.

Although the Romans copied much from Mycenae, Greece, Egypt, Assyria, and other cultures, they were responsible for two outstanding engineering developments. The first was the Roman arch (Figure 1.2). Through the use of the arch, the Romans built structures of unprecedented height with minimum material. Despite their classic beauty, Grecian buildings were constructed simply, with post and lintel or slab laid on pillar. The maximum span between columns was limited by the tensile stress developed in simple bending of the stone lintel.

The second development was the Roman road (Figure 1.3.)

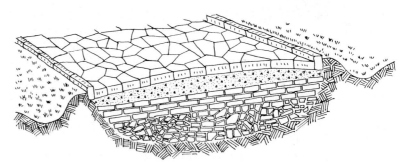

Figure 1.3 The Roman Road.

The Romans built their roads much as we do, with a compacted subbase, crowned surface, and drainage. These roads were usable in good and bad weather by humans, animals, and carts. As Rome flourished and its empire grew, it was serviced by a system of roads such as the world had never seen. They were built as highways for the Roman Legions, but the Roman roads gave people goods, material, and mobility that previous civilizations had not had. At one time, the road system extended 50,000 miles, which would be a lot in some countries today.

Medieval engineering fell short of the achievements of Greece and Roman times, with one exception. In the Middle Ages, the Catholic Church dominated society. Not surprisingly, the greatest medieval accomplishments were associated with the church. The Gothic arch and the flying buttress were an improvement over the Roman arch. They permitted the construction of the great cathedrals of the Middle Ages with arches spanning longer distances and with less mass (see Figure 1.4).

In the Far East—in India, China, and Japan—engineering had a separate but similar development. These achievements were also accomplished with sophisticated engineering techniques. Little survives today because these peoples used wood and unfired brick instead of stone and marble as building materials, and their lands did not have the preserving dry climates of Egypt and the Middle East. The Great Wall of China (Figure 1.5) was originally built in the reign of the emperor Shih Huang-ti (246–209 B.C.) to defend China's northern frontier. It runs 1500 miles from the coast of northeast Hopei to west Kansu. It has an average height of 22 feet and width of 23 feet. The Chinese also constructed large canals. The Grand Canal, 860 miles long from Hangchow to Tientsin, dates from the fifth century B.C..

Figure 1.4 The Flying Buttress. (*The Horizon Book of Great Cathedrals,* published by American Heritage Publishing Co., Inc., New York, 1968.)

Figure 1.5 The Great Wall of China. (*The Archaeological Journal of 1888, Vol. XLV,* New York Public Library.)

By the beginning of the Renaissance, new engineering achievements were observed; again, these achievements were connected with commerce and trade. New lands were discovered across the sea, demanding improvements in the ships needed to carry goods and people. New empires were built around exploration and colonization. The Renaissance also marked the appearance of two of the greatest scientific minds of all history, Leonardo da Vinci (1452–1519) and Galileo Galilei (1564–1642).

Leonardo da Vinci was a superb artist, and he was also an architect, musician, designer, inventor, anatomist, botanist, geologist, astronomer—a truly great experimental scientist for his age. Unfortunately, he lived hundreds of years before the world could make use of his genius. For 20 years, Leonardo served the Duke of Milan. He painted, sculpted, and designed court pageants; he also worked on fortifications, locks, canals, and the diversion of rivers. Hydraulic and aeronautical engineering was a major part of his life. His remaining works are among the art treasures of the world, but very few of his engineering ideas were realized.

Galileo was an experimental physicist, not an engineer. He is often credited with founding modern experimental science. He is best known for his work on dynamics, but among other things he is the inventor of the thermometer. At the age of 20, in 1584, Galileo correctly conceived the principle of the isochronous pendulum. In 1590 he crowned his achievements in experimental physics by discovering and proving the law of falling bodies. He also invented a hydrostatic balance used to determine specific gravity, and he improved the telescope, which led to his discoveries in astronomy. His work on motion and acceleration formed the basis for the laws of motion that Sir Isaac Newton (1642–1727) later formulated. These contributions became the foundation for all dynamics.

1.3 THE RISE OF SPECIALIZATION IN ENGINEERING

Over the years, engineering grew into two divisions, *military engineering* and *civil engineering*. Military engineering remained concerned with the technology of war, but as our knowledge of science and our civilization grew, civilian engineering became as important as military engineering. It became too much for one person to know all about engineering. Specialization was a natural development. Specialists in civilian engineering were restricted to static structures, such as highways, sanitation systems, canals, dams, bridges, and buildings.

In 1747 in France, the first true engineering school was founded: the École Nationale des Ponts et Chaussées, or the National School of Bridges and Highways. It grew out of the Corps des Ponts et Chaussées, or the National Bridge and Highway Commission. Its graduates were interested in engineering as a subject, and there were many things to learn. Statics, dynamics, descriptive geometry, the strength of materials, hydraulics, and fluid pressure were studied systematically. As knowledge was learned, it was passed on to others. The first schools in the United States to offer an engineering education were the United States Military Academy at West Point in 1817, and the Rensselaer Polytechnic Institute in 1825.

By the second half of the nineteenth century a new specialty, *mechanical engineering*, was recognized, and the need developed for more engineers who were interested in mechanical things. Specialists in mechanical engineering were concerned with engines and machinery, essentially dynamic structures. It was a natural development, after the steam engine gave us a new source of power, a source not tied to wind or water.

Early mechanical engineers were also concerned with electric generation, but by the late nineteenth and early twentieth century *electrical engineering* split from mechanical engineering. The experiments of Alessandro Volta and Michael Faraday were known, but electricity remained pretty much a novelty. In 1872, however, Z. T. Gramme developed the direct-current generator and motor, making electricity of practical use. Electronics and electronic engineering are outgrowths of the research of James Clerk Maxwell of Scotland and Heinrich Hertz of Germany and the development of the vacuum tube by Lee De Forest in the United States.

Chemical engineering grew out of the development of industrial processes involving chemical reactions in metallurgy, textiles, food, and munitions. By 1900, the use of chemicals had created a whole new industry whose function was the mass production of chemicals in industry and manufacturing. The design and operation of the plants of this industry became the function of the chemical engineer.

There are other specializations, and the splintering process continues today, but chemical engineering, civil engineering, electrical engineering, and mechanical engineering remain the four main disciplines of engineering.

1.4 SPECIALIZATION IN ENGINEERING TODAY

Civil engineers are concerned with civil works: bridges, dams, reservoirs, water-supply systems, sanitation and waste systems, canals, highways, transportation systems, harbors, and airports. They are also interested in the properties of construction materials, the characteristics of soils and rivers, and any effects that natural phenomena, such as earthquakes, may have on structures and structural materials.

Mechanical engineering is the science and art by which mechanical systems and components are conceived, designed, developed, and controlled. Mechanical engineers are concerned with the transformation of energy: chemical to thermal, thermal to mechanical, kinetic to potential, etc. This broad spectrum encompasses aeronautics and astronautics, design, environmental engineering, heat and mass transfer, thermodynamics, internal combustion, solar energy, and manufacturing processes.

Electrical engineering was originally concerned with the generation and utilization of electrical power, but now that is only one of many parts of a very much expanded profession. Today the field includes electronics, design of solid-state devices, computers and computer science, logic and information theory, energy conversion, communication systems, system control, and system engineering.

Chemical engineers design the processes for our chemical industries. Petroleum products, plastics, explosives, paints, fertilizers, and pesticides are all examples of goods produced in our chemical industries. Chemical engineers are closely related to chemists, but they are interested in the practical aspects of chemical processes such as fluid flow, thermal processes, controls, and physical properties as well as the chemical transformation of material. They are also interested in systems, system engineering, and plant design.

There are many other more specialized disciplines: nuclear, industrial, biological, environmental, geophysical, agricultural, metallurgical, mining, sanitary, and transportation engineering are a few. The Accreditation Board for Engineering and Technology lists 98 separate undergraduate curricula. Some new disciplines in engineering to be recognized include food, computer, human, rehabilitation, and safety engineering. Although many are combinations and options, we have indicated the variety of specialization available in engineering today. Needless to say, there are many interdisciplinary activities.

PROBLEMS

1.1 Research one of the following historical engineering developments.

    a. Archimedes' screw.

    b. Hero's aeolipile

    c. Triremes of Greece.

    d. Aqueducts of Rome.

    e. James Watt's steam engine.

    f. Automobiles.

    g. Airplanes.

    h. Spinning wheels.

    i. Gunpowder.

    j. Pneumatic tires.

    k. Photography.

    l. Alloying of metals.

    m. Refrigeration.

    n. Cyclotron.

1.2 Research one of the following names from engineering history.

    a. Leonardo da Vinci.

    b. Galileo Galilei.

    c. Michael Faraday.

    d. James Clerk Maxwell.

    e. Lee De Forest.

    f. Charles Steinmetz.

    g. John Smeaton.

    h. Theodor von Karman.

    i. Thomas Newcomen.

    j. Thomas A. Edison.

    k. Osborne Reynolds.

    l. Charles A. Parsons.

    m. Lord Kelvin.

    n. Orville and Wilbur Wright.

    o. Frank Whittle.

    p. Henry Ford.

    q. Werner Siemens.

    r. Stephen Timoshenko.

    s. Alexander Bell.

    t. Alessandro Volta.

    u. Nikola Tesla.

    v. Nikolaus Otto.

    w. John A. Roebling.

    x. Jacques and Joseph Mongolfier.

    y. Robert H. Goddard.

    z. Lord Rayleigh.

1.3 Research one of the following inventions.

    a. The Whitney cotton gin.

    b. The Edison electric lamp.

    c. The Bell telephone.

    d. The Francis turbine.

    e. The Bessemer converter.

    f. The Westinghouse compressed air brake.

    g. The Diesel engine.

    h. The De Forest vacuum tube.

    i. Macadam paving.

    j. Carrier air conditioning.

    k. Nobel dynamite.

    l. The Gramme generator.

1.5 WHAT IS GRAPHIC LANGUAGE?

People communicate with one another through speech, written prose, symbols, and graphics; more simply stated, they talk, write, and draw.

Many primitive societies never advanced to a written language. All communication and history were oral. Unfortunately, not many schools emphasize oral expression as a part of their curricula, so your ability to speak to others is largely a product of your home and your environment, not a product of your educational experience.

Oral expression is the first form of human communication. Children learn to express themselves before the age of two, and, strangely, parents can understand them. You have heard baby talk. The parents learn to speak a child's language. It may be incomprehensible to others, but the parents and the child communicate.

In your education, your courses in English composition and literature have furthered your ability to communicate with written words. Reading and writing have long been two of the basic skills taught in our educational systems. Indeed, reading and writing skills will be useful throughout your engineering education.

Symbols are an important and separable communication means, and your courses in mathematics and chemistry have taught you how to communicate with symbols. Can you imagine algebra without symbols or simple inorganic chemistry without symbols? The language of symbols is universal and not a matter of tongues. Symbols are an effective way to communicate complex thoughts with signs.

Graphics is the fourth form of communication, and it is as much a language as the other three. All forms of graphics are particularly important to engineers and to engineering. Although we could mix words and symbols with our graphics, in this text we will concentrate on the development of the various forms of *engineering graphics* and leave the mixing of words, symbols, and graphics until later.

Engineering graphics is the language used by engineers to communicate the ideas and information necessary for the construction of engineering devices and systems. This language includes engineering drawings, charts, sketches, layouts, and graphs. Graphics in engineering has three major objectives.

1. To analyze and reproduce a design.
2. To communicate information about the design.
3. To record a history of the design and all changes in that design.

Engineering graphics includes formal drawings and informal sketches, all graphs and charts, and sometimes the relationships of nonphysical ideas, if these relationships can be communicated graphically. Most engineers will gain a working knowledge of formal engineering drawings through some limited course work, such as the course in which you are using this book.

Informal drawing or sketching—"talking with a pencil and paper"—is another use of graphics. Engineers use this form of graphic communication all the time, and they usually learn it without formal instruction. Engineering drawing has been formalized, but "talking with a pencil and paper" has not. This does not mean that it is less important than engineering drawing.

Engineering graphics is the bridge where ideas are translated to real things. In this frame of reference, engineering graphics plays a powerful role in modern technology. It would be difficult to visualize the existence of today's modern society without engineering graphics. Indeed, without graphics, much of modern industry would cease to exist.

Through the use of discussion and examples, you will be introduced to this graphic language. If your goal is to become an engineer, you should learn how to read and write the language of engineering graphics with some proficiency.

1.6 A HISTORY OF GRAPHICS IN ENGINEERING

Graphic communication has existed since the beginnings of civilization. In fact, our written alphabet was developed from hieroglyphics, which were an early form of graphic communication. It is reasonable to assume that some formal method of communicating engineering plans must have been developed by the engineers of ancient Egypt, Greece, and Rome, although artifacts that have been passed on to us from those civilizations do not include engineering plans. It is known that drafting instruments were used in Greek and Roman times. A set of bronze instruments, including rules, compasses, and dividers, was unearthed during the excavation of Pompeii.

Perhaps the oldest technical drawing in the world is the ground plan of the ziggurat at Ur, cut in a stone slab. The idea of drawing to scale and in detail to show others how technical things were done came early, but there was very little advancement in technique for thousands of years.

Egyptian tomb drawings show much detail about every form of life and occupation in early Egypt, and some can be described as technical drawings. If you were an engineer in ancient Egypt

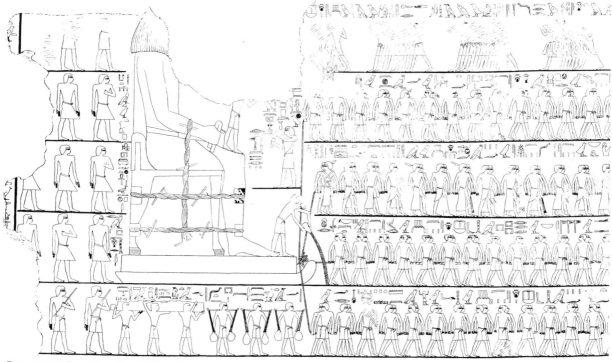

Figure 1.6 Egyptian Tomb Drawing. (*The Archaeological Survey of Egypt,* El Bersheh, Part I, 1893. New York Public Library.)

you might have been asked to move a statue. From the walls of an ancient tomb, we have a drawing (Figure 1.6) that shows how this was done. Needed were 72 workers or slaves, in pairs, pulling four sets of ropes, one leader, a water-pourer to reduce the frictional drag, three water carriers, three material carriers and a foreman. This is graphic communication about labor-intensive engineering as it existed in 1250 B.C.

During Roman times and through the Middle Ages, paper was in short supply and expensive. Since there were no reproduction methods, all copies had to be handmade and the engineer probably worked from original drawings. It is quite likely that ancient engineering drawings were simply accurate plans to lay out the work to be constructed. We call this the *direct method* of viewing. The drafter is directly in front of or directly over the work. The drawing shows what any observer would see. This "observer's view" was the only means of graphic communication used until the beginning of the nineteenth century.

The use of graphic language to communicate ideas received a huge boost from Leonardo da Vinci. The best evidence of his genius is in his sketches. His legacy to engineering was his recorded ideas and experiments in notebooks; he was one of the first experimental scientists to record his works, and over 7000

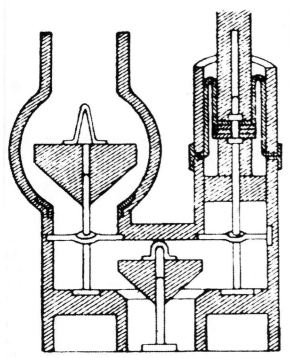

Figure 1.7 Rolling membrane compressor with self-acting values proposed by Leonardo da Vinci in 1492. (*The Codex Antlanticus,* Leonardo da Vinci, New York Public Library.)

Figure 1.8 Sketch of an odometer by Leonardo da Vinci.

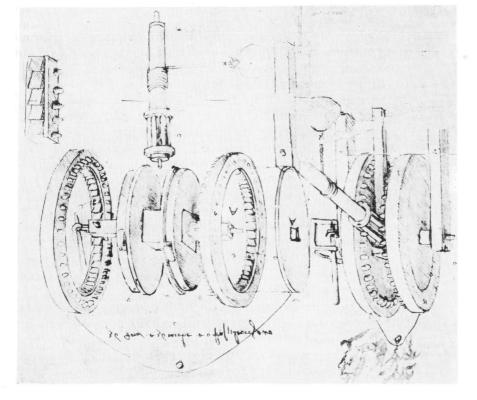

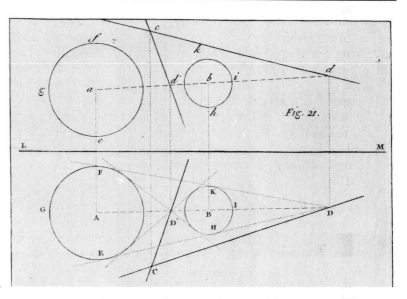

Figure 1.9 An orthographic projection of Gaspard Monge.

pages of his notes still exist. He used isometric drawing, cross-hatching, sections, and pictorial sketching, much as we do today, but he did not use projective geometry. Figures 1.7 and 1.8 are taken from his notebooks. The first is a rolling seal for pressurized chambers, very difficult in 1500 A.D. The second is a sketch of an odometer.

Many of his designs were too far advanced to be constructed in Leonardo's own time, but models of his sketches have been built, and they have worked. It is astounding that engineers can read the graphics of Leonardo da Vinci after 450 years!

Gaspard Monge (1746–1818) originated the *projection method* or descriptive geometry. He was a French mathematician employed as a designer by the French government. He worked out the theory of descriptive geometry as an easy, graphic means of working problems in the design of fortifications—tedious problems involving long mathematical calculations (see Figure 1.9). He invented graphic solutions and completed the plans in such a short time that at first they were not accepted. For nearly 30 years, this graphic process was kept as a military secret, finally being published in the book entitled *Descriptive Geometry* in 1795. Monge showed how to obtain compound angles from projections, how to determine the intersection of two cones, and how to project perspective.

The latest developments in the history of engineering graphics have been the introduction of the modern computer and the introduction of the dry-copying process or xerography. These two developments have streamlined engineering graphics. Through the computer it has become possible to eliminate much of the more formal and elaborate phases of engineering graphics. Engineers need not be intensely involved in the details of engineering drawings, which can be made swiftly and competently by well-trained drafting technicians. The use of pens and ink, which a generation ago was widespread, has all but vanished. Most reproductions are made from penciled drawings on white paper. Reproductions are clear, accurate and easily made.

Our text reflects this latest evolution in engineering graphics. We will concentrate on the use of graphics in engineering to solve engineering problems, and not graphics as subject matter by itself.

1.7 ENGINEERING SKETCHES

Sketching is the simplest form of engineering drawing. It is used to develop ideas quickly and to communicate these ideas to others. For sketching, little is needed beyond a pencil and paper. Tools or straightedges are not needed for sketching. When sketching is done without any aids it is called freehand sketching. If sketching becomes elaborate or more formal it becomes engineering drawing, which is quite another thing and used for another purpose. It is important to differentiate between the two. You should practice and develop your sketching abilities and techniques, but to make elaborate drawings where only sketches are needed is a waste of your time and ability.

Very little can be done to develop artistic talent in engineering drawing, but we can show some simple techniques that will help you include graphics as one of your engineering resources. It is also possible to make better engineering sketches with a few simple drawing aids such as a compass, a couple of triangles, and a scale. Indeed, these are the only drawing aids that most engineers have. It is technique, and not aids, that makes a good engineering sketch.

1.8 DESCRIPTIVE GEOMETRY

The geometric methods of Monge, which we know today as descriptive geometry, were based on the principle of *orthographic projection*. Objects are shown by views, called the front, top, side, auxiliary, or oblique views, or combinations of these views. The complexity of an object determines the number of views which should be used to describe it. Sometimes pictorial views are shown. We will develop these different views in full detail in Chapters 3 and 4, but it is sufficient to state here that any two projections will completely describe a three-dimensional object if those two views are at right angles to each other. Since each view is a flat, two-dimensional picture or drawing, and both can be placed on one side of one piece of paper, these orthographic views represent a simple means of describing a three-dimensional object.

Our task in descriptive geometry will be to determine distances between lines, between lines and planes, and between planes, if the two views are known. These geometric problems in the description of three-dimensional space have the name *descriptive geometry*.

1.9 ENGINEERING DRAWINGS

Engineering drawings are the plans from which engineering machines, structures, or systems are constructed. They consist of (1) detailed drawings, which show an engineering component in detail, the materials from which it is to be made, its dimensions, and other information such as who designed it; (2) assembly drawings, which show how the components are assembled; and (3) pictorial drawings. The first two use orthographic projection. Pictorial drawings convey a perspective that the first two kinds of plans do not convey. *Layout* is a general term for a planning drawing for a system. Charts and graphs can convey functional information and data.

The means to communicate technical and scientific ideas and concepts must be done in a manner that leaves little or no room for error or misjudgment. There are many *conventions* in this engineering "language," and engineers must study and become familiar with them. A convention is a commonly accepted practice, rule, or method. In our daily life there are many conventions, such as red and green lights meaning stop and go or nodding your head for yes or no. An engineer using an accepted drawing convention knows that other engineers or engineering drafters will read and understand that convention and understand consistently.

The following problems are included in this chapter as an additional way to introduce you to engineering graphics. You may or may not have much success with them, but we suggest you try them. They will help you realize the significance of the material in this book.

SAMPLE PROBLEM 1.4

A two-dimensional object can be represented with one continuous line that has no sharp corners. Try to sketch it from the following description.

Its overall length is about 8 units and its overall width is 2 units. For convenience, assume that one length unit is $\frac{1}{4}$ in. Start by drawing a 5-unit-long vertical line. Then locate a point 1 unit directly to the right of the top of this line. Then draw a half circle of radius 1 unit with this point being the center of curvature. Begin the half circle at the top of the straight line. You should now have what looks like an upside-down and reversed J. From the free or right end of the half circle draw a 6-unit line vertically down. The bottom end of this line should be 1 unit below and 2 units to the right of the bottom of the original line, the first line you drew. Now locate a point $\frac{3}{4}$ unit to the left of the bottom of this second straight line and construct about this point a half circle with radius $\frac{3}{4}$ unit. This half circle connects smoothly with the second straight line and is concave upward forming a J-like shape. Then from the left end of the second half circle draw a straight line 5 units long vertically up. Locate a point $\frac{1}{2}$ unit to the right of the top end of this line and construct a third half circle of radius $\frac{1}{2}$ unit about that point to connect in a smooth manner with the end of the third straight line. Then add a fourth and last straight vertical line of 4 units to the right end of the third half circle. This last line goes vertically down.

What you have just drawn is a common device. See the statement at the end of this chapter after Problem 1.42.

PROBLEMS

1.5 Prepare a verbal description of the shapes shown. Note that each shape can be thought of as a rectangle, a right triangle, and an equilateral triangle. After your description is complete, have someone use your verbal description to sketch the shape.

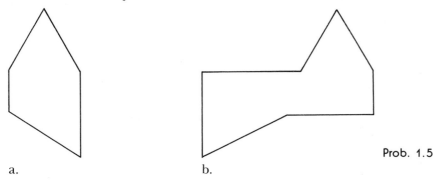

Prob. 1.5

a.   b.

1.6 Describe in words the object that you think each of the sketches is meant to describe.

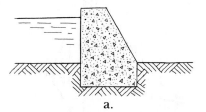

a.

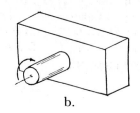

b.

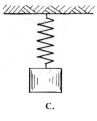

c.

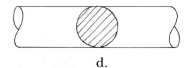

d.

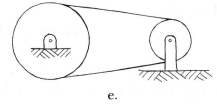

e.

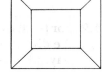

f.

Prob. 1.6

1.7 Describe in words what you think each of the drawings is meant to represent.

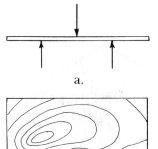

a.

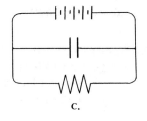

b.

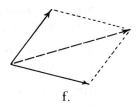

c.

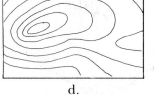

d.

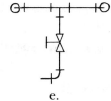

e.

f.

Prob. 1.7

1.8 Points *P* and *Q* are on two faces of a tetrahedron *ABCD*. The sides of the four triangles that make up the tetrahedron are all 100 mm long. Face *ABC* is resting on a horizontal surface. Point *P* is at the geometric center of face *ABD*. Point *Q* is 60 mm from *A* toward *C* and 20 mm from the line *AC*. What is the shortest distance from *P* to *Q* measured on the lateral surface of the tetrahedron?

Answer: 71 mm

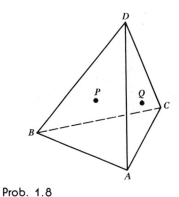

Prob. 1.8

1.9 For the tetrahedron *ABCD* described in Problem 1.8, what is the direct distance between points *P* and *Q*? Stated another way, what is the length of line *PQ*?

Answer: 42 mm

1.10 Determine graphically the distance *AB* for the three-dimensional object shown. Use the uneven scales on the sides of the object.

Answer: 3.8 units

1.11 For the object described in Problem 1.10, determine the distance *AC*.

Answer: 5.8 units

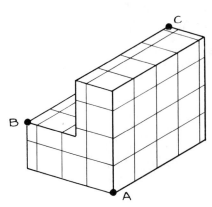

Prob. 1.10

1.12 A cross section of an object is the plane surface that would be generated if the object were cut in two. Sketch a vertical cross section of a coffee cup freehand. What do you do about the handle?

1.13 A cross section of an object is the plane surface that would be generated if the object were cut in two. Sketch a vertical cross section of a transparent tape holder through the axis of the roll of tape. Show the roll of tape in place.

1.14 a. Prepare a sketch that locates a 25-mm diameter hole through the center of a 75-mm cube.

　　 b. How would you verify the size and location of the hole after the hole is in the block?

　　 c. Alter your sketch, if necessary, to do (b).

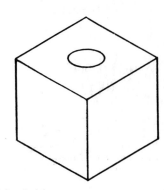

Prob. 1.14

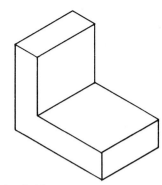

Prob. 1.15

1.15 An L-shaped bracket is to be made with all its surfaces flat and perpendicular or parallel.

    a. Prepare a drawing that includes dimensions of the surfaces that can be used to make the bracket.

    b. How would you verify how well this part meets your description of the part?

    c. Alter your drawing to include (b).

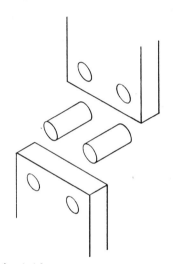

Prob. 1.16

1.16 Two flat plates are to be connected by two dowels.

    a. How would you dimension the plates and the dowels so that they can be assembled by hand.

    b. How would you verify that each part is the proper size after they are made.

    c. Alter your dimensions to include (b).

1.10 GRAPHICS IN ENGINEERING ANALYSIS

Engineering analysis is the study of physical quantities and the physical behavior of quantities involved in engineering. Most of engineering analysis involves modeling and mathematics, which go together. Physical behavior is modeled through mathematics.

Graphics can be very useful and very informative in understanding physical and mathematical relations, particularly when only two or three variables are involved. Two-dimensional space can represent how one quantity varies as a function of another. Three-dimensional space can be used to describe the functional behavior of three variables. All of what is included in this book will simply introduce you to graphic techniques and demonstrate how they are used.

You cannot begin an engineering analysis without understanding that your mathematics will be only as valid as your physical model. Modeling is the key to engineering analysis, but modeling is very much a function of judgment and experience. You will be able to solve mathematical problems in engineering long before you have the ability to formulate them—to develop the model.

After you have a grasp of the meaning of mathematical and physical modeling, you will find many concepts in mathematics that can be better understood with the aid of graphics. Algebraic functions and their behavior, the solution of algebraic and particularly transcendental equations, and the concepts of the derivative and the integral are a few.

We will cover these topics in Chapters 8, 9, and 10, but, to stimulate your thoughts on using graphics in analyzing problems, consider some of the next problems. Again, you may or may not be successful, but try them.

SAMPLE PROBLEM 1.17

If A can dig a hole in 6 h and B can dig that same hole in 3 h, how long would it take to dig the hole if A and B worked together?

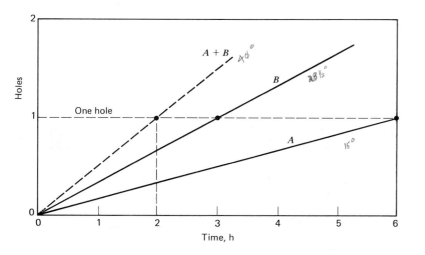

$$\frac{1}{3} + \frac{1}{6} = \frac{1}{2}$$

Solution

This is a classic rate problem, which can be solved easily with graphics. If A digs one hole in 6 h, two holes in 12 h, three holes in 18 h, etc., the line labeled A represents the rate of A's digging. The line labeled B represents B's digging. If they dig together, their rates are added. In the graph, their combined digging is represented by the line labeled $A + B$. That line crosses the one hole ordinate at 2 h. A simple algebraic check will verify that 2 h is the correct answer. What could be simpler than this graphic solution?

PROBLEMS

1.18 Sketch a thermometer that has both Celsius and Fahrenheit scales.

1.19 Sketch a scale (a straight line) that has meters on one side of the scale and feet on the other side for lengths of 0 to 20 ft.

1.20 Make a scale that will show both the diameter and the area of a circle for diameters from 6 to 20 mm. A scale is a single straight line; the diameters are marked off on one side of the line, and the corresponding areas are labeled on the other side of the line.

1.21 Make a scale that will show the function e^x for x from 2 to 7. A scale is a straight line. The function e^x and x are marked off on either side of the line.

1.22 If B can dig a hole in 3 h and A can fill it in 6 h, how long would it take to dig the hole if A and B are working at the same time, one working against the other?

Answer: 6 h

1.23 The tortoise and the hare begin a footrace of 400 m. The hare hops off at 2 m/sec, which is 10 times faster than the tortoise can crawl. After hopping for 2 min, the hare is so far ahead that it stops, becomes disinterested in the race and falls asleep. The tortoise, however, lumbers slowly along, passes the sleeping hare, and wins the race. When and where does the tortoise pass the hare? Graph the race of the tortoise and hare.

Answer: at 240 m, after 20 min.

1.24 a. Find several solutions to the equation $\sin x = 0.5$.

   b. Find several solutions to the equation $\sin x = 1/x$.

   c. Compare (a) and (b). Are they similar problems? How did graphics help you?

Answer: (a) $\dfrac{\pi}{6}, \dfrac{5\pi}{6}, \dfrac{13\pi}{6}, \dfrac{17\pi}{6}, \ldots$

   (b) $0.354\pi, 0.882\pi, 2.050\pi, 2.966\pi, \ldots$

1.25 a. Find several solutions to the equation cos $x = 0.75$.

b. Find several solutions to the equation cos $x = \tan x$.

c. Compare (a) and (b). Are they similar problems? How did graphics help you?

Answer: (a) 0.723, 5.560, 7.006, ...
(b) 0.666, 2.475, $2\pi + 0.666$, ...

1.11 WHAT IS ENGINEERING DESIGN?

The term *engineering design* is more difficult to define than *engineering*. The English word *design* is derived from the Latin *designare* ("to mark out"). *Design* is used broadly to describe a *process* of conceiving, analyzing, and choosing alternative solutions to meet and solve engineering problems. When it is used in conjunction with engineering as engineering design, it involves the specification and detailing of the machines, structures, and systems used in engineering. It requires a wide range of knowledge, experience, skills, and techniques. Engineering design is a central part of all engineering.

The design process involves assembling all the related facts and information, weighing and sorting them, and selecting a solution to the problem. Inherent in the process of design are decisions. Design is often used synonymously with decision making.

Conception is a part of this design process. If no solution exists, we will need to create one. Creating is an activity that is peculiar to engineering as contrasted to science. Innovation, invention, creativity, and brainstorming are all part of conception.

Analysis is a part of design. The difficulty in making this statement is that, once stated, it can be misinterpreted. Analysis is not all there is to design, but design without analysis is little more than art. The best achievements in design are thoroughly analyzed, and the best designers are very good at analytical techniques.

There can be alternate solutions for a design. This is an eye-opening revelation to many young engineers. In science and mathematics, problems have unique solutions. There is one right answer and there are many wrong answers. How can there be more than one right answer to a problem? In mathematics, mathematics is the criterion for judgment. In science, science is the criterion for judgment. The laws of science and mathematics are clear and precise. If we used science and mathematics as criteria for design, our design solutions would be unique, but we do not use science and mathematics as criteria. We use economics, weight, size, aesthetics, marketability, reliability, and, today, environmental factors as criteria for design. As a result, our solutions change with the criteria. The most economical design may not be the smallest, or the lightest, or the most aesthetic. The most economical design may be, and often is, the least reliable. It may be the most polluting. Probably more projects have been eliminated for their prohibitive cost than for any other reason. Working with such intangibles can be a difficult challenge, but it can therefore be one of the most rewarding experiences in engineering.

The designs that are developed in response to the stated needs must be communicated to those responsible for production and to those who have the need. It is part of the designer's task to see that this is done accurately, economically, and compatibly with the environment. The engineer communicates the design in three ways: orally, graphically, and through writing reports. We will concentrate our activity on graphic communication, but we cannot and will not neglect the other forms of communication. For good reason, the competent engineer must develop all three of these skills.

The following problems are typical of engineering design problems. They have more than one solution. Your only limits are your creativity, your experience, and your knowledge of what will and will not be acceptable.

SAMPLE PROBLEM 1.26

A bracket is needed to transfer the effect of the 80–N force to the solid wall *A*. Design a bracket that will do this. The load to be applied to the bracket is eccentric.

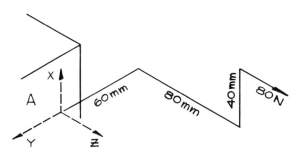

Solution

This problem can be solved in several ways. There is no one solution. First, there are several means of attaching the bracket to the wall. In *(a)*, the bracket is welded. This would be permanent, but it is only usable if the bracket could be welded to the wall. Glue is a possibility. In *(b)* the bracket has been drilled and bolted to the wall. If you are going to bolt the bracket, why not use two, as in *(c)* or *(d)*? Which is better, *(c)* or *(d)*?

A bracket that would do the task is shown in *(e)*. It would be cut from sheet metal and bent.

A different design would be the twisted bracket in *(f)*. It has two advantages. The first advantage is that it can be made from a strap rather than sheet metal and therefore would be less expensive. A second advantage is that the 80–N load would be simpler to apply. Do you see how economic factors influence design?

The second design has advantages, but it is flexible. It will deflect more than the first design, given equivalent dimensions.

Consider the third design, *(g)*, which uses a stiffening web. Note that the web is in the plane formed by the 80–N force and the z-axis. It is a stiffer design, probably stronger, but it will cost more. Do you see how economic factors influence design?

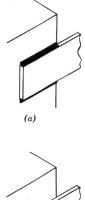

(a)

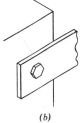

(b)

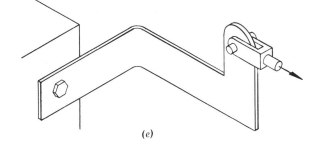

(e)

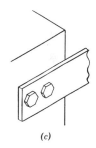

(c)

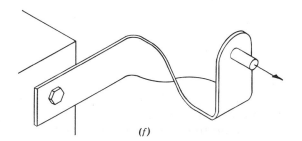

(f)

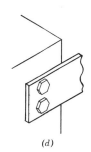

(d)

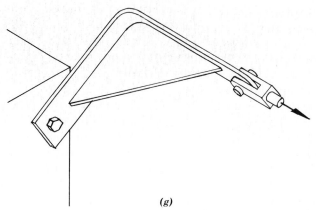

Sample Prob. 1.26 *(continued)*   (g)

PROBLEMS

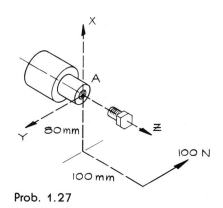

Prob. 1.27

1.27 A bracket is needed to transfer the effect of the 100–N force to the shaft *A*. It is intended to bolt the bracket at *A*. Design a bracket that will do this. Sketch it.

1.28 Some handle is needed to transfer the effect of the 100–N force to the shaft. The handle is intended to be removable. Design something that will do this. Sketch it.

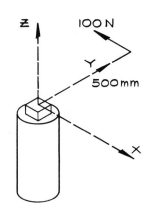

Prob. 1.28

1.29 Design a gusset plate that will transfer the effect of the 10–kN force to the two rivets *A* and *B*. Sketch it.

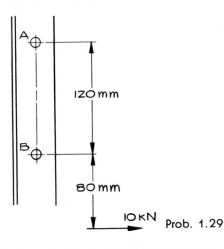

Prob. 1.29

1.30 A tensile load of 50 lb is to be supported at points *A* and *B*. A 12-in. cylindrical obstruction is located as shown. Design something that will transmit the force from *A* to *B*. Sketch it.

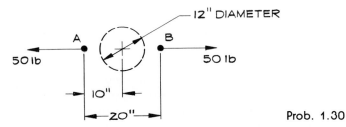

Prob. 1.30

1.31 Link *AB* and link *DC* are to be joined so that one slides along the other. There must be no resistance to this sliding motion, but when the two links are joined there should be no tendency to buckle or bend. Sketch a means to do this.

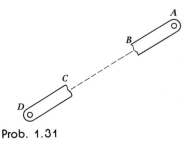

Prob. 1.31

1.32 Shaft *A* turns on an axis that is skewed 10 deg from the axis of shaft *B*. Design something that will transmit motion from *A* to *B*.

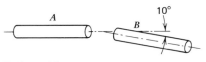

Prob. 1.32

Prob. 1.33

1.33 Design a footing and support for a pole 250 mm in diameter and 15 m high.

1.34 Design a container for 500 gal (1.763 m³) of hot water.

1.35 Design a container for 6 ft³ of household waste (food, paper, hardware, etc.).

1.12 GRAPHICS IN ENGINEERING REPORTS

The end result of an engineering design may be something as simple as a bracket or a mechanical fixture, or it may be as large and as complex as a refinery or a supersonic aircraft. If it is simple, the engineering designer can always fondle the completed object and see what his or her engineering skills have accomplished.

In a more complex project, an engineering report is usually one of the milestones the work produces, and there are many, many reports. These reports serve two major purposes: they communicate the results of an engineering project to others and they serve as a permanent record of the project. The size and form of the report are related to the subject matter. Reports vary from a few pages to multivolume productions.

Written prose and graphics make up engineering reports. Plans, spatial analysis, analytical models, data, and functional relationships require graphics for communication to others. Chapter 12 of this book discusses in some detail the makeup of engineering reports.

Note that in each of the following problems it is both natural and necessary to utilize graphics in your communication.

SAMPLE PROBLEM 1.36

Obtain a flat plastic ruler 45 cm (18 in.) long. Support one end as a cantilevered beam. Note the unloaded position of the free end of the cantilevered ruler with overhanging lengths of 20, 25, 30, 35, and 40 cm. Now place a mass on the free end of the rule and note the loaded position of the cantilevered ruler for the same overhang. Report on your results.

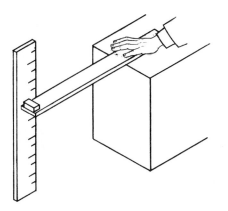

Objective. The objective of this study is to observe the relation between deflection and the loaded length of a cantilevered beam.

Data

Beam Length, cm	Unloaded Deflection, mm	Loaded Deflection, mm	Column 2 —Column 3, mm
20	720	711	9
25	717	699	18
30	711	681	30
35	703	655	48
40	690	618	72

Results. Two new erasers were selected for the mass placed at the end of the beam. They were heavy enough to deflect the beam measurably, but not extensively. The deflection is plotted as a function of beam length in the graph.

It is evident that the deflection is a function of the beam

length within the bounds of this simple experiment. On rectangular graph paper it is evident that the deflection is proportional to some power of the beam length. When the logarithm of the deflection is plotted against the logarithm of the beam length it appears that

$$\ln y = 3 \ln l - 6.79$$

where y is the deflection of the beam due to the added mass and l is the length of the beam.

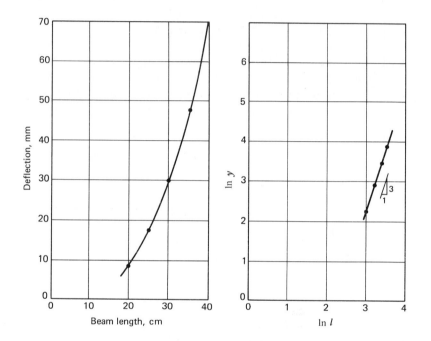

PROBLEMS

For the following six problems, prepare a two-page report that consists of a page of written description and a page of graphics.

1.37 Investigate the cooling rate of a cup of hot coffee.

1.38 Repeat the experiment of Problem 1.37, but this time drop one ice cube into the hot liquid and report its effect. Does it make any difference if you put the ice cube in at the beginning or after a short delay?

1.39 Investigate the gear ratios for a 10-speed bike. Present how you measured the ratios and the 10 ratios you determined graphically along with necessary description and conclusions.

1.40 Run in place for 1 min. Cease exercising and immediately start counting your pulse. Count the cumulative number of beats and record the total after every 15 sec for up to 5 min. What was your initial pulse rate?

1.41 With a piece of string and any small solid mass, make a pendulum. Observe and record the frequency of the pendulum for various string lengths.

1.42 Vary the load on a cantilever or simply supported beam and report how the deflection of the beam varies with the load in addition to how you conducted the investigation. Use a beam you can easily obtain, such as a thin strip of wood, a paper straw, a piece of corrugated paperboard, etc.

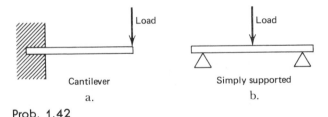

Cantilever
a.

Simply supported
b.

Prob. 1.42

Answer to Sample Problem 1.4. *If you followed the written description correctly, you drew a common paperclip.*

2
SKETCHING
AND DRAWING

2.1 SKETCHING

Sketching is informal drawing. A good sketch has three basic characteristics: it is prepared quickly, it is simple, and it is easily understood.

All you need in the way of supplies is something with which to write or draw and something on which to write or draw. In sketching only two tools are required: a paper and a pencil. Your eyes will estimate lengths and angles successfully. Your eyes are much more accurate than you realize.

In addition to communicating something quickly and clearly to someone else, sketching is an obvious method of communicating with yourself. A sketch helps you think. A sketch can extend your memory or help clarify a spatial situation. A pencil and paper can be very useful in developing a physical or spatial concept.

In this chapter we will suggest some hints on sketching techniques. Some may work for you and some may not. You will have to adapt our suggestions to your own abilities and training. We hope that these suggestions will give you a nudge to improve your engineering sketching ability, but we cannot make you an artist. Sketching must be learned through practice and use. The proposed techniques are the result of observing ideas and techniques from both art and engineering. Our discussion is limited to sketching individual shapes, and our suggestions are intended to build on one another.

2.2 ENGINEERING DRAWINGS

Formal engineering drawings are made with a variety of mechanical aids to gain accuracy, communicate details, and prepare a permanent record. Engineering drawings are usually prepared by professional drafters, although engineers may also be called on to prepare formal drawings of their work. Often the preparation of drawings will be an integral part of the engineer's process of thinking through a design problem.

Formal engineering drawings will be discussed in detail in Chapter 7.

2.3 TOOLS

All aspects of graphics require tools. At the very least, you will need a pencil and a sheet of paper. The points and lines that you use to make sketches, engineering drawings, and graphic analyses must be made with some writing instrument on a writing surface. A wide variety of drawing guides and measuring devices are available to help you with accuracy and style. This brief discussion of tools focuses on what you will need as an engineering student and a practicing engineer. A drafting technician, one whose vocation is graphic communication, will need and use a much larger variety of tools.

Virtually all of your graphics work can be done with a very few drawing guides. The basic graphic tools of an engineer are shown in Figure 2.1. An engineer can handle all typical graphic communications and analysis situations with these six tools.

The various writing instruments available today include many types of wood pencils, mechanical pencils, and pens. Most of the work you will want to do will require a pencil, although you may need a pen for dark, uniform, permanent lines. You should spend a little time determining what style and pencil hardness works best for you. Choose a pencil that will be soft enough to give dark, clear lines that can be easily seen and copied, but hard enough to prevent smudging of your work by hands, arms, and other sheets of paper.

Figure 2.1 Basic tools—pencils, compass, dividers, triangles, scales, protractor.

Paper is the most commonly used writing surface for engineering graphics. Hundreds of papers are available—plain and ruled, colored, transparent paper for easy copying, and vellum for permanent records.

Some other media are also used. The device used to make a sketch could be a felt-tip pen, chalk, or your finger in some dust. The drawing surface could be a scrap of brown kraft paper (butcher paper), a chalkboard, a napkin, the floor of a workshop, or the back of an envelope. The surface need only be as appropriate as the situation and its permanency require. Have you ever looked through your handbag or pockets for a scrap of paper to help explain an idea or concept? We all draw pictures in the air with our hands as we are explaining something. Such gestures are a part of oral expression, but they really are a visual aid. The glass front on a cathode-ray tube serves as a drawing medium for computer-aided graphics.

Circles and arcs of a circle are common shapes in engineering. For these, a compass is an important tool. A compass with a mechanical radius adjustment can be much more accurate than one with a friction hold. Dividers provide the easiest and fastest way to accomplish the common chore of transferring lengths, and a friction-hold divider has the advantage of quick adjustment. Triangles can be used as straightedges, to establish common angles, and to draw parallel and perpendicular lines. Many other guides such as templates, French curves, parallels, and drafting machines are available, but you can do most engineering with a simple compass, a pair of dividers, and several triangles.

A scale is also a primary tool. Scales are used to measure and lay off lengths. A protractor to measure and lay off angles may also be needed. Keep in mind, when you obtain scales for your personal use, that you will probably need both English (conventional) and SI (metric) scales, at least at the beginning of your professional career.

Angular and linear measurements in accurate sketches and mechanical drawings are usually made with the aid of protractors and scales. The accuracy of the drawing is related to how carefully these tools are used and how large the drawing is to be —that is, to what scale the drawing is made. For greater accuracy, the scale is enlarged. With careful work, your inaccuracy should be no greater than one line width.

Since the primary goal of a sketch is to prepare a clear representation of something quickly, your primary measurement aid should be your eyes. Measurement using your eyes is more accurate than you realize. With practice, measurement with your eyes can be developed into a real skill. Draw some random lines and angles, estimate their lengths and angles, and then measure the lengths and angles for comparison.

Your outstretched hand can also be a good reference. What are the widths of your fingers, the distances between the ends of your fingers and thumb, and angles between your fingers and thumb? Measure and remember all these, since you will always have this reference scale with you.

2.4 SKETCHING LINES

A sketch of an object is simply a combination of curved and straight lines that define the shape of that object. You have undoubtedly heard the saying that someone "cannot even draw a straight line." The statement is derogatory, but try it! Drawing a straight line is not easy. This section contains some suggestions relative to sketching straight and curved lines for you to consider and possibly to alter and improve your techniques.

To construct a straight line, first gently grasp your pencil close to its point (about 25 mm from tip). Next, rest the pencil point and your hand on the paper. Mark or imagine two points where the straight line should begin and end. Start at one point and aim for the other. Either use a light, quick sweeping motion, keeping your eye as much or more on your destination point than on your pencil, or inch along lightly, as shown in Figure 2.2. As you draw from one position on the paper to another, move your hand, pencil, and arm together, as if they were one instrument. Do not plant your hand firmly on the paper and move only your fingers.

Your first try will probably not be satisfactory. You may make a crooked line, or your line may miss the second point. In one or more additional tries, you can draw a better line, using the first line as a guide. Last, make a dark distinct line for the one you wish to preserve.

It may be to your advantage to change the position of your paper. It may be easier for you to draw horizontally than vertically, perhaps even easier at a slight angle. Try several positions. Many right-handed people prefer to draw lines diagonally (8 o'clock to 2 o'clock). If this works for you, the surface on which you are writing can always be turned. All your straight lines can be drawn in the most convenient direction for you.

When drawing straight lines, learn to recognize whether any of them are to be parallel. Keep your eye on the first line while drawing a second line parallel to the first.

Curved lines vary in shape so much that no one method can be used to sketch them all. If there is any symmetry or pattern, place a few points on a straight line as guides. First sketch the curved lines a few times very lightly. Follow these light lines with one dark final line. Keep correcting yourself. Figure 2.3 shows an example.

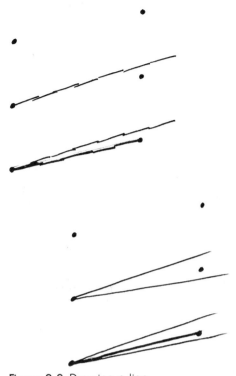

Figure 2.2 Drawing a line.

short, light strokes ⇒
heavy the line (2H)

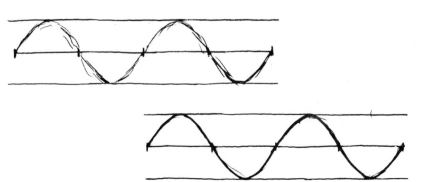

Figure 2.3 Drawing a sinusoidal curve.

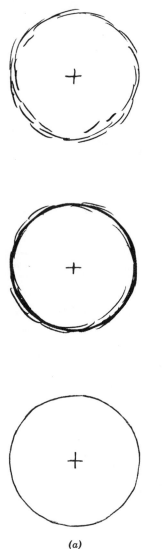

(a)

Figure 2.4 Sketching a circle.

To draw arcs of circles, you can use your hand and arm as a compass by pivoting on the ball of your hand, or your elbow, or at your shoulder. These will give you different radii.

To sketch a circle, locate the center point of the circle to be constructed. You may mark it, but you need not. Keep your eye on the center and rest your little finger very lightly on the paper. The pencil point should be close to the paper, but not touching the paper. Begin making a continual circular motion just above the paper. Slowly lower your pencil until it draws light circles on the page. Your circles will run together, making a fuzzy circle. Go over this fuzzy circle with a single dark circular line. Inch along if necessary. Your final circle will be quite accurate. Another technique for sketching a circle is to construct a square and to quarter the square, then cut the corners off with arcs, or build up the circle with several arcs. Figure 2.4 shows these procedures.

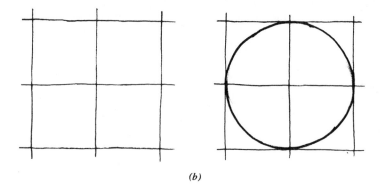

(b)

A circle viewed from an oblique angle is an ellipse. The top of a cylinder viewed as if it were sitting on a table in front of you is an ellipse. To sketch an ellipse, draw two lines that bisect each other, one vertical and one horizontal, as shown in Figure 2.5a. The lengths of these lines represent perpendicular diameters of the circle. Now sketch short perpendicular dashes through the endpoints. Proceed in a light continual motion as suggested for circles, concentrating on the center point and making the curve tangent to the four dashes. Last, draw a single dark line, as before.

In many cases, it is convenient to orient the two-dimensional axes differently (see Figure 2.5b). Once the four tangential dashes are established, the drawing process is the same as before.

You can also box in the boundaries of an ellipse in the same manner used for a circle. This box represents a square in which a circle will be scribed. Remember, both the circle and the square are viewed obliquely. After the box is located and the points of tangency identified, the ellipse can be sketched quickly and accu-

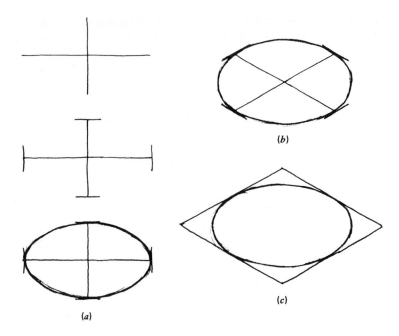

Figure 2.5 Sketching ellipses.

rately. This technique is demonstrated in Figure 2.5c. A common mistake when drawing ellipses is an improper orientation of the axes, which results in an incorrect ellipse. Another common error is to make the ends of the ellipse too rounded or too pointed.

If the object you are sketching has a cylindrical element, such as an axle or shaft, that is perpendicular to a circular surface, the element should always appear perpendicular to the major or largest axis of the ellipse (see Figure 2.6).

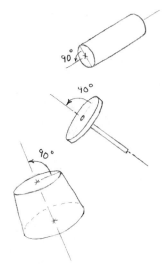

Figure 2.6 Ellipse-axis orientation.

One of the most useful aids you can use to improve your sketching clarity is a grid. Light crosshatched lines, such as those shown behind this paragraph, give a useful guide for establishing lines, curves, and distances. A great variety of papers with grids are available; some are prepared with lines that will not show when copies are made. Relatively dark-lined grids can be placed under and seen through plain paper on which you are preparing a sketch.

2.5 SLIPPING TRIANGLES

Parallel lines can be drawn accurately and quickly by "slipping triangles," as illustrated in Figure 2.7. By guiding one side of a triangle along a straight surface, a series of parallel lines can be established with either of the other two sides of the triangle. The straight surface could be one edge of another triangle. In using this technique with two triangles, one slips on the other—hence the term *slipping triangles.*

The same technique can be used to establish a series of perpendicular lines. In Figure 2.8*a*, observe that when the hypotenuse of a triangle is guided along a straight surface, the lines established by the other two sides are parallel and perpendicular to each other. Note that only perpendicular lines can be established by "slipping" the hypotenuse of the triangle.

A second method of establishing orthogonal (perpendicular) lines is included in Figure 2.8*b*. The hypotenuse of one triangle is oriented with the given line. The second triangle is then held against one of the other two sides of the first triangle. The first triangle is then rotated through 90 deg by placing the third edge on the fixed edge of the second triangle. The hypotenuse of the first triangle has then been rotated through 90 deg.

It is best to have two different triangles, one a 45–45–90–deg triangle, and the other a 30–60–90–deg triangle. Practice using a pair of triangles for drafting exercises. Triangles are convenient and are the only mechanical drawing aid that most engineers have.

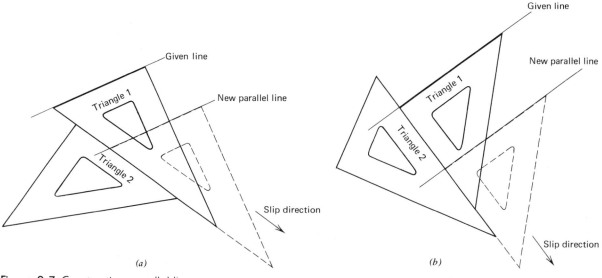

Figure 2.7 Constructing parallel lines.

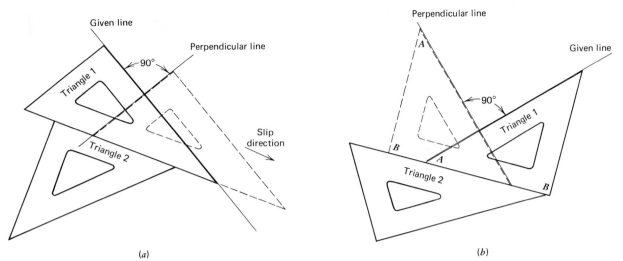

Figure 2.8 Constructing perpendicular lines. (*a*) Hypotenuse slip. (*b*) Hypotenuse rotation.

2.6 SHAPES

The shape of an object can be described or built up using combinations and intersections of basic surface shapes.

Plane.
Cylinder.
Cone.
Sphere.
Double-curved surface.

Dark lines in a sketch usually represent a sharp edge resulting from the intersection of two surfaces or the limits of a body (e.g., a circle is used to represent a sphere).

Most engineers do not have the ability, nor should they take the time, to shade their sketches, but the effect of light can and should be used in obvious situations. Deep holes and slots can be shown very clearly with shading or darkening. Since the area to be darkened is clearly defined, no special abilities are required.

We don't recommend that you use surface shading, if you do not possess the ability to draw it well, but in many cases some indication of surface shape is necessary and simple to accomplish. Try the following. Think of groups of perpendicular lines on several surfaces—for example, the four lines in Figure 2.9a. How do these lines appear from different directions and angles?

Now, if the page were rolled up to form a cylinder, two of the lines would appear curved while the other two would remain straight. A double-curved surface could then be described with

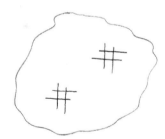

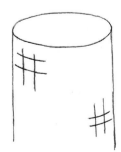

Figure 2.9a
Shape emphasis lines.

Figure 2.9b Showing surface shape.

curved perpendicular lines. These perpendicular shape-emphasis lines, or a series of parallel lines, can be used as an occasional suggestion of surface shape or can completely cover the surface, as the situation dictates. See Figure 2.9b.

As another example, consider four different graphic representations of the pulley shown in Figure 2.10. A photograph, an artist's sketch, and two engineering sketches using two degrees of shape-emphasis lines are shown. Draw your own conclusions concerning the concentration of the shape-emphasis lines.

Figure 2.10 Representations of a pulley.

The representation of two intersecting plane surfaces is commonly required in the sketching of frames and machines. If the intersection is a sharp, well-defined angle, it should be represented with a straight line. If the surface transition from the two planes is curved with a relatively small radius, it can be sketched several ways, as shown in Figure 2.11. Shape-emphasis lines may not be practical if the radius between the planes is relatively small.

Engineering sketches must communicate not only shape but also shape change. Often the deformations and relations of interest would not be apparent if the sketch were made to scale. Overdoing small deformations and relations can greatly clarify a sketch. Several examples are included in Figure 2.12.

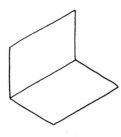
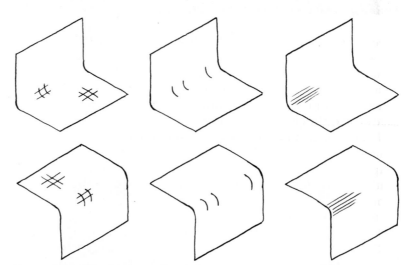

Figure 2.11 Sketching small radii between intersecting planes.

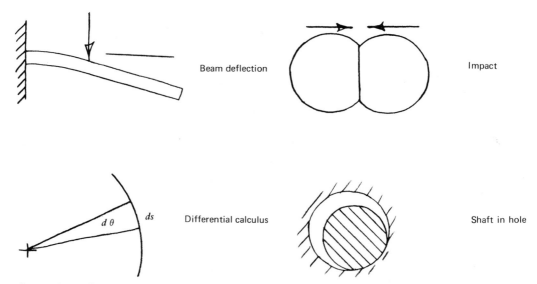

Beam deflection

Impact

Differential calculus

Shaft in hole

Figure 2.12 Overdoing deformations and relations for clarity.

2.7 SYMBOLS

Specifying the shape of an object usually doesn't include specifying all the information one may want to communicate. Specific dimensions, materials, finish, motion, etc., all can be presented with a variety of symbols. Many standard symbols used in mechanical drawing by drafting technicians can and should be used in sketching.

These conventions (i.e., the conventional or accepted way something is presented) will be discussed in some detail in chapters to follow. You need only to learn what conventions are available. If conventions are needed, they can always be obtained from references. As one example, sometimes it is desirable to remove a portion of an object mentally or to discontinue it, making *breaks*. Four conventional breaks are shown in Figure 2.13.

In addition to using the standard conventions for specifying materials, line quality and character can help communicate what the object is. Figure 2.14 includes some examples.

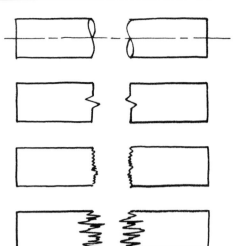

Figure 2.13 Representation of breaks.

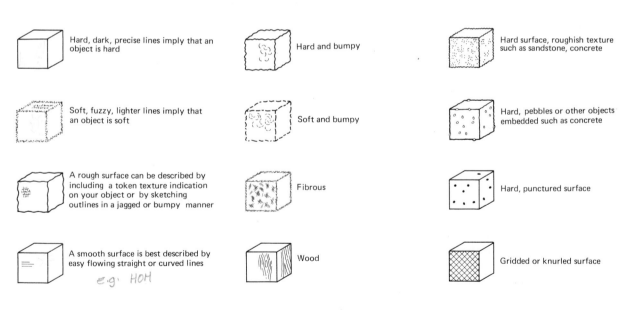

Figure 2.14 Sample conventions.

Hard, dark, precise lines imply that an object is hard

Soft, fuzzy, lighter lines imply that an object is soft

A rough surface can be described by including a token texture indication on your object or by sketching outlines in a jagged or bumpy manner

A smooth surface is best described by easy flowing straight or curved lines

e.g. HOH

Hard and bumpy

Soft and bumpy

Fibrous

Wood

Hard surface, roughish texture such as sandstone, concrete

Hard, pebbles or other objects embedded such as concrete

Hard, punctured surface

Gridded or knurled surface

contouring

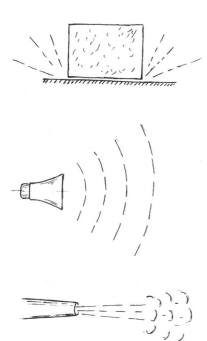

Lines added for emphasis can be very useful in sketches. Emphasis lines are not used in formal drawings. Note how such things as light, speed, direction, sound, and impact are shown in Figure 2.15. Cartoonists make heavy use of emphasis lines. To show that one object or surface has hit another, add a few raylike dashes close to the point of contact. These can be interpreted as noise or dust that may occur on impact. Light can be represented by rays, in the direction they would come from the source. To indicate that an object is traveling, draw a few lines from behind the object on the side away from the direction it is headed. If you wish to emphasize the history of the motion, these lines trace the path an object has traveled. A few light lines or wiggly lines on each side of an object can indicate vibration. Curved lines that abbreviate waves indicate sound. To show fluid flow, such as air being blown from a tube, draw several straight lines in the direction the air is being blown and then a few "half clouds."

A very important symbol used by engineers is the arrow. It is used to define dimensions, but the most important use to the engineer is to symbolize a *vector quantity.* A vector is a physical quantity that has both direction and magnitude. The direction is symbolized by the direction of the arrowhead, and the magnitude is symbolized by the length of the arrow. Vectors are used in a formal way for analysis and informally to help understand and communicate the situation. Figure 2.16 contains some isolated examples.

Figure 2.15 Symbols influenced by cartoonists. *shading*

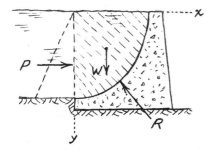

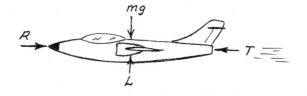

Figure 2.16 Vectors included in sketches help clarify the situation.

Construction lines used during the preparation of a sketch can also improve clarity. Note in Figure 2.17 how the additional lines that are not part of the specific shape description help the readability of the sketches.

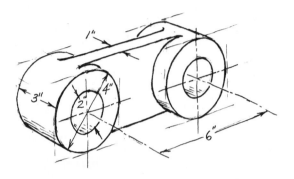

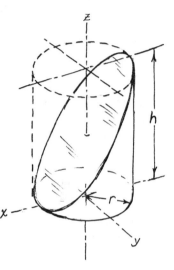

Figure 2.17 Use of construction, layout, and centerlines.

Understanding the operation of a device can be greatly improved by adding the path of selected points on the device, and curved and straight arrows help indicate rotation. Several examples are shown in Figure 2.18. Indicating several positions of a device as shown in Figure 2.19 also helps communicate operational details.

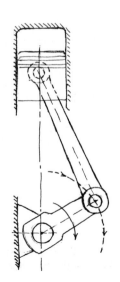

Figure 2.18 Use of arrows to indicate movement.

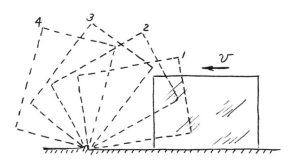

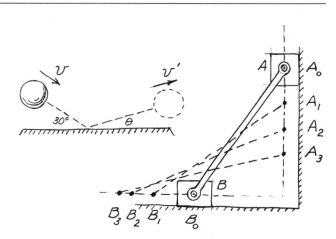

Figure 2.19 Use of positions for clarification.

2.8 LETTERING

Engineers do not write; they letter or print. Virtually all handwritten prose authored by engineers to be read by others is prepared in printed form. Notes and calculations prepared by engineers for use by themselves and others typically are a mixture of words, letters, and numbers. Engineers do not actually write equations; they print or letter equations to ensure clarity. Figure 2.20 presents a typical analysis that an engineer would prepare. Note that the example includes sketches of shapes, symbols, numbers, letters, and printed words. The work is clear because all the elements making up the analysis are clear, including the printed letters and words.

Informal sketches and formal drawings usually must include information that cannot be communicated as a picture or shape. A complete sketch or mechanical drawing will include dimensions, letters, notes, and specifications. Drafting technicians must perfect their lettering technique, but even this skill isn't as necessary as it used to be. The advent of various types of lettering guides, paste-on letters, and typewriters that can be used for the preparation of drawings make hand-lettering skills unnecessary even for drafters.

The one rule you should follow for the preparation of notes and specifications is *clarity*. If you are capable of printing relatively quickly and legibly, then use this graphic tool. Try to develop a clear, efficient style. If your work is clear and consistently

Determine the resultant force due to the pressure of the water on the vertical door

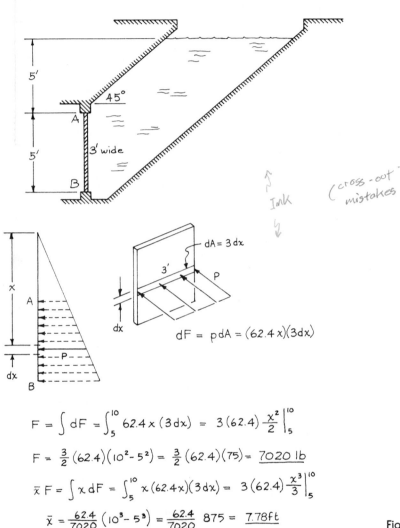

$$F = \int dF = \int_5^{10} 62.4\,x\,(3\,dx) = 3\,(62.4)\,\frac{x^2}{2}\Big|_5^{10}$$

$$F = \frac{3}{2}\,(62.4)\,(10^2 - 5^2) = \frac{3}{2}\,(62.4)\,(75) = \underline{7020\ \text{lb}}$$

$$\bar{x}\,F = \int x\,dF = \int_5^{10} x\,(62.4x)\,(3\,dx) = 3\,(62.4)\,\frac{x^3}{3}\Big|_5^{10}$$

$$\bar{x} = \frac{62.4}{7020}\,(10^3 - 5^3) = \frac{62.4}{7020}\ 875 = \underline{7.78\text{ft}}$$

Figure 2.20 Typical analysis problem.

clear, it will be all that you will need as an engineer. If it is not, the use of guide or grid lines may benefit your lettering technique with very little additional effort.

There is no preferred lettering style, but most engineers do develop a style of their own. They want to be able to identify their work and they want others to be able to identify their work. Again, clarity and consistency are the characteristics required.

PROBLEMS

2.1 Copy the following lines freehand, estimate their length, and then check your accuracy with a scale.

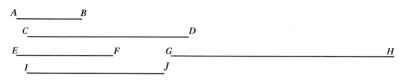

2.2 Estimate the following angles and check with a protractor.

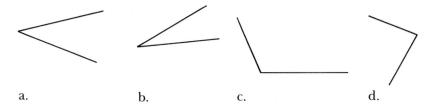

a.   b.   c.   d.

2.3 Estimate the diameter of the following circle with your dividers; find the center of the circle and the radius. Copy the circle with a compass, center at *A*.

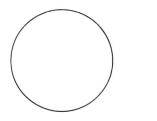

2.4 Copy freehand the following shapes.

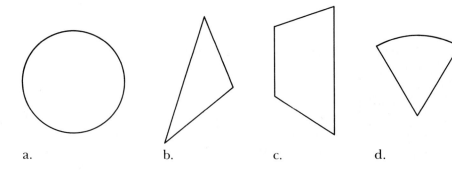

a.   b.   c.   d.

2.5 Slipping triangles, for each of the three cases, draw line *AB* parallel to line *CD*. Lines *AB* and *CD* have the same length.

2.6 Slipping triangles, for each of the three cases in Problem 2.5 draw line *AB* perpendicular to line *CD*. Make the length of line *AB* twice *CD*.

a.

+*C*

+*D*

+*A*

+*A*

C
+

D
+

b.

c.

+
D

+
C

+*B*

2.7 Slipping triangles, crosshatch the shapes.

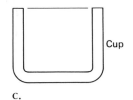

Cup

a.

b.

c.

2.8 Slipping triangles, crosshatch the shapes.

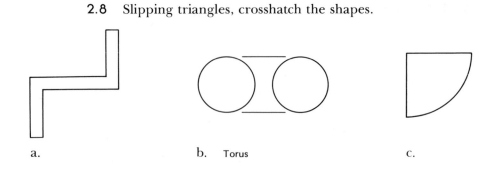

a.   b.   Torus   c.

2.9 In the box, sketch a circle that just fits within the box.

a.   b.

2.10 In the box, sketch an ellipse that just fits within the box.

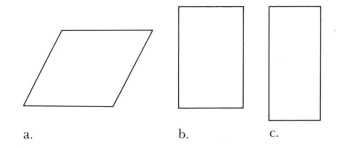

a.   b.   c.

2.11 Sketch the cosine wave $y = a \cos x$ from $x = 0$ to $x = 3\pi$.

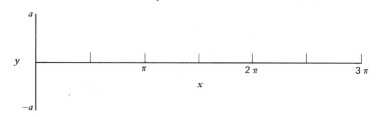

2.12 Sketch the function $y = e^x$ from $x = -2$ to $x = 2$.

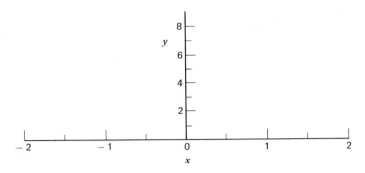

2.13 Sketch the function $y = \csc x$ from $x = 0$ to $x = 3\pi$. Complete the vertical axis.

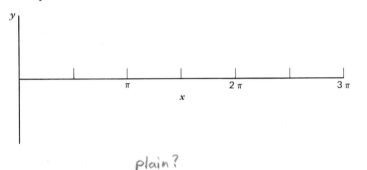

plain?

Sketch the function on a plane sheet of paper. Then plot the functions on graph paper and compare the results.

2.14 $y = x^2$ $(-3 \le x \le 3)$

2.15 $y = \sqrt{x}$ $(0 \le x \le 9)$

2.16 $y = \sin \theta$ $(0 \le \theta \le 360° \text{ deg})$

2.17 $y = x^3$ $(-3 \le x \le 3)$

2.18 $x^2 + 4y^2 = 4$

2.19 Using motion emphasis lines, sketch the action of the situations on the following page.

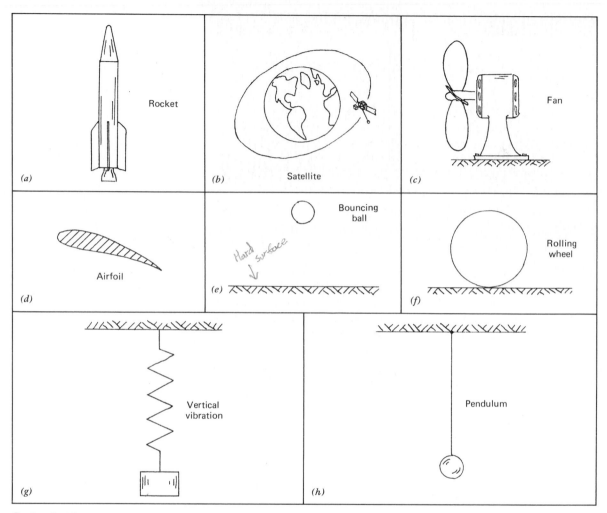

Prob. 2.19

2.20 Sketch 8 cylinders. By adding something symbolic to each, portray the following qualities (but do not letter them).

a. Made of plastic.   e. Hard surface.

b. Made of ice.   f. Covered with dust.

c. Covered with silk.   g. Sticky surface.

d. Smooth surface.   h. Soft surface.

Now exchange your 8 cylinders with someone else. Have that person label your cylinders *a–h*. You should do the same for his or hers. How many surface textures were you able to convey accurately to the other person?

2.9 PICTORIAL VIEWS

Drawings and sketches utilized by engineers can be placed into two general categories. The first and largest category is *multiview drawings* that can accurately communicate complex forms and relationships. They will be discussed in detail in Chapter 7. *Pictorial drawings* are the second category. As the name implies, they show the object in a manner that approximates how we see the object or how it would look if we took a photograph of it. While pictorial drawings do not contain as much detailed information as do multiview drawings, they provide a clear method of communicating overall appearance and relationships to engineers and people working with engineers.

The three mutually perpendicular axes shown in Figure 2.21 provide us with a common base to discuss the three basic pictorial views. Assume that the x–y plane is horizontal and the z–axis is vertical, with $+z$ being in the up direction. We can view an object in any of the three planes, x–y, y–z, x–z (Figure 2.22). Assume that you are viewing this object in the x, y, or z directions and your eye is a relatively long distance from the object. The front view is what you see when you look along the y-axis, the top view is what you see when you look vertically down, and the side view is what you see when you look down the x-axis.

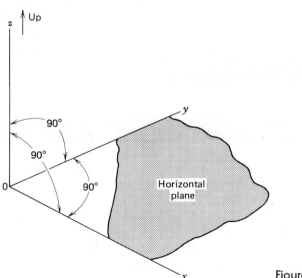

Figure 2.21

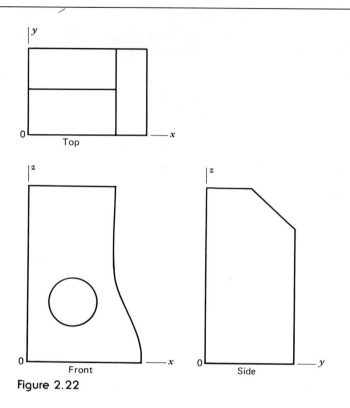

Figure 2.22

The object has half of the top cut off at a bevel, one side is curved, and a circle is scribed on the face. While Figure 2.22 gives all the details of what the object looks like, this multiview representation does not give a clear idea of how the object really appears if we are looking at it. Multiview drawings convey much information, but they are hard to read. Our discussion of pictorial views will use this same object. As we proceed, it will become clear what the advantages and the disadvantages of pictorial views really are.

2.10 OBLIQUE VIEWS

An oblique view is constructed by placing two of the mutually perpendicular axes in the plane of the paper and representing the third axis as going back behind the paper at an oblique angle. The resulting axis system is shown in Figure 2.23. Any line in the object parallel to the x-axis is drawn parallel to that axis. Any line parallel to the z-axis is drawn parallel to the z-axis. Any line parallel to the y-axis is drawn parallel to the y-axis. The object shown in Figure 2.22 is now drawn in oblique view in Fig-

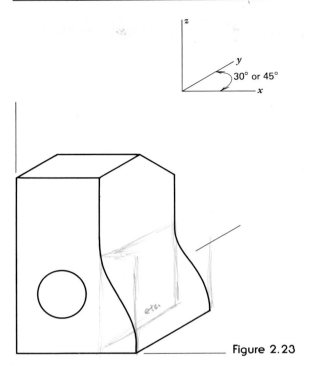

box im

Figure 2.23

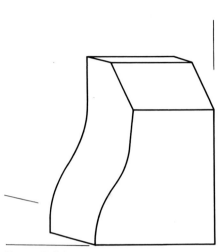

Figure 2.24

ure 2.23. Any distance that is measured parallel to one of the three axes is scaled directly onto the drawing.

The angle between the horizontal axis, in this case the x-axis, and the oblique axis, in this case the y-axis, is usually taken as 30 or 45 deg. But other angles can be chosen to help the clarity of your presentation. Notice how distorted the object appears in Figure 2.23. The scale associated with the oblique axis, in our case the y-axis, can be arbitrarily reduced to help the clarity of presentation.

Note that irregular shapes in planes parallel to the plane of the paper are not distorted. The circle scribed on the face of the block remains as a circle, and the curved side is not altered. This is a basic advantage of oblique views. Circles and curves remain undistorted when drawn in planes parallel to the plane of the paper.

A second oblique view of our object is presented in Figure 2.24. To decrease distortion and help the viewer better understand the shape of the object, it has been turned around so that we see its back; thus we can see the beveled part of the top clearly. The scale along the oblique axis is taken as $\frac{1}{2}$ in this case, and the oblique axis is at a 15 deg angle with the horizontal axis. Experience will help you choose the axis angle and oblique scale to use.

3 axis; 120°

vert. is true

oth. forshortened

not true shape

2.11 ISOMETRIC VIEWS

Isometric views orient the three mutually perpendicular axes on the paper with equal angles between them. The 120 deg angles between the three axes are shown in Figure 2.25. Unlike the oblique view, none of the planes described by the axes are parallel to the paper. All are distorted. Any vertical distance, however, remains in true length, which is the only true representation in an isometric view.

Again, any line on the object that is parallel to one of the axes can be drawn parallel to the axis and scaled directly. The grid contained in Figure 2.25 becomes very useful in carrying out this process. The object we have been considering is shown in isometric view in Figure 2.25. Note that the scribed circle appears elliptical and the beveled top surface is not well represented. A clearer representation of the object can be obtained by choosing another orientation, as shown in Figure 2.26. The choice of orientation is up to you, but note the curved side in this figure and the beveled top all show clearly.

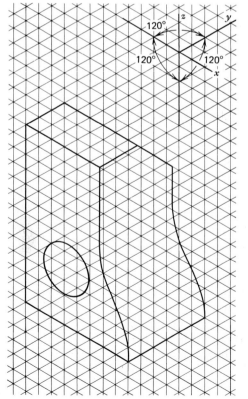

Figure 2.25

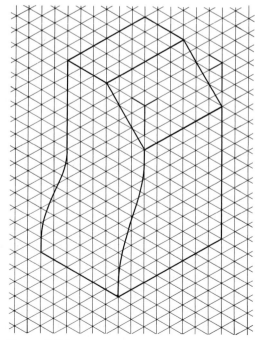

Figure 2.26

2.12 PERSPECTIVE VIEWS

Perspective drawings more closely approximate what the human eye sees than oblique and isometric views. Architects and illustrators take courses and study whole textbooks on the subject. Our goal here is simply to introduce to you what a perspective is.

When you observe parallel lines, they appear to vanish at a single point. This characteristic is utilized in perspective drawings. The three basic types of perspectives are shown in Figure 2.27. When one of the sides of the simple cube is placed parallel to the paper, the perspective is prepared using one vanishing point (VP). When the intersection of two sides of the simple cube is placed parallel to the paper, the perspective is prepared using two vanishing points. And when only a corner of the cube is thought to be in the plane of the paper, the three-point perspective is used. A two-point perspective of our sample object is shown in Figure 2.28. For this sketch, the vanishing points are arbitrarily chosen. Note that both vanishing points lie on the horizon.

There are well-established formal ways to determine the location of vanishing points and the scale of perspective drawings. We suggest you consult a textbook on the subject if you wish to go into further detail. The subject is much too involved to handle in a few paragraphs.

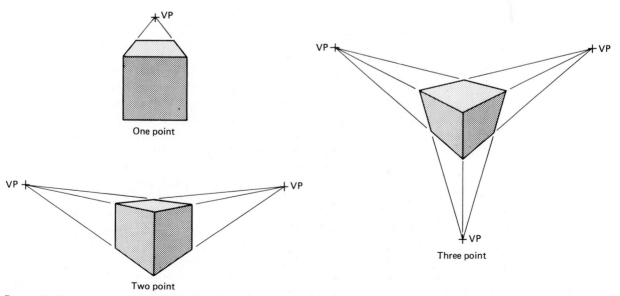

One point

Two point

Three point

Figure 2.27

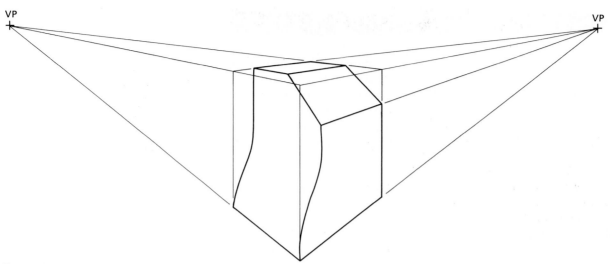

VP

VP

Figure 2.28

2.13 PICTORIAL CHOICE

In addition to sketching lines and basic shapes and understanding how to prepare the three basic pictorial views, your facility with informal graphic communication must include choosing which pictorial view to use and how to orient the object within the view. When we discuss engineering drawings in Chapter 7, you will see how multiviews are used to communicate detailed manufacturing instructions. Pictorial views are usually used to communicate appearance in addition to a variety of other details.

Multiviews and the three basic pictorial views of a simple L-shaped object with a circular hole are shown in Figure 2.29. As you progress from the multiview to the perspective (left to right in Figure 2.29) note how the presentations become closer to how your eye would actually see the object. As you move the other way in the figure (right to left) you gain the ability to communicate construction details, lengths, and angles.

Obliques and isometrics can be prepared rapidly, especially if you utilize grids. In general, perspectives are more time consuming to prepare. Oblique views exhibit considerable distortion but do have the advantage, as we have already discussed, of presenting circles and curves undistorted. The isometric offers both visual realism and ease in preparation. Before you make a sketch, take a moment to consider the alternatives: which pictorial is best and how should the object be oriented?

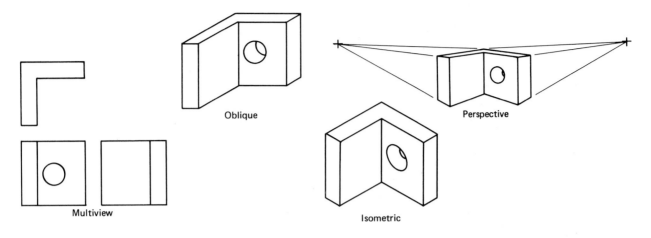

Figure 2.29

2.14 SKETCHING STEPS

Making a sketch is much like building a house. First, you must lay out and construct a foundation. Second, the house is framed. Third, the finish work is done. The details of the finished product are done last, not first.

The foundation of a sketch consists of choosing a view and orientation best to communicate the shape to be sketched and defining the space occupied by the object. The variety of views useful in sketching have been discussed. To illustrate the three sketching steps, consider an isometric sketch of a 3 ft × 5 ft trap door over a hole in a horizontal surface. The trap door is hinged at the intersection of the horizontal surface and a vertical wall. Figure 2.30 is the initial step that defines the space occupied by the door that is to be shown in a partially open position. Line 1 defines the hinge line, and lines 1, 2, 3, and 4 define the hole in the horizontal surface. Curves 5 define the paths of the corners of the door.

Now we are ready to frame in the door and its means of support, as shown in Figure 2.31. Line 6 is drawn parallel to line 3. Lines 7 represent ropes that support the door. Note how easy the framing job is once the foundation is defined.

The amount of finish depends on the purpose of the sketch. In Figure 2.32 the door and hole have been given depth, hinges have been added, the top of the door has texture, and the horizontal and vertical surfaces are suggested by breaking them

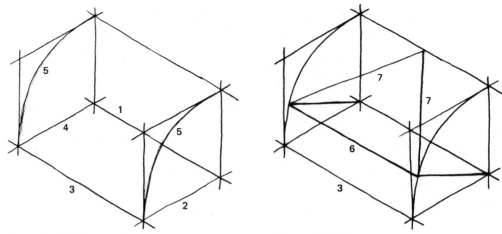

Figure 2.30 The foundation.　　　　　　　　　**Figure 2.31** The frame.

from the surroundings with wiggly lines. The wiggly lines are a convention. All important lines are darkened for emphasis, but none of the construction lines are erased. In fact, note how the construction lines add to the understanding of the object and its function.

Now let us make a sketch of a somewhat more difficult object, the small boat shown in Figure 2.33. Several warped surfaces add to the challenge. Again, we must first sketch a foundation as shown in Figure 2.34. To ease laying out the first sketch, the various tubes are given a rectangular cross section, and the front (bow) of the boat is not yet tipped upward.

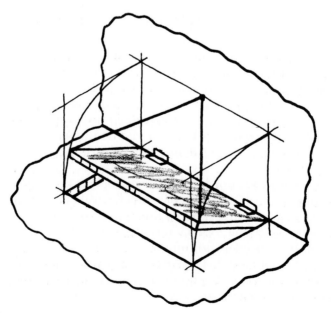

Figure 2.32 The finished product.

Length 8'
Width 4'
Tube diameter 1'

Figure 2.33 Sketch of a small inflatable boat.

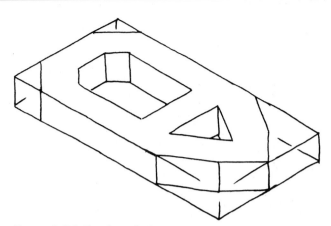

Figure 2.34 The foundation.

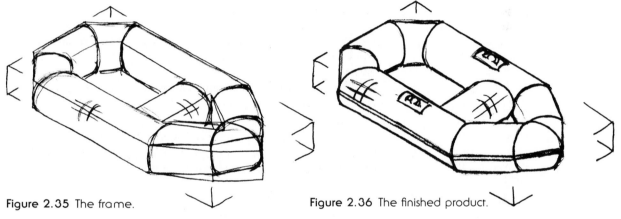

Figure 2.35 The frame.

Figure 2.36 The finished product.

The frame step (Figure 2.35) includes rounding out the tubes and tipping up the bow of the boat. The finish step (Figure 2.36) consists of adding details such as the fender strip around the outside and oarlocks, as well as darkening important lines. Again, note that all the construction lines remain and aid in your understanding of the depicted shape.

In making your sketches, deliberately build a foundation, frame in the shape, and then finish with the necessary details. Don't try to make the finished sketch first. During all three steps, sketch each step in much the same manner as we suggested for lines and curves: try something lightly and then correct it.

If you don't already use the boxing-in (foundation) technique illustrated here, try it. Boxing in is defining, in a simple way, the space (volume) to be occupied by the object being sketched. This foundation will provide reference points and lines from which you can frame the object.

PROBLEMS

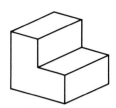

Prob. 2.21

Prob. 2.23

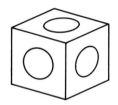

Prob. 2.25

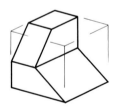

Prob. 2.27

2.21 Make an oblique sketch of the simple shape shown. It is made from a 1-in. cube.

2.22 Make a perspective sketch of the shape in Problem 2.21.

2.23 Make an oblique sketch of the shape shown (made from a 1-in. cube).

2.24 Make a perspective sketch of the shape in Problem 2.23.

2.25 Make an oblique sketch of the shape shown (made from a 1-in. cube).

2.26 Make a perspective sketch of the shape in Problem 2.25.

2.27 Make an oblique sketch of the shape shown (made from a 1-in. cube).

2.28 Make a perspective sketch of the shape in Problem 2.27.

2.29 Make isometric sketches of the following shapes.

    a. Oblique cylinder.

    b. Cylindrical shell.

    c. Oblique cone.

    d. Torus.

    e. Right five-sided prism.

2.30 Sketch an oblique view of a new automobile of your choice. You may refer to our drawing, a photograph or the real thing.

2.31 Sketch an isometric view of the automobile in Problem 2.30.

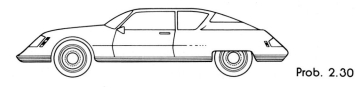

Prob. 2.30

2.32 Sketch an isometric view of a dump truck (or trailer, or bus).

2.33 Sketch an isometric view of a boat (or ship).

2.34 Sketch an isometric view of an open door (45 deg open). Sketch two additional positions, 30 and 90 degrees.

2.35 Sketch an isometric view of bicycle handlebars.

2.36 A strap of metal is twisted 90 deg. Sketch an isometric of the strap. There is a small hole in both ends.

Prob. 2.36

2.37 A cylinder is 100 mm high and 50 mm in diameter. Pass a plane through the cylinder, inclined at 45 deg. The plane contains the center of the cylinder. Now remove the top half of the cylinder. Make an isometric sketch of the remaining portion of the cylinder.

2.38 An equilateral tetrahedron, 80 mm on an edge, rests on a flat horizontal surface. Pass a horizontal plane midway through the tetrahedron and remove its top. Make an isometric sketch of the truncated tetrahedron.

Make isometric sketches of the following objects.

2.39 Pipe elbow.

2.40 U-tube manometer.

2.41 Simply supported I-beam.

2.42 Circular arch.

2.43 Bell-shaped shell.

2.44 Make a perspective sketch of the building in which you live.

2.45 Make a perspective sketch of a street intersection.

2.46 Make a perspective sketch of a room where you live and its contents.

2.47 For each of the following figures take a thin slice from the solid perpendicular to the z-axis. Sketch this element.

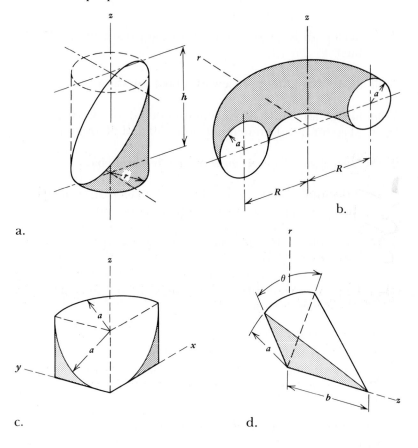

a.

b.

c.

d.

2.48 For each of the shapes shown in Problem 2.47, remove a cylindrical shell from the solid with the z-axis being the axis of the shell. Sketch the shell using a pictorial view that best shows the shape.

Make a pictorial sketch of a device that clearly shows both what the device looks like and how it operates.

2.49 Paperclip.

2.50 Zipper.

2.51 Cork screw.

2.52 Carpenter's hammer.

2.53 Belt buckle.

2.54 Scissors.

3

DESCRIPTIVE GEOMETRY

3.1 PROJECTION

Descriptive geometry is the geometry of describing three-dimensional objects. This chapter presents the fundamentals of descriptive geometry. They are the bases of several important tools in engineering communication and analysis.

The problem of describing three-dimensional objects using a two-dimensional medium, which is usually paper, is solved through the use of *projection*. The elements of projection are shown in Figure 3.1. Given a three-dimensional object and the location from which that object is viewed, your *point of sight,* a transparent plane can be inserted between your point of sight and the object. Your eyes are the point of sight. Using lines that go from points on the object to your eyes, you can draw what you see as a picture on that transparent plane. Projection is the process of recording on a transparent plane what we see as we view an object through that transparent plane. You are projecting how you see that object on the *projection plane.* The situation shown in Figure 3.1 includes *lines of sight* that diverge from your point of sight.

Descriptive geometry utilizes a specific and different situation, shown in Figure 3.2. All the lines of sight are parallel, which is another way of saying that your point of sight is at infinity, and the projection plane is perpendicular to all the lines

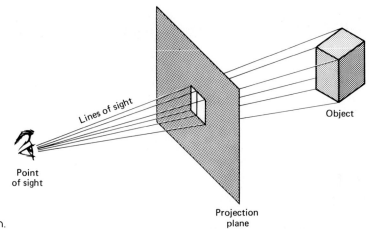

Figure 3.1 Elements of projection.

of sight, not just one line of sight. This particular form of projection is called *orthogonal projection* or *orthographic projection,* and it is the basis of descriptive geometry. In orthogonal projection, lines parallel to the projection plane are projected in true perspective; that is, these projected lines will be the same length as those of the object. This property is very important; taking the measurements of these projected lines will be the equivalent of taking the measurements of the object. Unless we state otherwise, when we discuss the projection of an object or points onto a projection or viewing plane, we are assuming (1) that it is orthogonal projection, (2) that all lines of sight are parallel, and (3) that the projection plane is perpendicular to our lines of sight.

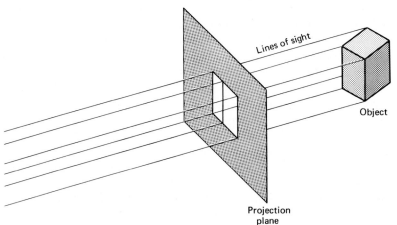

Figure 3.2 Orthogonal projection.

3.2 FUNDAMENTALS OF DESCRIPTIVE GEOMETRY

In descriptive geometry we will make use of three particular projection planes: (1) the *horizontal projection plane*, which is parallel to a horizontal surface; (2) the *frontal projection plane;* (3) and the *profile projection plane*. Both the frontal plane and the profile plane are vertical planes, one being perpendicular to your horizontal sight and the other parallel to it. Figure 3.3 shows these three planes with respect to lines of sight and with respect to each other.

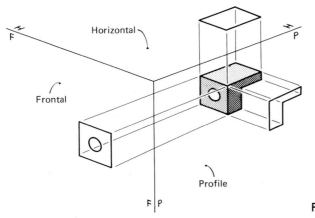

Figure 3.3 Horizontal, frontal, and profile planes.

For our discussion, and for problems in descriptive geometry, we will need to define points and lines. Points in space will be specified with capital letters, $A, B, C, \ldots$, and projection planes will be specified with capital letters or numbers such as $H, F, P, 1, 2, 3$. $H, F,$ and P refer to the horizontal, frontal, and profile planes, respectively. If a point lies in a particular plane, it will usually carry the nomenclature for the plane as a subscript, A_H, A_F, B_1, C_2, etc. Line segments will be called by the points that define or limit the line segment, such as line $A–B$, line $C–D, \ldots$, or more simply line AB, line CD, etc. Figure 3.4 shows three projections of line AB. Lines may also be called by lowercase letters $(a, b, c, \ldots, m, n, \ldots)$ or as the line of intersection between two planes. For example, line $H–F$ is the line of intersection shared by the horizontal and frontal planes, and line A–1 is the line of intersection shared by plane A and plane 1. This nomenclature differentiates a line designated by two endpoints and a line designated by the intersection of two planes, but a line is still a line.

Consider the horizontal projection plane H and the vertical projection plane F shown in Figure 3.5a. These two projection

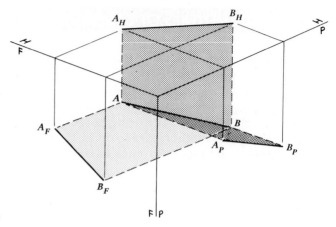

Figure 3.4 Projections of line *AB* in the *H*, *F*, and *P* planes

planes are perpendicular to each other. Point *A* is projected on projection plane *H* using the line of sight labeled *f*. Point *A* is also projected on projection plane *F* using the line of sight labeled *h*. The lines of sight *h* and *f* are also projected on their respective planes, forming a rectangle. Now, if we take the two projection planes in Figure 3.5*a* and unfold them about their intersection as in Figure 3.5*b*, they will be parallel to each other and in the same plane. The result is shown in Figure 3.5*c*. The line labeled *H–F* represents the intersection of the two projection planes. If you are looking at the *H* projection, this line represents the edge view of the *F* projection plane. If you are looking at the *F* projection, this line represents the edge view of the *H* projection plane. Figure 3.5*c* represents in two dimensions what is actually two views of point *A* on the perpendicular projection planes.

The horizontal projection plane and another vertical projection plane labelled 1 are shown in Figure 3.6*a*. Planes *H* and 1 are perpendicular to each other, and the respective projections of point *A* are made using lines of sight *f'* and *h'*. When these two views are unfolded (Figure 3.6*b*), the result is shown in Figure 3.6*c*, where the line labelled *H–1* represents the intersection of the two projection planes. Again, when you are looking at the *H* projection, line *H–1* represents the edge view of projection plane 1, and when you are looking directly at projection plane 1, line *H–1* represents the edge view of the *H* projection plane.

The two situations represented in Figures 3.5*a* and 3.6*a* are combined now in Figure 3.7. The two previous two-dimensional representations are combined to give the projections shown in Figure 3.8, which represents three views of point *A*. Projection planes *F* and 1 are both perpendicular to projection plane *H*, and the *H–F* and *H–1* lines represent the intersections of the perpendicular projection planes. The line from A_H to A_F is per-

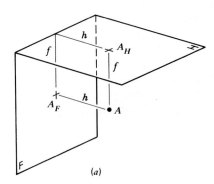

(a)

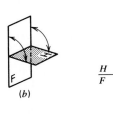

(b)

(c)

Figure 3.5

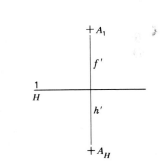

(a)

(b)

(c)

Figure 3.6

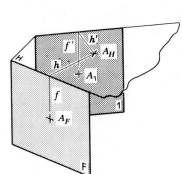

Figure 3.7

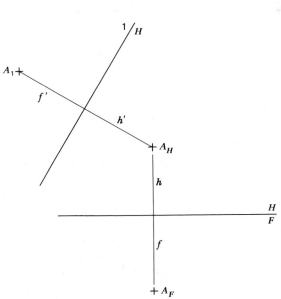

Figure 3.8

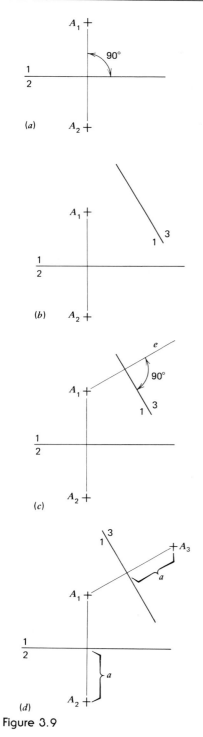

(a)

(b)

(c)

(d)

Figure 3.9

pendicular to the H–F line, and the line from A_H to A_1 is perpendicular to the H–1 line. Note also that the line labeled f is the perpendicular distance from point A to the H projection plane, and the line labeled f' is also the perpendicular distance from the H projection plane to point A; f and f' are the same length.

Figure 3.9a shows two views of a point A in planes 1 and 2 that are perpendicular to each other. Now, we wish to find the projection of point A in plane 3, where projection plane 3 is defined in Figure 3.9b. Since the projection appears in plane 1 as a line, it is also perpendicular to plane 1. This means that projection planes 2 and 3 are both perpendicular to projection plane 1. From what we have previously seen, we know that the projection of point A in plane 3 is somewhere along line e, shown in Figure 3.9c, and line e is perpendicular to the intersection of planes 1 and 3. We also know, as shown in Figure 3.9d, that the distance from projection plane 1 to point A would appear in plane 2 and plane 3. This distance is labeled a. We now have a way to construct a third projection of a point when we have two projections of that point on *orthogonal planes*. The term *orthogonal* means at right angles, so orthogonal planes are at right angles to each other. The two projection planes that are both perpendicular to a third plane are also referred to as the adjacent planes of the third plane.

Refer again to Figure 3.9. If plane 1 were horizontal, we would call it plane H. Then plane 2 would be vertical, and we could name it plane F or P. Also, plane 3 could be reidentified as plane 1 or any other number. Now, planes F and 1 are both adjacent to plane H. With this possible change in nomenclature, the projection process is unchanged.

We are now ready to develop the *two fundamental principles of orthogonal projection* for the representation of three dimensions in two.

1. *The projections of a point in space onto two orthogonal projection planes (two perpendicular planes) lie on a line that is perpendicular to the line of intersection of the two planes.*

2. *When two projection planes are both orthogonal to a third projection plane, the distance that a point lies from the third projection plane can be seen twice, as the distance along the projection line in either of the first two projection planes.*

To review the two fundamental rules of orthogonal projection, consider the five views of point A given in Figure 3.10. To start a problem using descriptive geometry, you must have, or be able to establish, two orthogonal views of the object in question.

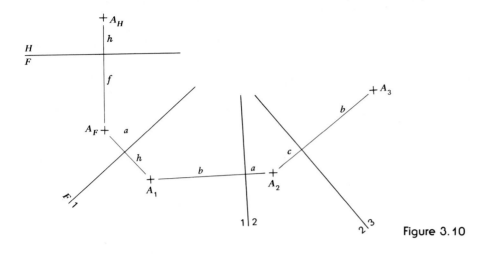

Figure 3.10

For this case, assume that the *H* and *F* views of point *A* are known. Then the problem is to find the projection of point *A* in plane 1, where projection plane 1 is perpendicular to projection plane *F*. Plane 1 is only seen as line *F*–1, since it is perpendicular to the *F* plane. The first step is to construct a line perpendicular to this line of intersection *F*–1 from point A_F. The projection of point *A* in plane 1 will lie on this line. The distance *h* represents the distance that point *A* lies behind the *F* plane. Recall the second fundamental principle of orthogonal projection, that distance *h* can also be viewed in projection plane 1. In a similar manner, you can project point *A* into projection planes 2 and 3. Note that planes *H* and 1 are perpendicular to *F*, planes *F* and 2 are perpendicular to plane 1, and planes 1 and 3 are perpendicular to plane 2.

You can follow this procedure for any number of points and through any number of orthogonal projection planes. Geometrically, two points determine a line and three points determine a plane. If the orthogonal projection involves many points, planes, and lines, and many projection planes, consider the projection of one point at a time, on any three projection planes where two planes are adjacent to a third. This is the basic structure of descriptive geometry, and all problems in descriptive geometry can be broken down to the projection of one point at a time.

SAMPLE PROBLEM 3.1

Project line *AB* successively into projection planes 1 and 2 in *(a)*.

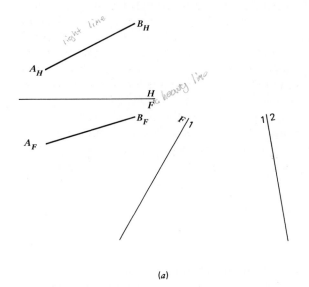

(a)

Solution

Line *F*–1 represents the intersection of the frontal plane and projection plane 1. When you are looking at the frontal plane, line *F*–1 represents the edge view of projection plane 1. When you are looking directly at projection plane 1, line *F*–1 represents the edge view of the frontal plane.

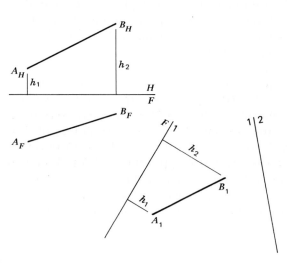

(b)

To project line *AB* into projection plane 1, first draw lines from A_F and B_F perpendicular to the line of intersection *F–1*, as in *(b)*. Now, locate points A_1 and B_1. Remember the second of the two fundamental principles of orthogonal projection: when two projection planes are both orthogonal to a third projection plane, the distance that a point in space lies from the third projection plane can be seen twice, as a projection line in either of the other two projection planes. In this case, the distances h_1 and h_2 can be seen in both projection plane 1 and the horizontal plane.

Similarly, line 1–2 represents the intersection of projection planes 1 and 2, as in *(c)*. In projecting line *AB* from plane 1 to plane 2, the frontal projection distances f_1 and f_2 can be seen in both the frontal plane and projection plane 2.

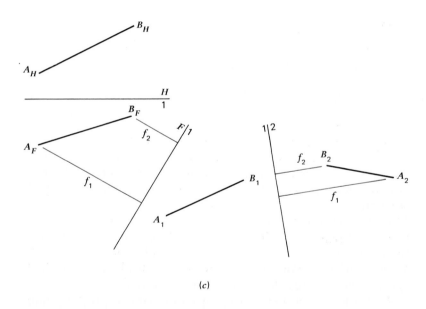

(c)

Note that these projections are the projection of frontal lines, and they represent the distance from points *A* and *B* to projection plane 1. Note also that projection lines from A_1 to A_2 and B_1 to B_2 are perpendicular to the line representing the intersection of planes 1 and 2.

SAMPLE PROBLEM 3.2

Project plane *PQR* successively into projection planes 1 and 2 in *(a)*.

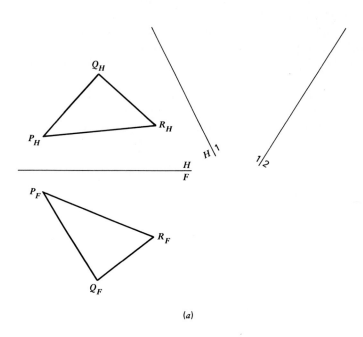

(a)

Solution

Line *H*–1 represents the intersection of the horizontal plane and projection plane 1. When you are looking at the horizontal plane, line *H*–1 represents the edge view of projection plane 1. When you are looking directly at projection plane 1, line *H*–1 represents the edge view of the horizontal plane.

To project plane *PQR* into projection plane 1, project points *P, Q,* and *R* into plane 1 individually, as in *(b)*. The frontal projection·distances f_1, f_2, and f_3 can be seen twice, in both the frontal plane and projection plane 1.

To project plane *PQR* into projection plane 2, project points *P, Q,* and *R* from plane 1 to plane 2, as in *(c)*. The horizontal projection distances h_1, h_2, and h_3 can be seen in both the horizontal plane and projection plane 2. Note that these projections are the projections of horizontal lines, and they represent the distance from points *P, Q,* and *R* to projection plane 1. Note also that projection lines from P_1 to P_2, Q_1 to Q_2, and R_1 to R_2 are perpendicular to the line of intersection of planes 1 and 2.

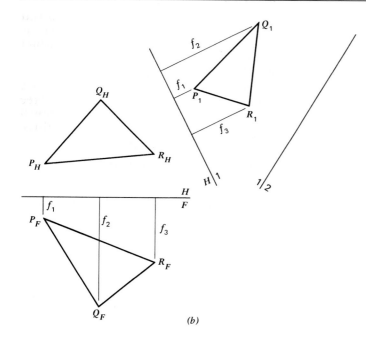

(b)

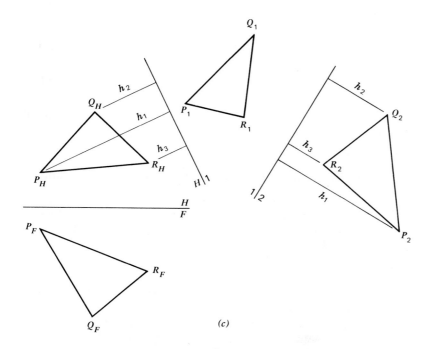

(c)

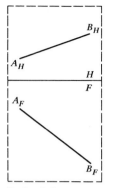

Prob. 3.3

PROBLEMS

3.3 Sketch an isometric drawing that shows line AB, the two projections of line AB, and the two projection planes, as they all appear in three-dimensional space. The dashed lines indicate the limits of the two projection planes.

3.4 Using only a pencil, sketch an isometric drawing showing line AB, the three projection planes, and the three projections of line AB as they all appear in three-dimensional space. The dashed lines indicate the limits of the three projection planes.

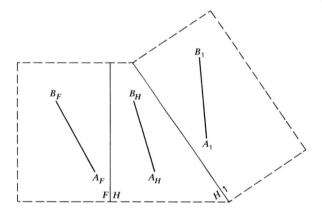

3.5 Refer to the three views of points A, B, and C.
  a. Which point is farthest from the F projection plane?

  b. Which point is highest?

  c. Indicate the horizontal distance between points A and C. Label the distance h.

  d. Indicate the vertical distance between points B and C. Label the distance v.

  e. Point D is 1 in. directly below point A. Show point D in all three views.

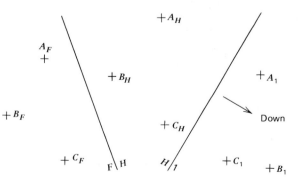

3.6 Trace this problem on a separate sheet of paper. Complete the 2 and 3 views of point A and line BC. After removing the portion of the sheet indicated, fold the paper to put the four projection planes in their proper spatial relationship with one another. Then satisfy yourself that you can locate the actual position in space of point A and line BC.

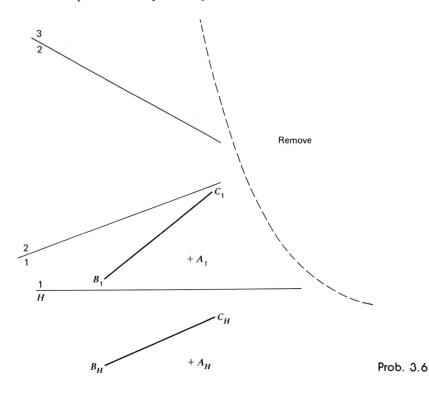

Prob. 3.6

3.7 Complete the three views of points A, B, and C.

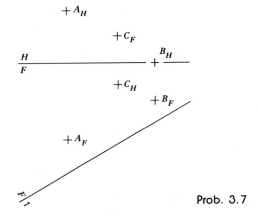

Prob. 3.7

3.8 Complete all views of the two intersecting planes *ABC* and *ABD*. Trace and cut out the object and then view it in space corresponding to the views.

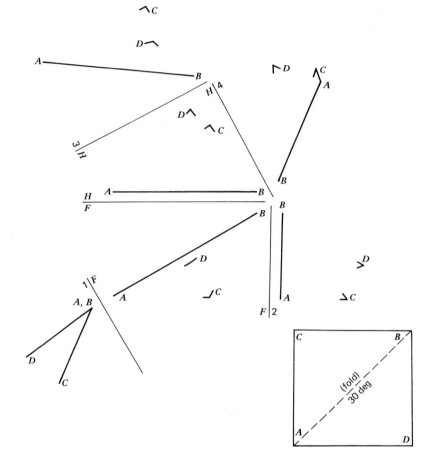

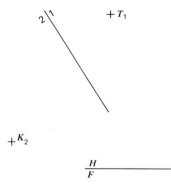

$+T_1$

3.9 Determine the *H*, *F*, 1, and 2 views of line segment *KT*.

$+K_2$

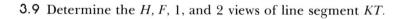

$+K_H$

$+T_F$

3.10 Determine the H, F, 1, and 2 views of line segment AB.

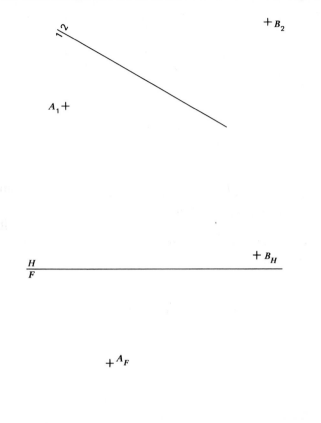

3.3 BASIC GEOMETRIC RELATIONSHIPS

Before beginning any discussion of geometric problems, we should review a few basic ideas about points, lines, planes, and their relationship. Most of this should have remained with you from plane and solid geometry.

In engineering, perpendicular and parallel lines and planes deserve a bit of special thought. It is self-evident that lines can be parallel to lines and planes can be parallel to planes, but it may not be as obvious that a line can be parallel to a plane. Perpendicularity is a special relationship that implies the existence of a 90 deg angle. Perpendicular lines are lines that form a 90 deg angle, and perpendicular planes are planes that meet at 90 deg. A line can also be perpendicular to a plane. Perpendicularity is important because it occurs so frequently in engineering. Buildings are constructed so that loads are carried vertically. Bodies fall perpendicular to the surface of the earth. To put it bluntly but succinctly, engineers deal with a square world.

You should have 10 basic geometric concepts.

1. Two points determine a straight line (Figure 3.11).

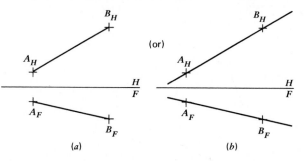

(or)

(a)   (b)

Note: (a) and (b) are interpreted identically.

Figure 3.11 The end of a line may or may not be defined by points.

2. A straight line appears as a straight line, however it is viewed (Figure 3.12).

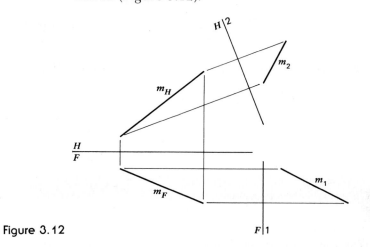

Figure 3.12

3. The true length or actual length of a line will show in a view where the projection plane and the line are parallel (Figure 3.13).

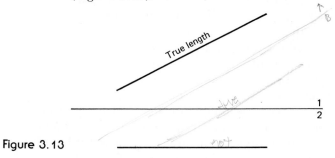

Figure 3.13

4. The orthogonal projections of parallel lines will also be parallel lines. In special cases, parallel lines may appear as one line or two points (Figure 3.14).

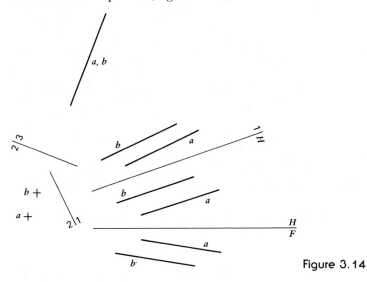

Figure 3.14

5. Intersecting lines have a point in common (Figure 3.15). _Same in all views_

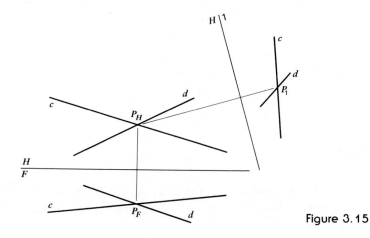

Figure 3.15

mais look || in 1 view

6. Skew lines are lines that are neither parallel nor intersecting (Figure 3.16).

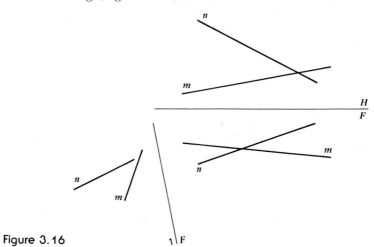

Figure 3.16

7. If two lines are perpendicular, they will appear to be at right angles in any view in which one or the other is in true length. In a view in which neither line is in true length, perpendicular lines will not appear to be at right angles to each other (Figure 3.17).

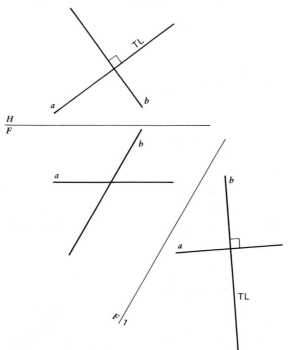

Figure 3.17

8. A plane is defined by three points, two intersecting lines, two parallel lines, or a line and a point, if that point does not lie on the line (Figure 3.18).

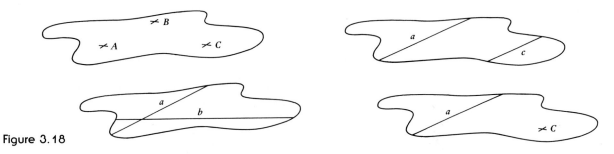

Figure 3.18

9. A line is parallel to a plane if the line is parallel to any line that lies in the plane (Figure 3.19).

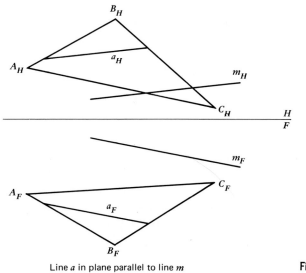

Line *a* in plane parallel to line *m*

Figure 3.19

10. A line is perpendicular to a plane if that line is perpendicular to two intersecting lines that lie in the plane. Perpendicular lines don't necessarily have to intersect (Figure 3.20).

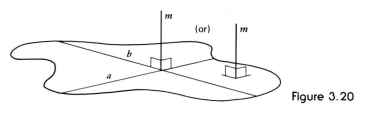

Figure 3.20

The basic concepts presented in Figures 3.11 to 3.20 are fundamental to working descriptive geometry problems. Study them carefully. Keep these 10 ideas in mind as you study the sample problems in this chapter.

SAMPLE PROBLEM 3.11

Three points A, B, and C define a plane (a). Point D_H is shown in the horizontal view. Find frontal view D_F if point D also lies on the plane.

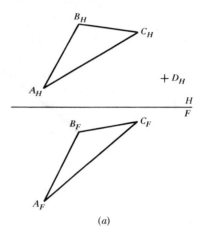

(a)

Solution

Two intersecting lines define a plane. In the horizontal view, construct intersecting lines AC and BD, as in (b). Point P is their point of intersection, and it is unique. Project point P into the frontal view. Both points P and D are on line BP. Point D_F is located by extending line BP to the projection line of D in the frontal view, as in (c).

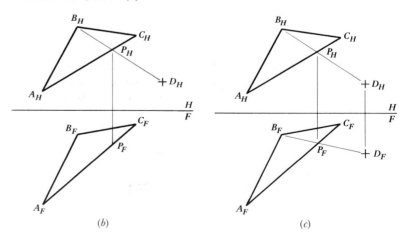

(b) (c)

SAMPLE PROBLEM 3.12

Lines a and b are skew lines. They do not intersect, and they do not lie in the same plane. Construct a plane that is parallel to both lines a and b and contains point P.

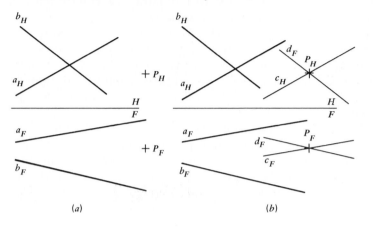

(a) (b)

Solution

Solving the problem is much simpler than it may appear at first reading. Construct line c parallel to line a through point P. Remember, parallel lines will be parallel in all views of those lines. Now, construct line d parallel to line b through point P. The problem is solved. Lines c and d define a plane, and lines a and b are both parallel to that plane because they are each parallel to a line in the plane.

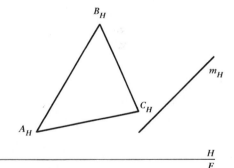

SAMPLE PROBLEM 3.13

Define a plane that contains line m and is also perpendicular to plane ABC, shown in (a).

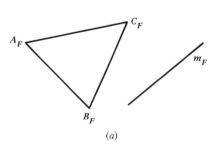

(a)

Solution

An infinite number of planes can contain line m. To define a particular plane that contains line m and is also perpendicular to plane ABC, we must establish perpendicularity with some line other than m. Call this line n. In order for m and n to define a plane, the lines would have to intersect.

Construct a frontal line CD in both views, as in *(b)*. Note that CD is in true length in the frontal view. Now, construct line n to intersect line m and be perpendicular to line CD in the frontal view. P is the point of intersection of lines m and n, and it can be projected into both views. The location of point P on line m is arbitrary.

Now, construct a horizontal line AE, as in *(c)*. In the horizontal view, line n should also appear as perpendicular to line AE. Since line n is perpendicular to both lines CD and AE in plane ABC, line n is perpendicular to plane ABC. The plane defined by lines m and n is therefore perpendicular to plane ABC.

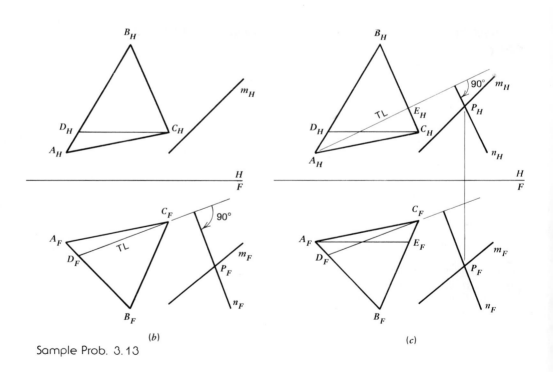

(b)   *(c)*

Sample Prob. 3.13

PROBLEMS

3.14 Intersecting lines r and s define a plane. Point D also lies in that plane and point D_H is known. Locate point D_F.

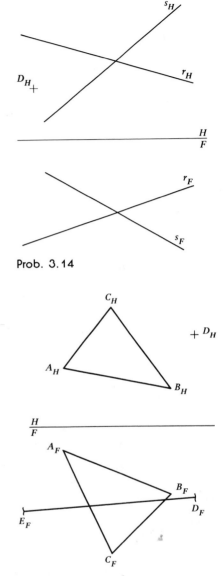

Prob. 3.14

3.15 Plane ABC and line DE are parallel. Locate point E_H.

3.16 a. Construct a line b perpendicular to the horizontal line a and through point P.

b. Construct a line c parallel to the horizontal line a and through point P.

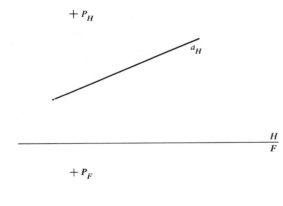

Prob. 3.15

Prob. 3.16

3.17 Construct a line through point D that is perpendicular to plane ABC.

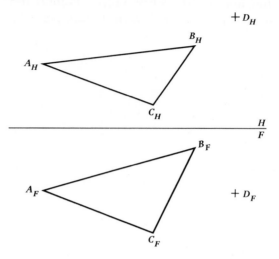

3.18 Point P and line m form a plane that is parallel to line n. Locate point P_F.

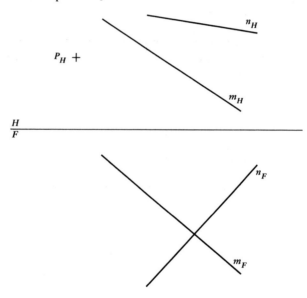

3.19 Point *A* and line *m* form a plane that is parallel to the plane formed by point *B* and line *n*. Locate points A_F and B_H.

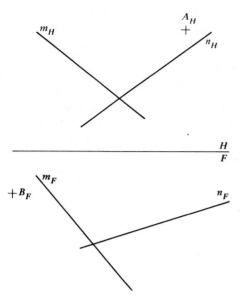

3.20 *A, B, C*, and *D* define a plane parallelogram. Construct a line *m* that is perpendicular to plane *ABCD* and passes through its geometric center.

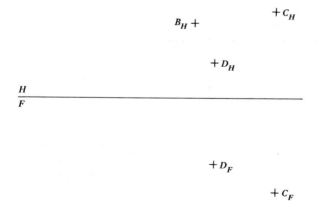

3.4 VISIBILITY OF LINES

Sometimes a line may be all or partially obscured by a plane or a solid. Visibility refers to the problem of distinguishing lines that are in full view and lines that are obscured from view. For example, in Figure 3.21 a block is represented two ways: (1) an isometric view that completely and clearly describes the shape and (2) three orthogonal views. In the orthogonal views are important lines that cannot actually be seen but are included for clarity. These lines are called *hidden lines*. The conventional way to represent hidden lines is by using dashed lines. The orthogonal views in Figure 3.21 would be inadequate to represent the shape if the hidden lines were not included. Establishing the visibility of lines when making views of objects and understanding or reading the visibility of lines when using views of objects are essential graphic tools.

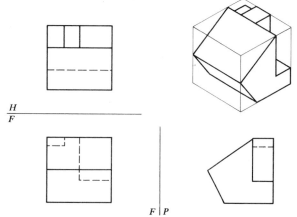

Figure 3.21

The visibility of lines can be determined directly from two orthogonal views. Consider the two skew lines *m* and *n* shown with the *H* and *F* views in Figure 3.22. In the *H* view they appear to intersect, but we can see that they really do not because the apparent intersections in the frontal view and the horizontal view are not the same point. If they did intersect, the point of intersection would be a unique point, and if it were established in one view it could be projected to all views. Lines *m* and *n* are skew lines—lines that are neither intersecting nor parallel. The apparent intersection in the horizontal view is labeled *A;* it represents two points, one on line *m* and one on line *n*. One of the two points is directly above the other. In the horizontal view they appear as one point, but they appear as two different points in the frontal view. Since we know that the *H–F* line represents the

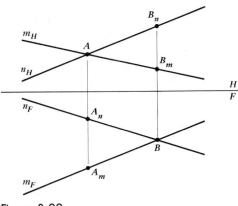

Figure 3.22

edge view of the horizontal plane when looking at the frontal view, we can see that point A_n is closer to the horizontal plane than is A_m. If lines m and n were part of an object we were viewing, we would conclude that in the horizontal view point A_n is closer to our eye than A_m and we could then use this fact to establish the visibility in the horizontal view. Likewise, in the frontal view the apparent intersection labeled B actually represents two points, B_m and B_n, shown in the horizontal view. If we were establishing the visibility of an object containing lines m and n in the frontal view, we could use the fact that B_m is closer to our eye than B_n. This reasoning is based on the fact that we always assume that the viewing plane is between our point of sight, which is our eye, and the object of interest.

In your professional life, you will probably read more drawings than you prepare. The visibility of lines is a direct application of the two principles of orthographic projection. You can use these principles to establish the geometric location of two points in space. It requires no special aptitude for spatial perception.

SAMPLE PROBLEM 3.21

The figure (a) presents three views of two intersecting planes. Plane ABC and plane ABD have a common line AB. These three views are complete with the exception of showing correct visibility of the two planes. Complete the views by determining which lines are hidden and which lines are not.

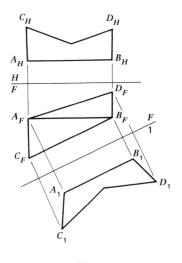

(a)

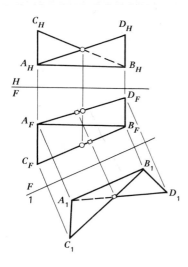

(b)

Solution

The three views are repeated in (b). The apparent intersection of lines AD and CB in the horizontal view is identified in the frontal view with small circles. We now can see that the point on line AD is closer to the H projection plane than is the point on CB. Therefore, in the general location of the apparent intersection in plane H, AD is closer to our eye and will cover up line CB, as shown in the horizontal view with the dashed line.

The apparent intersection of lines AD and BC in plane 1 is identified with a small square, and these two points are also identified with small squares in the frontal view. The point on line CB is closer to projection plane 1 than is the point on line AD. Therefore, in the area where there is an apparent intersection in projection plane 1, line CB is closer to our eye than is line AD. The hidden portion of line AD must be dashed. Actually, identifying points in views as we did with the small circles and squares is not necessary, but it illustrates how you can read the views to establish visibility.

SAMPLE PROBLEM 3.22

Three views of a pyramid-shaped object are shown in (a). The object is a tetrahedron, a solid bounded by four triangular-shaped planes. In this figure, visibility has not yet been established. Complete the three views showing correct visibility.

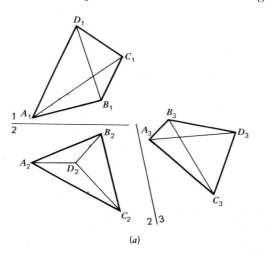

(a)

Solution

To establish visibility in plane 1, consider the apparent intersection of AC and DB in that view identified with a small circle. This circle again identifies two points, one on line DB and the other on line AC, as shown in *(b)*. From view 2 we can see that the point on line DB is closer to projection plane 1. Therefore, line DB should be solid and line AC should be dashed.

An identical process can be used to establish the visibility in view 3, shown in *(b)*. An apparent intersection of BC and AD is identified in view 3, and view 2 shows that BC is closer to our eye than AD. Therefore, AD should be dashed. In view 2 there are no apparent intersections to work with, but by extending line DC so that it does apparently intersect line AB, we can establish visibility as before. Two points represented by this apparent intersection in view 2 are identified in view 1 in *(c)* with small circles.

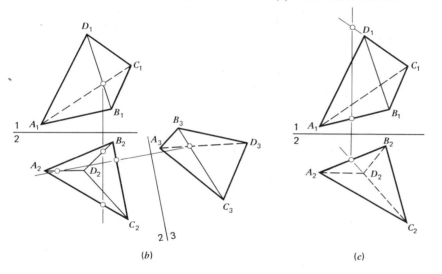

(b)   (c)

Of these two points, the one on line AB is closer to projection plane 2 than is the point on line DC. We conclude that the extended line DC is much farther from our eye, indicating that DC should be dashed. Because we know the shape of the object, it follows that AD and BD in view 2 must also be dashed. Part *(d)* shows the completed views. Note how much easier it is to read the views when correct visibility is shown.

The visibility in view 2 can actually be established without going through this formal process. Since it can be seen that point D is much farther from plane 2 than is plane ABC, you can conclude directly that ABC is closer to your eye than is point D; therefore, the visibility that has been established for view 2 must be true. This approach does require some aptitude for spatial perception. On the other hand, the formal construction process will always work.

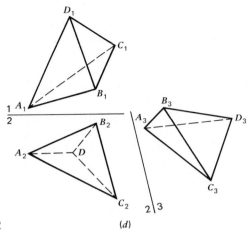

Sample Problem 3.22

(d)

PROBLEMS

3.23 Lines AB, BC, and AC define a triangular-shaped plane. Points D and E are the two ends of a line that does not intersect plane ABC.

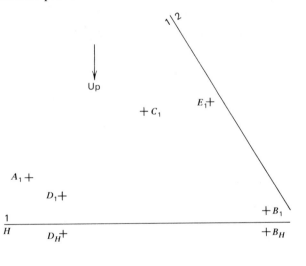

a. Complete the three views and show correct visibility in all views.

b. Label the vertical distance (rise) between points A and C as v.

c. Label the horizontal distance (run) between points A and C as h.

d. Which is the highest of the five points?

e. Which of the five points is farthest from plane 2?

3.24 Lines AB, BC, and AC define a triangular-shaped plane. Points D and E are the two ends of a line that does not intersect plane ABC.

a. Complete the three views and show correct visibility in all views.

b. Label the vertical distance (rise) between points C and D as v.

c. Label the horizontal distance (run) between points C and D as h.

d. Which is the lowest of the five points?

e. Which of the five points is closest to plane 1?

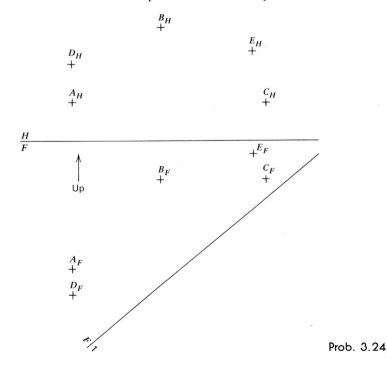

Prob. 3.24

3.25 Establish visibility using dashed lines in the three views of the tetrahedron.

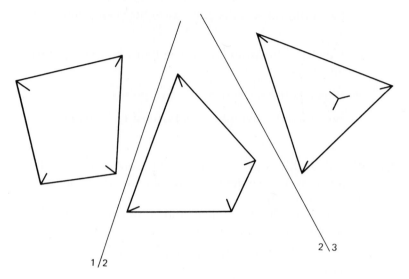

3.26 Complete the two views of the tetrahedron. Show visibility using dashed lines.

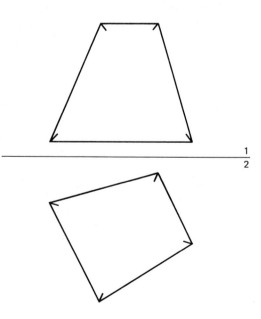

3.27 In both views, show visibility of the two intersecting planes.

3.28 Lines *AA'*, *BB'*, and *CC'* intersect the corners of plane *ABCD*. Show the correct visibility in both views.

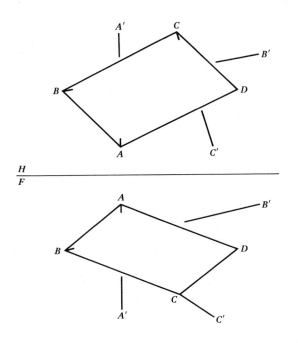

3.29 Lines a, b, and c are the centerlines of small diameter rods. Complete the three views treating the three rods as solids. Omit the usual dashed lines for hidden lines.

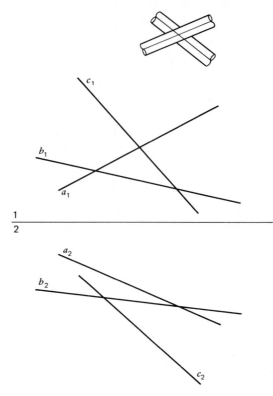

3.30 The intersection points of lines p, q, and r with the surface of the cylinder can be seen in the H view. Complete the F view and the 1 view showing intersection points and visibility. Omit dashed lines for that portion of the lines internal to the cylinder.

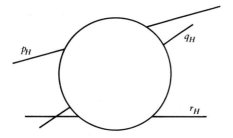

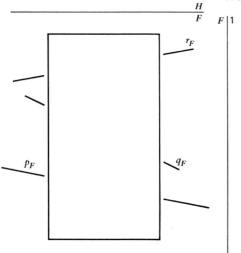

3.5 THE THREE BASIC PROBLEMS

All the geometric shapes used in engineering can be described with points and lines. Even curved lines and warped surfaces can be described by a series of points. Naturally, the number of points needed to describe a warped surface would depend on the accuracy desired in the description.

Most of the problems in descriptive geometry reduce to two fundamental questions: what is the true distance between two points, and what is the true perception of a shape? Descriptive geometry and orthogonal projection exist and are used to answer these two questions, and they are used to communicate the answers between engineers.

Descriptive problems may be involved or simple; they may be obvious or disguised. You have the tools to draw projections of shapes and objects provided you have available two adjacent (orthogonal) views. Whatever the situation, there are only three basic problems in descriptive geometry: what is the true length of a line, what is the point view of a line, and what is the true shape of a plane surface? The two fundamental rules of orthogonal projection are all that you need to solve these problems.

3.6 TRUE LENGTH OF A LINE

The true length (TL), or actual length, of a line will show in any view that is parallel to that line.

In Figure 3.23a line AB is a frontal line. It appears in the frontal view in true length because line AB is parallel to the frontal plane. We use the frontal projection to measure the length of line AB. Note that point B is lower than point A, so the projection of line AB in the horizontal plane is *not* in true length. In Figure 3.23b line AB is horizontal. It appears in the horizontal view in true length, and it is parallel to the horizontal plane. We use the horizontal projection to measure the length of line AB.

In Figure 3.23c line AB is not in true length in either the horizontal or frontal view. To find the true length of line AB shown in Figure 3.23c, a projection plane that is parallel to line AB must be established. Projection plane 1 in Figure 3.23d is defined by drawing the H–1 line parallel to the H projection of line AB. Remember that when you are observing the H view, line H–1 represents the edge view of plane 1 and is parallel to line AB. Therefore, the projection of line AB in plane 1 is a true-length view of line AB. A view established parallel to the projection of

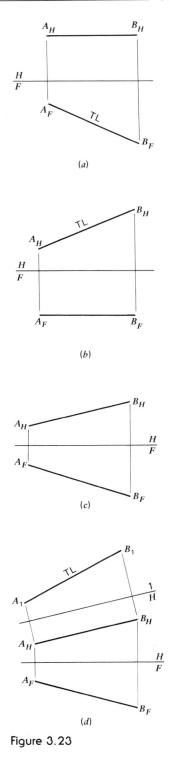

Figure 3.23

line AB in the F plane, which would be perpendicular to the F plane, can also be used to obtain the true length of line AB. As a check on your work, remember that the true-length view is always the longest projection of the line.

3.7 POINT VIEW OF A LINE

When a straight line is viewed looking down the line, it appears as a single point, for a line has no lateral dimension, only length. In order to view a line as a point, the viewing plane would have to be perpendicular to the line. This means that the point-view plane would also be at a right angle to the plane in which the line appears in true length. We will use this idea to solve the second basic problem.

Begin with any two orthogonal views, one of which contains the line in true length. The orthogonal planes H and 1 of Figure 3.23d are repeated in Figure 3.24a. Line AB is viewed in true length in plane 1. If a projection plane is established perpendicular to the true-length view of line AB in Figure 3.24b, the projection plane would also be perpendicular to projection plane 1. This step is done in Figure 3.24b by establishing the line of intersection of planes 1 and 2 perpendicular to the true-length view of line AB. Then, using the two fundamental rules of orthogonal projection, the point view of AB can be established in projection plane 2. Note that everywhere in the H view line AB is the same distance from the line of intersection H–1, which establishes the position of the point that is the point view of line AB in plane 2.

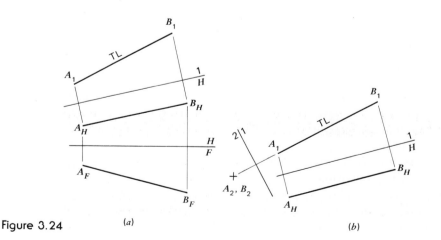

Figure 3.24   (a)   (b)

When you see the point view of a line you will also see the edge view of any plane that contains that line. Consider plane *ABC* described with the *H* and *F* projection planes in Figure 3.25*a*. Plane *ABC* contains line *AB*. To have an edge view of plane *ABC*, we need only find the point view of any line in plane *ABC*. In Figure 3.24*b* the point view of line *AB* is obtained. Point *C* is also projected into planes 1 and 2 resulting in an edge view of plane *ABC* in plane 2. Note that Figures 3.24 and 3.25 are actually the same problem, the only difference being that the object in one is a line and in the other is a plane.

A variation of the problem is shown in Figure 3.26. Plane *ABC* is defined with the *H* and *F* views. A horizontal line *DB*, which is contained in plane *ABC*, is constructed parallel to the horizontal plane. Since line *DB* is parallel to the *H* plane, it will be in true length in the *H* view. Now, by taking the point view of line *DB*, which is in true length in the *H* view, the edge view of the plane is obtained in adjacent plane 1. Note that this is *not* the same plane 1 as in Figure 3.25. This procedure requires one less view.

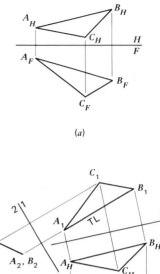

(a)

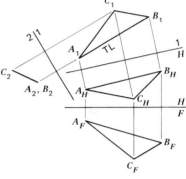

(b)

Figure 3.25

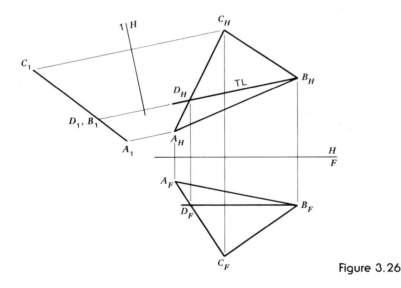

Figure 3.26

3.8 TRUE SHAPE OF A PLANE

The third basic problem of descriptive geometry is to obtain the true shape (TS) of a plane surface. To see the true shape of a plane surface, that plane and the projection plane in which the plane is viewed must be parallel. We will use this idea and extend what we already have.

Begin with any two orthogonal views of a plane surface, one of which contains the edge view of the plane. Figure 3.25*b* is repeated in Figure 3.27*a*. The edge view of the plane surface is seen in plane 2. Now, establish projection plane 3 parallel to plane *ABC*. Projection plane 3 is defined by drawing its line of intersection with plane 2 parallel to the edge view of plane *ABC*. The projection of plane *ABC* into plane 3, as in Figure 3.27*b*, is the true shape of the plane.

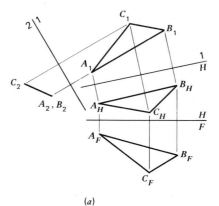

(*a*)

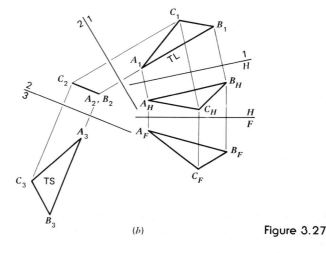

(*b*)

Figure 3.27

The true length of a line requires one additional view beyond the two views given originally. The point view of a line requires two additional views. The true shape of a plane requires three additional views. The point view of a line is obtained from the true-length view, and the true shape of a plane is obtained from a point view of any line in the plane. All problems in descriptive geometry will be variations of these three basic problems.

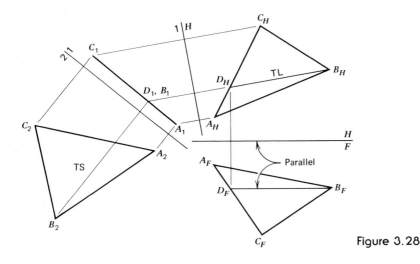

Figure 3.28

The procedure of Figure 3.26 can also be used to find the true shape of plane surface *ABC*. In Figure 3.28, simply establish a new projection plane 2, orthogonal to projection plane 1 and parallel to plane *ABC*. The projection of plane *ABC* into plane 2 is the true shape of the plane. Note again that projection planes 2 and 3 in Figure 3.27 are not the same as projection planes 1 and 2 in Figure 3.28. The three-step process beginning with the true length, then finding the point view, and finally obtaining the true shape is also used here, the difference being that the true-length of line *DB* is established in plane *H* rather than in an adjacent plane.

At this point, we should state that there are other geometric techniques to determine the true length and point view of a line and the true shape of a plane. We have not shown you everything, only the minimum necessary to understand descriptive geometry. If you become interested in techniques, the other ways to solve these geometric problems may interest you.

3.9 ELEMENTARY DESCRIPTIVE GEOMETRY

Descriptive geometry is basically the geometry of points, lines, and planes and their representation. Although there are instances of warped surfaces, most curved surfaces are spherical or cylindrical. If we restrict ourselves to the geometry of points, lines, and planes, elementary descriptive geometry is quite straightforward, and you will gain a working understanding of most geometric problems that you will see in engineering practice.

Sample Problems 3.31 to 3.39 are quite elementary. They are direct applications of the three basic problems and are meant to be read as an integral part of the text, not as a supplement. Read them carefully; Chapter 4 is built with these problems as a foundation.

SAMPLE PROBLEM 3.31

Construct a perpendicular line from point *C* to point *D* on line *AB* in (*a*) and determine its length.

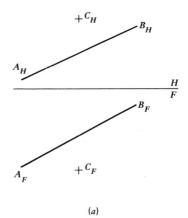

(*a*)

Solution

Perpendicular lines will appear as perpendicular lines in any view in which either line appears in true length. Since line *AB* is not in true length in either the frontal or the horizontal projection plane, in (*b*) place a new projection plane 1 parallel to line *AB* and project line *AB* and point *C* into plane 1. Now, construct perpendicular *DC*. It will appear to be perpendicular to line *AB* in plane 1, so it should be constructed at right angles to line *AB*. Although line *CD* is perpendicular to line *AB*, *CD* is not in true length.

To find the true length of line *CD* you will need a second projection plane. In (*c*) place a new projection plane 2 perpendicular to the view of line *AB* in plane 1 and project lines *AB* and *CD* into plane 2. In plane 2 line *AB* appears as a point. Line *CD* is in true length.

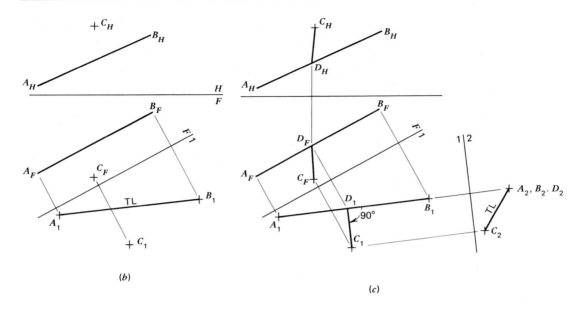

(b)

(c)

Now, project point D back into the frontal and horizontal planes. Note that lines CD and AB do not appear as perpendicular in these views.

SAMPLE PROBLEM 3.32

Determine the minimum distance between two skew lines, line AB and line CD in (a). *(perpendicular only)*

Solution

The minimum distance between two skew lines is the length of a line perpendicular to both line AB and line CD. To measure its length, the perpendicular must be viewed in true length. This view would be possible if either line AB or line CD is viewed in point view. Neither line AB nor line CD is in true length in either of the given views.

Selecting line AB arbitrarily, in (b) place a new projection plane parallel to line AB and project both line AB and line CD into plane 1. Line AB will be viewed in true length in plane 1. In (c) place a second projection plane, plane 2, at right angles to the projection of line AB in plane 1 and project both line AB and line CD into plane 2. Line AB will appear as a point. Perpendicular PQ from line AB to line CD will be in true length in plane 2. Now, project perpendicular PQ back to planes 1, H, and F, as in (d). Remember that line AB is in true length in plane 1, and perpendicular PQ must also appear at a right angle to line AB in

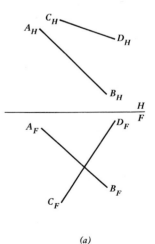

(a)

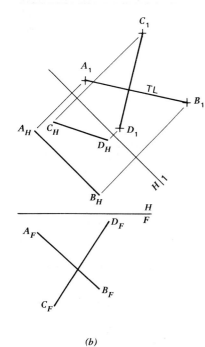

(b)

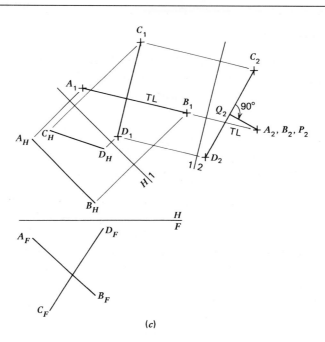

(c)

plane 1. It will not appear at a right angle to line *CD*.

This technique of finding the perpendicular distance between two skew lines is sometimes called the *line method*. In Sample Problem 3.35, a more involved but more versatile technique is used to solve the same problem.

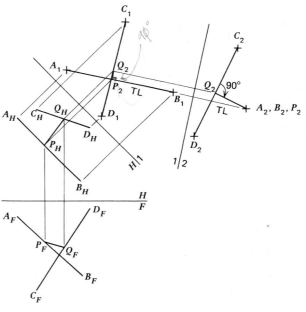

Sample Problem 3.32

(d)

SAMPLE PROBLEM 3.33

The angle between planes is called the dihedral angle. Determine the dihedral angle between plane ABC and plane ABD shown in (a).

Solution

Two planes that are not parallel intersect in a line. In this case line AB is common to both plane ABC and plane ABD. The angle between these planes can be measured by viewing line AB in point view.

Line AB is not in true length in either the horizontal or the frontal view. In (b) place a new projection plane 1 parallel to line AB, and project line AB and points C and D into plane 1. Line AB will appear in true length in plane 1. In (c) place a second projection plane, plane 2, at right angles to the projection of line AB in plane 1 and project line AB and points C and D into plane 2. Line AB will appear as a point. Since this line is common to both planes ABC and ABD, these planes will appear in edge view in plane 2. The dihedral angle between them can be measured, as 85 deg.

Note that in the frontal view plane ABC partially obscures plane ABD, and in plane 1 plane ABD partially obscures plane ABC.

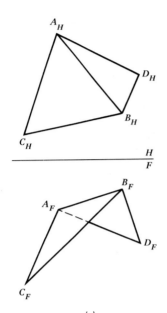

(a)

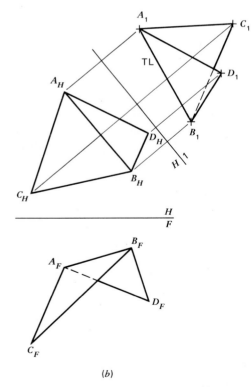

(b)

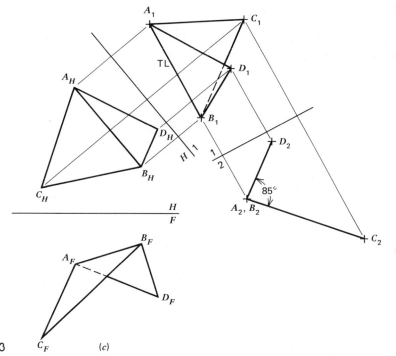

Sample problem 3.33

(c)

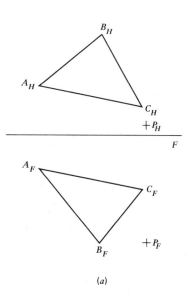

(a)

SAMPLE PROBLEM 3.34

Find the perpendicular from point P to plane ABC in *(a)*. Show this perpendicular in all views.

Solution

The perpendicular distance between a point and a plane can be determined when the plane is viewed in edge view. In *(b)* construct line AD as a frontal line. It will appear parallel to the line of intersection H–F in the horizontal view and in true length in the frontal view. Now, in *(c)* establish plane 1 perpendicular to line AD, and from the frontal view project plane ABC and point P into plane 1. Point Q is where the perpendicular intersects plane ABC. The true length of the perpendicular can be viewed in plane 1.

Finding the length of the perpendicular is comparatively simple. Projecting the perpendicular back to the horizontal and frontal views may take a little more thought, the problem being where to locate point Q. You will recall that if two lines are perpendicular they will appear to be at right angles in any view in which one or the other is in true length Line AD lies in plane ABC, and it is in true length in the frontal view. Perpendicular PQ must be at right angles to line AD in the frontal view. Construct a line from point P perpendicular to line AD and project point Q back to the frontal plane and then to the horizontal plane, as in (c). The frontal view of PQ can also be established by recognizing that $P_F Q_F$ must be parallel to the F–1 line since $P_1 Q_1$ is in true length.

Note that point Q is outside the area bounded by ABC. Is this a problem?

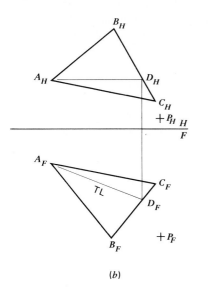

(b)

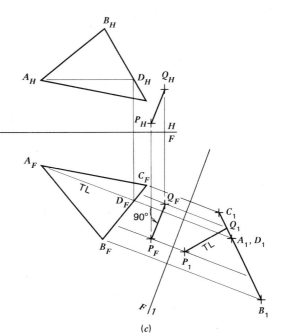

(c)

SAMPLE PROBLEM 3.35

Sample Problem 3.32 is repeated (a). Determine the minimum distance between two skew lines, line AB and line CD.

Solution

The minimum distance between two skew lines is the length of a line perpendicular to both line AB and line CD. To measure its length, the perpendicular must be viewed in true length. This view is possible if either line AB or line CD is viewed in point view, or it would be possible in a particular view in which the projections of line AB and line CD were viewed as parallel lines. Sample Problem 3.32 is the first alternative. Think about the second alternative. Model the circumstances (e.g., with two pencils) when skew lines are viewed as parallel lines. We are going to create this situation.

Neither line AB nor line CD is in true length in either of the given views (a). Selecting line AB arbitrarily, in (b) we will con-

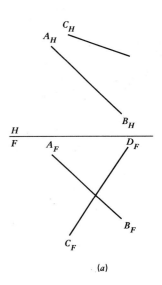

(a)

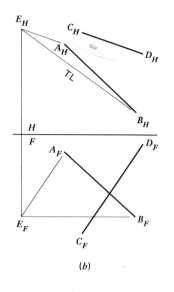

(b)

struct a plane *ABE* parallel to line *CD*. To do this, we must construct a plane that (1) contains line *AB*, and (2) contains a line parallel to line *CD*. Remember that a line is parallel to a plane if it is parallel to any line in the plane and that two intersecting lines define a plane.

To construct plane *ABE* construct horizontal line *BE* and construct line *AE* parallel to line *CD*. It is important that we construct a horizontal line. We will see why in Sample Problem 3.36. Lines *AE* and *BE* intersect at point *E*, which can be determined in both frontal and horizontal views. Note also that line *BE* is in true length in the horizontal view.

Now, establish projection plane 1 perpendicular to the projection of line *BE* in the horizontal view, as in (c), and project plane *ABE* and line *CD* into plane 1. They should appear as two parallel lines. Plane *ABE* will be in edge view since line *BE* in plane *ABE* will be in point view. The perpendicular distance *a* can be measured directly.

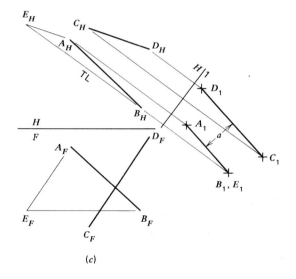

(c)

To locate the perpendicular takes a little more thought. First, it is in true length in plane 1. We are seeing the perpendicular distance from a point to a plane. Second, if we wish to locate that unique perpendicular, we will find it by establishing plane 2 parallel to the projections of lines *AB* and *CD* in plane 1. In plane 2 the projections of lines *AB* and *CD* will both be in true length, and the perpendicular between them will be in point view. Remember, the perpendicular was in true length in plane 1, which means that it will be in point view in plane 2. But where is it? The perpendicular is located where lines *AB* and *CD* appear to intersect. In (d) the perpendicular is projected back to plane 1

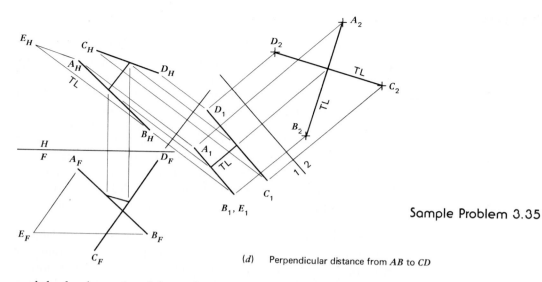

Sample Problem 3.35

(d)   Perpendicular distance from *AB* to *CD*

and the horizontal and frontal planes. This technique of finding the perpendicular distance between two skew lines is called the *plane method*. Note that the connecting line is the same line found in Sample Problem 3.32.

The plane method is more versatile than the line method described in Sample Problem 3.32. Sample Problem 3.36 is an extension of this problem, but it answers an entirely different question: what is the shortest horizontal distance between two skew lines?

SAMPLE PROBLEM 3.36

What is the shortest horizontal distance between the two skew lines, line *AB* and line *CD* of Sample Problem 3.35?

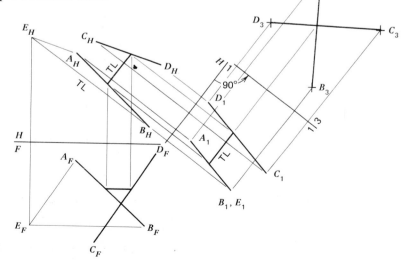

(e)   Shortest horizontal distance from *AB* to *CD*

Solution

In Sample Problem 3.35, a horizontal line *BE* was constructed to form plane *ABE*. Line *BE* is in true length in the horizontal view and in point view when projected into plane 1. Plane 1 is a vertical plane, as is the frontal plane. Very simply, we have rotated our viewing perspective to see the projections of lines *AB* and *CD* as parallel lines, but we still view them from the perspective of a vertical plane. That is, horizontal lines are parallel to the horizontal projection plane, and vertical lines are at right angles to the horizontal projection plane. To maintain this perspective, we constructed plane *ABE* to be parallel to line *CD*, with line *BE* as horizontal.

Using the same construction procedures of Sample Problem 3.35—(*a*), (*b*), and (*c*)—in (*e*) we establish projection plane 3 orthogonal to the horizontal plane. The apparent intersection of lines *AB* and *CD* in plane 3 will be the point view of the shortest horizontal connector between lines *AB* and *CD*. Remember, the shortest horizontal connector would appear in true length in projection plane 1. It is the point where lines *AB* and *CD* appear to intersect in projection plane 3. As a check, the connector should appear as a horizontal line both in projection plane 1 and in the frontal view, and it does.

In a similar manner, we could locate the shortest connecting line at a 20 deg slope, 30 deg slope, etc. We would simply orient a new projection plane 4 to be at right angles to whatever sloping connector is desired. In (*f*) the shortest connector at a 20 deg slope is located. Again, we can determine this because we con-

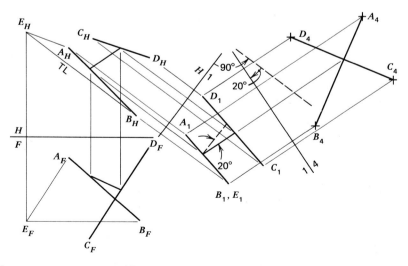

Sample Problem 3.36   (*f*) Shortest connector at 20-deg slope

structed a horizontal line to begin our solution and constructed projection plane 1 to be a vertical plane perpendicular to the horizontal plane.

The true-length connector appears at a 20 deg slope to the horizontal in projection plane 1.

SAMPLE PROBLEM 3.37

Sample Problem 3.31 is repeated *(a)*. Construct a perpendicular line from point C to point D on line AB and determine its length.

Solution

Two intersecting lines define a plane. Since perpendicular DC will intersect line AB at point D, constructing the perpendicular would be a simple matter if we could view the plane that contains point C and AB in true shape.

In *(b)* construct a frontal line from point C to line AB. Note that point E is on line AB but is beyond B from A. Line CE will appear in true length in the frontal view. Now, establish plane 1 at right angles to the projection of line CE in the frontal view. Project point C and line AE into plane 1. Since line CE will appear in a point view, the plane containing AE and C will appear in edge view.

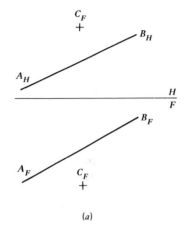

(a)

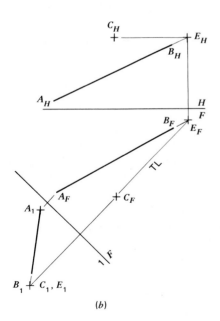

(b)

In *(c)* establish plane 2 parallel to the projection of plane *ABC* in plane 1. Project point *C* and line *AE* into plane 2. Line *AE* will be in true length, and the plane containing line *AE* and *C* will be in true shape. Now, construct perpendicular *CD*. Last, project it back to the other views, as in *(d)*.

Note that this technique could also be used to construct a line at any specified angle to line *AB*. Since plane *ABC* is viewed in true shape in plane 2, all angles between lines in plane *ABC* will be seen in true measure.

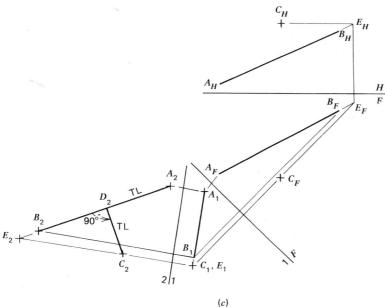

(c)

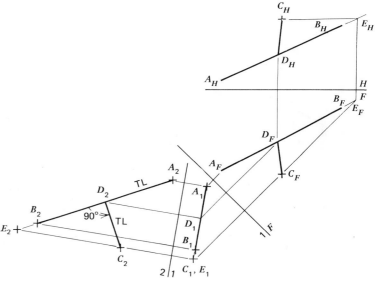

Sample Problem 3.37

(d)

SAMPLE PROBLEM 3.38

Determine the angle between the two intersecting lines a and b shown in (a).

Solution

Lines a and b do intersect since the intersecting point is P, and P appears as a unique point in both the horizontal and frontal views. The angle between lines a and b can be measured in a view in which the plane containing lines a and b appears in true shape.

In (b) construct horizontal line AB. The angle between lines a and b is angle APB. In the horizontal view, line AB will appear in true length. Now, establish plane 1 at right angles to the projection of line AB in the horizontal view. Project both lines and points A, B, and P into plane 1. Since line AB will appear in point view, the plane containing lines a and b will appear in edge view.

In (c) establish plane 2 parallel to the projection of the plane containing lines a and b. Project lines a and b into plane 2. Both lines will appear in true length, and the angle between them will appear in true measure — as 93 deg.

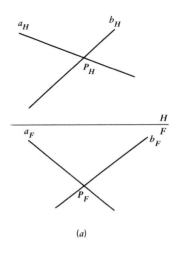

(a)

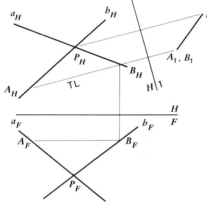

(b)

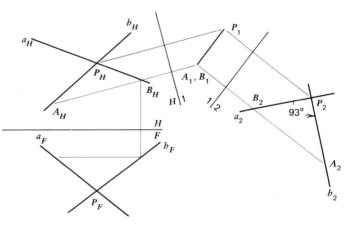

(c)

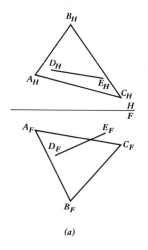

(a)

SAMPLE PROBLEM 3.39

Line *DE* intersects plane *ABC*. Determine the point of intersection and the angle of intersection between the line and the plane *(a)*.

Solution

The true measure of the angle of intersection can be seen only when line *DE* is in true length and plane *ABC* is in edge view. In all other views the angle will be distorted. Think about these two statements before continuing.

In *(b)* construct frontal line *AR*. This line will appear in true length in the frontal view. Now, establish plane 1 at right angles to the projection of line *AR* in the frontal view. Project line *AR*, line *DE*, and plane *ABC* into plane 1. Since line *AR* was in true length in plane *F*, it will appear as a point in plane 1, and plane *ABC* will be in edge view. Point *P* will be the point of intersection of line *DE* and plane *ABC*. Line *DE* will *not* be in true length, and the angle will *not* be seen in true measure. It is distorted.

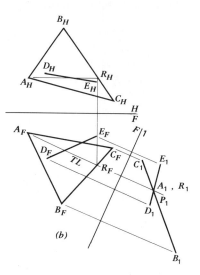

(b)

Now, establish plane 2 parallel to the projection of plane *ABC* in plane 1, as in *(c)*. Project line *DE* and plane *ABC* into plane 2. Plane *ABC* will be in true shape, but line *DE* is, again, not in true length. The sole purpose of this construction is to prepare for the next view.

In *(d)* establish plane 3 parallel to the projection of line *DE* in plane 2. In plane 3 line *DE* *will* be in true length, and plane *ABC* will again be in edge view. Note that both plane 3 and plane 1 are orthogonal to plane 2. Plane 3 is the view desired. The angle is in true measure, at 24 deg. Note the distortion of the angle in plane 1.

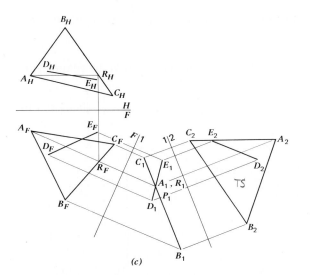

(c)

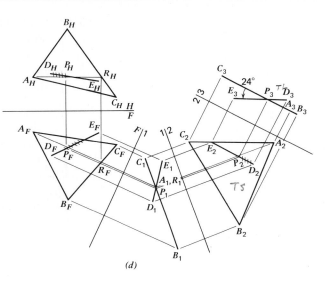

(d)

In planes *H*, *F*, and 2, part of line *DE* has dash marks on it—our way of communicating that part of line *DE* is hidden, but not a common convention.

To determine the angle between a line and a plane requires three successive adjacent views. This problem is the most lengthy of our elementary examples, but it is not more involved than any previous problem. Each successive orthogonal projection is treated one projection at a time.

PROBLEMS

3.40 Determine the point of intersection and the visibility of line *LM* passing through the plane triangle.

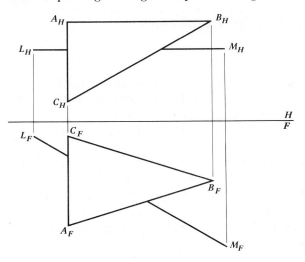

3.41 Find the perpendicular distance from Q to plane ABC. Show it in all views.

3.42 Find the perpendicular distance from R to plane ABC. Show it in all views.

Answer: $\frac{3}{4}$ in.

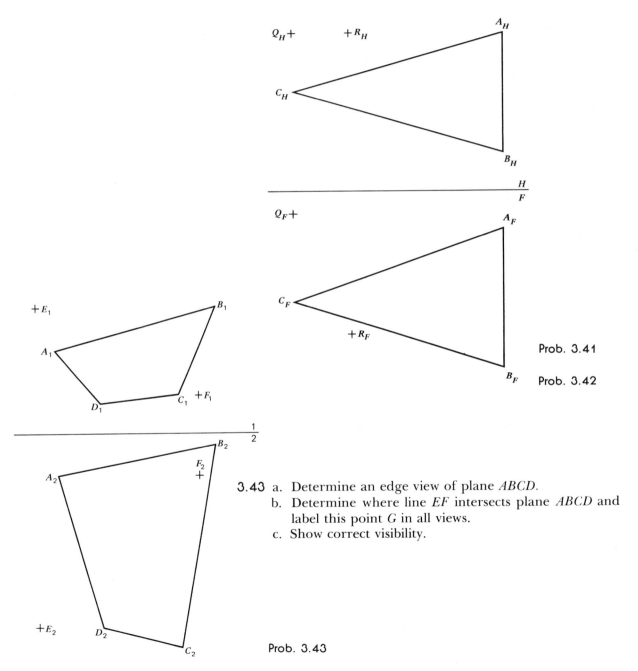

Prob. 3.41

Prob. 3.42

3.43 a. Determine an edge view of plane $ABCD$.
b. Determine where line EF intersects plane $ABCD$ and label this point G in all views.
c. Show correct visibility.

Prob. 3.43

3.44 a. Determine an edge view of plane ABO.
    b. Determine where line CD intersects plane ABO and label this point E in all views.
    c. Show correct visibility in all views.

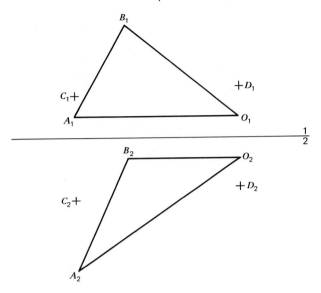

3.45 Locate a point E on plane ABC such that DE makes a 30-deg angle with plane ABC. Show point E in all views.

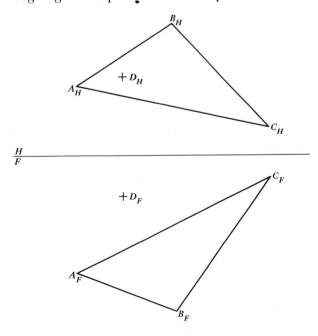

3.46 Lines m and n are two skew lines in space. Point A and line m form a plane that is parallel to the plane containing point B and line n. Locate points A_F and B_H and find the distance between these two planes.

Answer: $\frac{13}{32}$ in.

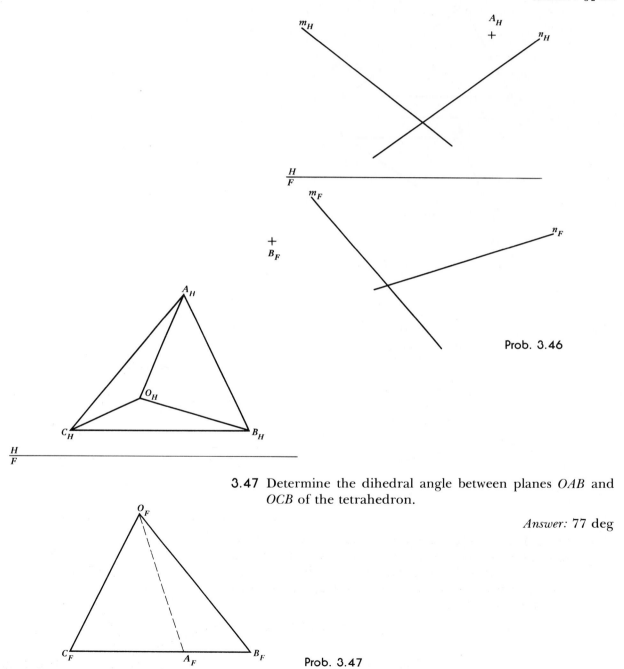

Prob. 3.46

3.47 Determine the dihedral angle between planes OAB and OCB of the tetrahedron.

Answer: 77 deg

Prob. 3.47

3.48 Determine the dihedral angle between planes OAC and OAB of the tetrahedron in Problem 3.47.

3.49 The angle between the two planes ABC and BCD is 50 deg, but the planes appear to be coincident in the horizontal view. Locate point D and show plane BCD in the frontal view.

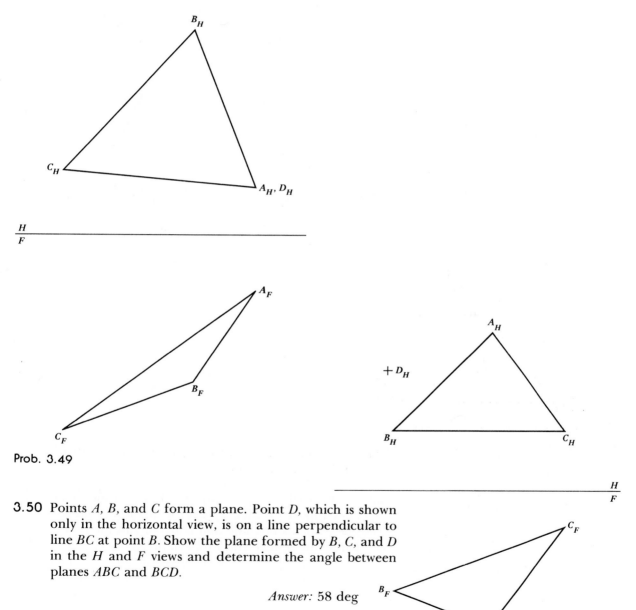

Prob. 3.49

3.50 Points A, B, and C form a plane. Point D, which is shown only in the horizontal view, is on a line perpendicular to line BC at point B. Show the plane formed by B, C, and D in the H and F views and determine the angle between planes ABC and BCD.

Answer: 58 deg

Prob. 3.50

3.51 Cables *AB* and *CD* are shown in frontal and profile views. Determine the minimum vertical clearance between the cables and show it in both views (scale 1:200).

Answer: 5.2 m

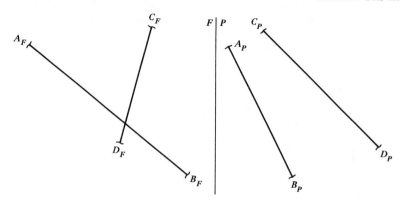

3.52 Determine the shortest connector between lines *AB* and *CD* using the 3 view to get the true length of line *AB*. Show the connector in the 1 and 2 views.

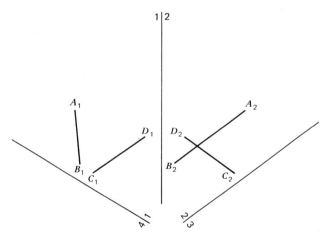

3.53 Work Problem 3.52 using the 4 view, which makes the lines appear parallel. To get the 4–1 line, a plane is used that contains one line and is parallel to the other. Compare your results with Problem 3.52.

3.54 Cables *AB* and *CD* in Problem 3.51 are repeated. Determine the minimum clearance between the cables and show it in all views (scale 1:200). Use the point view of one of the lines in your solution.

Answer: 3.5 m

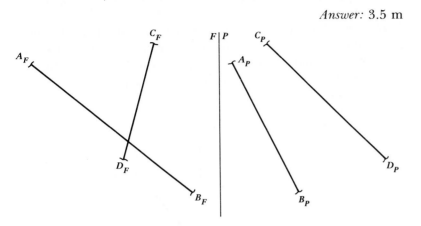

3.55 Given lines *AB* and *CD* in the horizontal and frontal views, find the shortest horizontal connector and show it in all views. What is the true length of the connector (scale 1:100)?

Answer: 41 in.

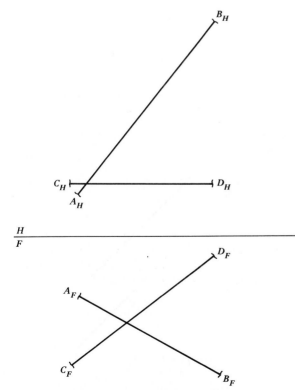

3.56 Work Problem 3.54 using a view that makes the lines appear parallel. Remember, to get this special view you find the edge view of a plane that contains one of the lines and is parallel to the other.

3.57 Given lines AB and CD in the horizontal and frontal views, find the shortest horizontal connector and show it in all views. What is the true length of the connector? In the frontal plane AB and CD appear to be parallel.

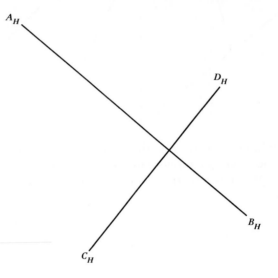

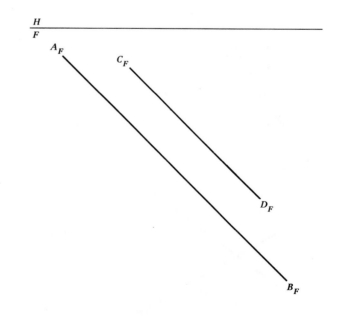

3.58 Find the shortest horizontal connector, *GH*, between skew lines *AD* and *BC*. Point *G* lies on *BC*. Point *H* lies on *AD*. Project *GH* to the *H* and *F* views, and measure its true length.

Answer: $\frac{7}{8}$ in.

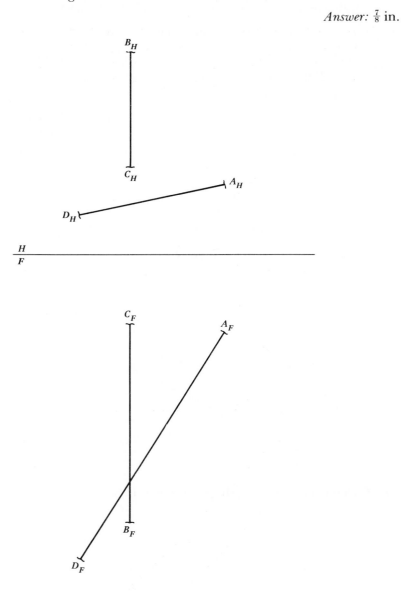

3.59 For the two lines in Problem 3.55, find the shortest vertical connector and show it in all views.

3.60 Two skew cables are drawn to scale (1:500). Determine the following. Show connectors in the H and F views.

  a. The shortest distance between the two cables.

  b. The shortest connector from m to n having a grade of −30 percent.

  c. The shortest vertical connector.

Answer: (a) 3.8 m
(b) 4.1 m
(c) 5.6 m

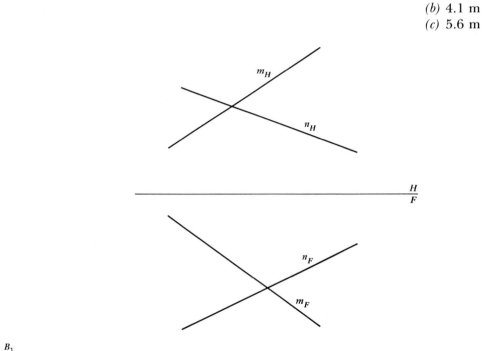

3.61 Determine the true shape of plane ABC.

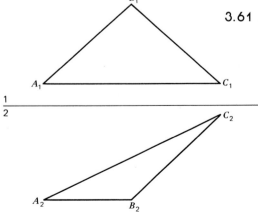

3.62 For the two skew cables in Problem 3.60, determine the shortest connector from *n* to *m* having a grade of 20 percent. Show the connector in the *H* and *F* views.

3.63 Determine the true shape of plane *ABC*.

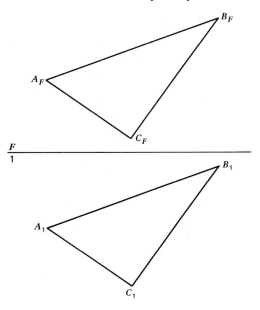

3.64 Find the angle between plane *ABC* and line *AD*. Show the angle where it is measured.

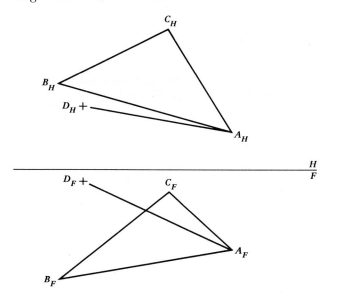

3.65 Complete the H and P views showing correct visibility and the point of intersection of line m and plane A. What is the angle between the line and the plane?

Answer: 18 deg

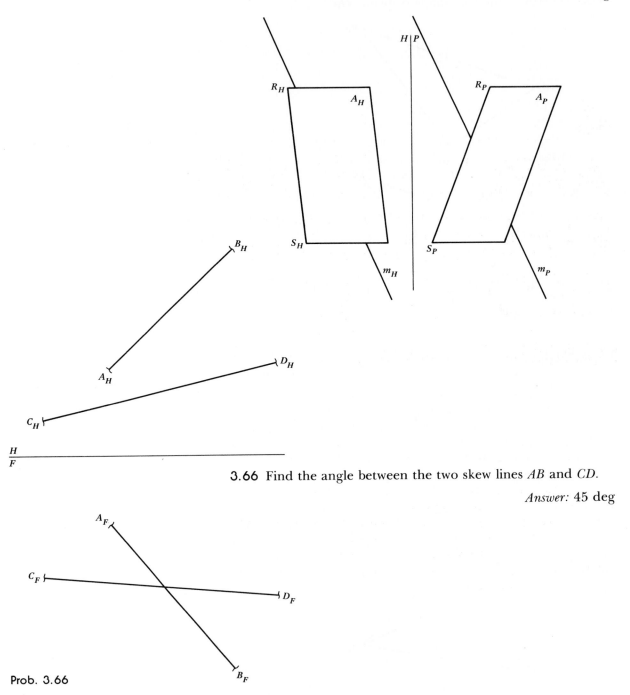

3.66 Find the angle between the two skew lines AB and CD.

Answer: 45 deg

Prob. 3.66

4
APPLICATIONS OF DESCRIPTIVE GEOMETRY

Real problems that involve lines and planes are everywhere. As we have said, engineers live in a square world. Engineers try mightily to build engineering structures with parallel lines, parallel planes, and right angles, but this cannot always be. Flow lines, gravitation, economics, and aesthetics all affect the square world of engineering. As a consequence, in many instances we need to determine distances, shapes, sizes, and positions of physical systems. In addition, as engineers, we must formulate these problems, carry out the analyses, and communicate the results to ourselves and to others. Orthogonal projection and descriptive geometry are the means for analysis and communication. This chapter presents some of the more important and common applications of descriptive geometry.

only need 2 views

3ʳᵈ view for clarity

right 2 views ino redundant

4.2 READING ORTHOGONAL VIEWS

Graphic communication includes both making drawings and sketches and interpreting drawings and sketches, which is the same as reading. Reading orthographic drawings is analogous to reading a sentence. A sentence contains words that are combined in a specific way to communicate a thought. Orthographic drawings combine points and lines in a specific way to describe a shape. The dilemma of working with a two-dimensional space to describe three-dimensional objects is amplified when we recognize that we handle objects in our minds in their original three dimensions. If we were to mention a common device to you, you would picture it as you would see it. You would not picture it in two or three orthographic projections.

You now have the vocabulary—points, lines, letters, and numbers—and the grammar of orthogonal projection. All that is left is practice!

To read or interpret orthographic views of a three-dimensional object, you must visualize a pictorial of the object. This pictorial must be constructed; it does not simply happen. Experience helps, because with experience the construction process becomes shorter, but visualization is a talent that some people have and some people do not have.

The three-dimensional construction of a pictorial is the result of your visual memory. You must remember a pictorial of one line or plane while you are constructing the pictorial of another. If you can remember and picture several lines or planes while you are constructing a fourth or fifth, the visualization process becomes very quick indeed.

It may be pedantic to try to cite formal steps in reading orthographic views, but there are certain things that you should do first and certain things that you should do later.

First, establish the overall dimensions of the object—the length, width, and height. It is obvious that an object will fit into a box. What are the dimensions of that box? Construct that box in your visual memory. To help yourself, make an isometric sketch of the box.

Second, picture any dominant features, such as surfaces, shapes, holes, or intersections.

Third, picture details. Picture each detail by itself, separate from other details, and follow it from one view to the next.

Last, review and check lines and points. If you are sketching the object to aid your visualization, label each point. Remember the meaning of points and lines in descriptive geometry. A line is an edge view of a plane or the intersection of two planes. A point is the end view of a line.

A simple object is represented in the three views in Figure 4.1*a*. Although the specific shape may not be immediately clear, it is obvious that the object is made from a cube. By making an isometric sketch of a cube and placing the three views on the appropriate sides of the cube, as shown in Figure 4.1*b*, the shape of the object becomes clear. For this example, the views on the sides of the cube also complete the isometric representation.

Figure 4.2*a* presents three views of another simple shape. These views can be placed on the appropriate surfaces of an isometric cube, as shown in Figure 4.2*b*. These views can then be more easily used to interpret the shape than can the original orthogonal views (Figure 4.2*a*). A possible shape is shown in Figure 4.2*c*.

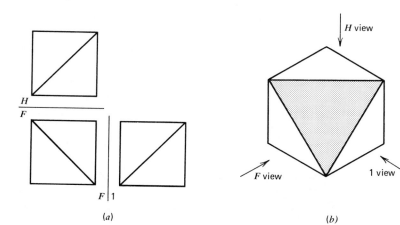

(a)

(b)

Figure 4.1

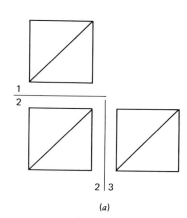

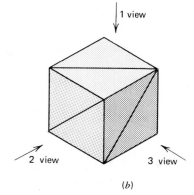

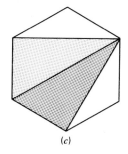

(a)

(b)

(c)

Figure 4.2

Three orthogonal views of a bracket are shown in Figure 4.3. We hope you easily read these views and understand the shape represented, but note how the pictorial view clarifies the shape. Figures 4.4 and 4.5 are two additional examples of correct orthogonal views that are clarified with pictorial sketches.

Figure 4.3

Figure 4.4

Figure 4.5

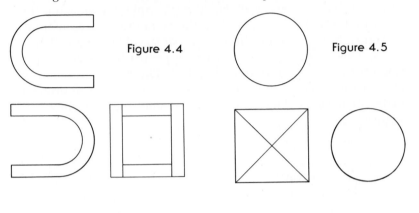

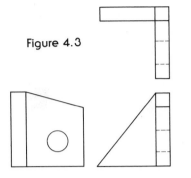

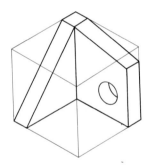

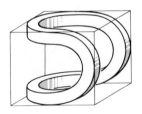

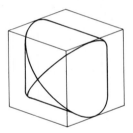

PROBLEMS

4.1 Sketch the third orthographic view and an isometric view of the two shapes shown with two orthogonal views. All surfaces are planar.

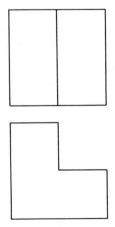

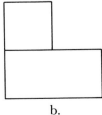

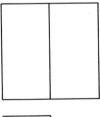

a.

b.

4.2 Sketch a pictorial view of the object represented with three orthogonal views. All surfaces are planar.

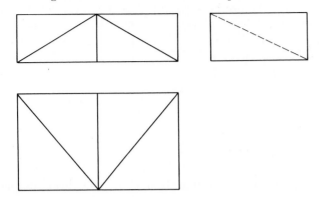

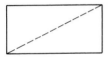

4.3 Sketch a pictorial view of the object represented with three orthogonal views. All surfaces are planar.

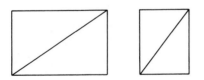

4.4 Sketch an isometric view of the object represented with two orthogonal views.

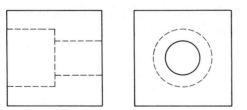

4.5 Sketch an oblique view of the object represented with two orthogonal views.

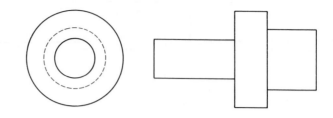

4.6 Sketch pictorial views that best show the three shapes represented with two orthogonal views. Will one type of pictorial view be best for all three cases?

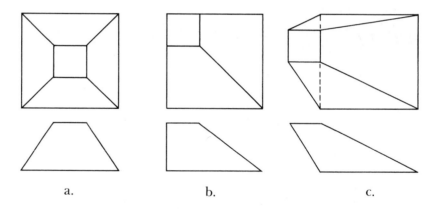

a.   b.   c.

4.7 For the 2-in. cube with three 1-in. diameter holes centrally located, make an oblique sketch.

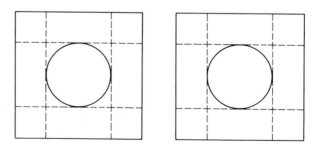

4.8 Make an isometric sketch for the cube in Problem 4.7.

4.9 For the shape shown in full scale, sketch an oblique view.

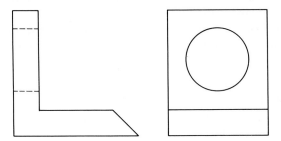

4.10 Sketch an isometric view of the shape in Problem 4.9.

4.11 Sketch a perspective view of the shape in Problem 4.9.

4.3 INTERSECTIONS

A very important application of descriptive geometry is the determination of the intersection of surfaces. In engineering, surfaces rarely exist alone and unviolated by holes, seams, or cuts. Flat surfaces are simply a small and finite part of a larger geometry. Intersections of surfaces are commonplace.

Surfaces intersect other surfaces. In many cases, these intersections are clear, simple, and easily determined. Some flat surfaces do meet other flat surfaces at right angles, and these intersections are aesthetically simple. In other cases they are not so simple, and determining a line of intersection between oblique and/or curved surfaces can be a considerable problem.

The intersection of surfaces breaks down to four basic problems.

1. A line and a plane.
2. A line and a cone.
3. A line and a cylinder.
4. A line and a sphere.

The surfaces of planes, cones, and cylinders can all be developed, which means that planes, cones, and cylinders can all be generated by a straight line; they have that common characteristic. Actually, the first problem, the intersection of a line and a plane, far overshadows the other three in importance. The surface of a sphere is a warped surface, but the procedures used to determine the intersections of a line and a cone and a cylinder can also be applied to a sphere. A sphere is a very special geometric figure. The last three really are special cases, and they will be considered in order of their importance.

The intersection of a line and a plane is the most important intersection problem and the basis of four other intersection subproblems:

a. A plane and a plane.
b. A line and a solid.
c. A plane and a solid.
d. A solid and a solid.

Each of these four subproblems can be broken down into an application of the intersection of a line and a plane, which is why they are all considered subproblems of the first.

4.4 INTERSECTION OF A LINE AND PLANE

In Figure 4.6a line m intersects plane ABC. An edge view of the plane will clearly show the point of intersection, as in Figure 4.6b. This approach is simple and direct. When intersections become involved and confused, it is always possible to use an edge view of one plane to determine its point of intersection with a line. This is tedious, however, particularly if several lines are involved.

In Figure 4.6c, the concept of a *cutting plane* is introduced. In the frontal view, consider a plane perpendicular to the frontal plane that contains line m. This plane cuts plane ABC. Since it is seen only in edge view, in the frontal view, line m and the line of intersection of the cutting plane and plane ABC are coincident, which is not true in the horizontal view. The line of intersection of the cutting plane and plane ABC is line $1_H 2_H$. The intersection of line m and plane ABC is the point where line $1_H 2_H$ crosses line m.

As an alternate solution, consider a plane perpendicular to the horizontal plane that contains line m, as in Figure 4.6d. This plane also cuts plane ABC, and it is seen only in edge view in the horizontal view. In the frontal view, line $3_F 4_F$ is the line of intersection of the cutting plane and plane ABC. The intersection of line m and plane ABC is the point where $3_F 4_F$ crosses line m. We will use the concept of a cutting plane for many of the other intersection problems.

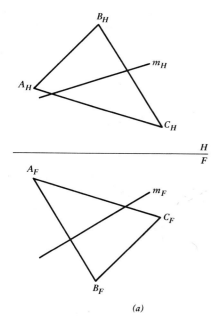

(a)

Figure 4.6 Intersection of a line and plane.

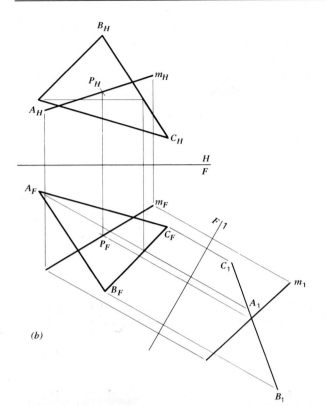

(b)

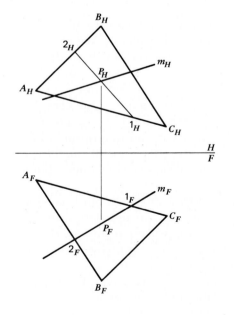

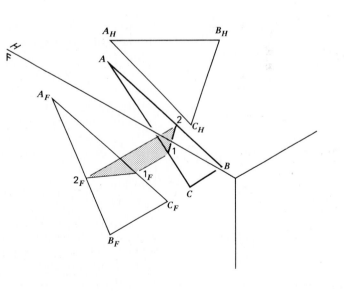

(c)

Figure 4.6 (cont'd)

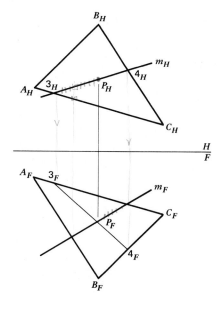

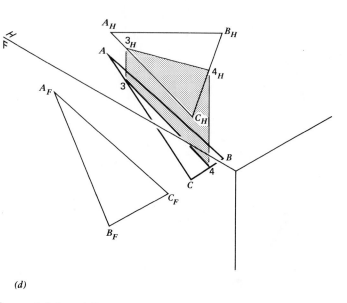

(d)

Figure 4.6 (cont'd)

then
visibility

Possible that

pl. not int. → look at

apparent ln. int.   NB

4.5 INTERSECTION OF TWO PLANES

We will now use the intersection of a line and plane to determine the line of intersection of two intersecting planes. We will consider this problem in detail. It is more convenient to discuss later problems involving solids if we are conversant with the intersection of two planes, rather than of a line and a plane, thinking about one point of intersection at a time. Quicker solutions are obtained if intersections can be found between planes.

It is important to remember that you can determine a line of intersection by finding an adjacent edge view of one of the two planes. This approach may not always be convenient, but a solution is always possible.

In Figure 4.7a planes *ABC* and *QRS* intersect, but their line of intersection is not known. In fact, no further description is given for either plane. Using the cutting-plane technique developed in the last section, consider two cutting planes, one containing line *QR* and a second containing line *AC*.

In the horizontal view, consider a cutting plane that is perpendicular to the horizontal plane and contains line *QR*. The cutting plane is seen only in edge view, and line *QR* and the line of intersection of the cutting plane and plane *ABC* are coincident, which is not true in the frontal view. Line $1_F 2_F$ is the line

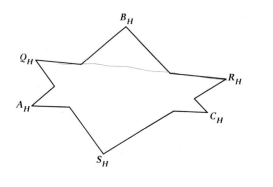

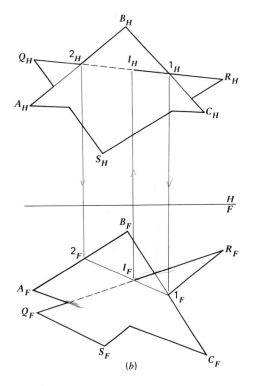

(a)

(b)

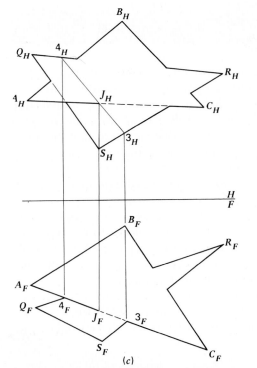

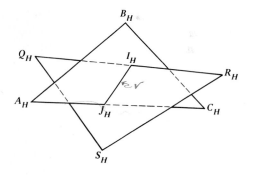

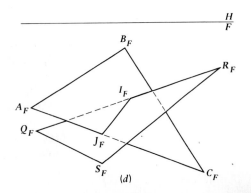

(c)

(d)

Figure 4.7 Intersection of two planes.

of intersection of the cutting plane and plane *ABC;* point I_F, where line 1_F2_F crosses line Q_FR_F, is the point of intersection of line *QR* and plane *ABC*. All this is shown in Figure 4.7*b*. The visibility of line *QR* is shown in both views.

In the frontal view, consider a cutting plane that contains line *AC* and is perpendicular to the frontal plane. The line of intersection of the cutting plane and plane *QRS* is line 3–4, which is coincident with line *AC* in the frontal view but not in the horizontal view. Point J_H, where line 3_H4_H crosses line A_HC_H, is the point of intersection of line A_HC_H, and point *J* is the point of intersection of line *AC* and plane *QRS*. The visibility of line *AC* is also shown in both views of Figure 4.7*c*.

Line *IJ* is the line of intersection of the two planes. The remaining visible and hidden lines are completed (see Figure 4.7*d*). Only two cutting planes were used to determine the line of intersection, one containing line *QR* and one containing line *AC*. But what about the four other lines? Was it random luck that brought us to line *QR* and line *AC*? How can we tell which lines to use? The answer lies in an earlier reconnaissance of the problem. Line *BC* is below and in back of plane *QRS*. Line *AB* is above plane *QRS*. Line *QS* is below plane *ABC*. Line *RS* is in front of and above plane *ABC*. Line *QR* is the only line to intersect plane *ABC*, and line *AC* is the only line to intersect plane *QRS*. It is all a matter of visibility.

4.6 INTERSECTION OF SOLIDS BOUNDED BY PLANES

The next problem we might consider involves the intersection of a plane and a solid, but this problem is passed over in favor of considering the intersection of two solids. The intersection of a plane and a solid can be easily found with one adjacent view with the plane in question in edge view. Finding the intersection of two solids bounded by several planes is a more complex problem. If two solids do intersect, there can be several lines of intersection, and you may not be able to determine intuitively how many different lines there are.

It is worth repeating a word of caution. Any problem with intersecting planar solids can be broken down to the basic element, the intersection of one line and one plane. When a problem looks complicated or becomes complicated, slow down and consider it one line and one plane at a time.

The three-sided prism in Figure 4.8*a* is known to intersect a four-sided prism, but the intersection is not known. From the given two orthogonal views, it is obvious that there will be more

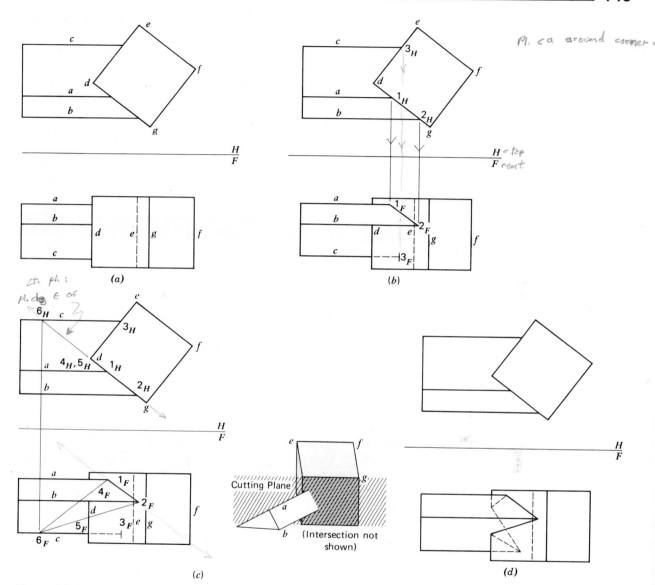

Figure 4.8 Intersection of planar solids.

than one line of intersection, but it is not obvious just how many there will be.

To find the intersection between these two solids, begin by finding obvious points of intersection. Lines a and b intersect the front face, and line c intersects the back face of the four-sided prism. These points all can be found. Locate them as points 1, 2, and 3, respectively, as shown in Figure 4.8b. One line of intersection, line 1_F2_F, is now known. It is also obvious that line d of the four-sided prism pierces two faces of the three-sided prism. Call these piercing points 4 and 5. Unfortunately, line d is in point view when viewed in the horizontal plane, so the locations of points 4 and 5 are not immediately apparent.

Points 4 and 5 can be found by using a cutting plane coincident with the face containing parallel lines d and g. The isometric sketch in Figure 4.8c shows the cutting plane. Note that with the cutting plane we can locate point 6 on line c. This point is all that is needed to locate points 4 and 5, and with them all the lines of intersection can be drawn, as shown in Figure 4.8d.

4.7 THE GENERAL INTERSECTION PROBLEM

The concept of a cutting plane can be generalized. If two intersecting solids S_1 and S_2 are cut with plane P, as shown in Figure 4.9a, curves of intersection C_1 and C_2 result (Figure 4.9b). The intersections I_1 and I_2 of curves C_1 and C_2 are two points on the intersection of the solids. To find the curve of intersection for the solids, many cutting planes would be needed.

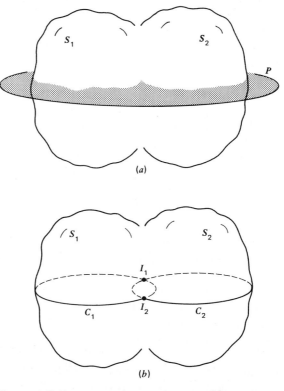

Figure 4.9 The general intersection problem.

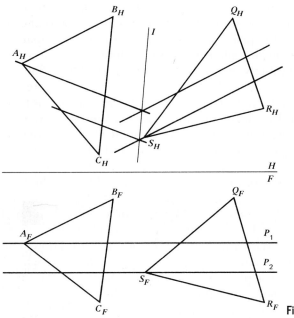

Figure 4.10 Intersection of limited planes.

In Figure 4.10 the two planes are visibly separated and are called *limited planes*. An intersection exists, but it could be outside the bounded areas of the planes.

In order to find the line of intersection, pass either two horizontal or two frontal cutting planes through both planes ABC and QRS. In Figure 4.10 two horizontal cutting planes are used. They appear as horizontal lines in the frontal view. The lines of intersection of the cutting planes with each of the planes are seen in the horizontal view. These lines will intersect unless the planes ABC and QRS are parallel.

Finding the intersection of limited planes is a direct application of the general problem represented in Figure 4.9.

The judicious choice of the cutting plane can greatly simplify intersection problems. For the intersection of a line and cylinder, a cutting plane parallel to the centerline of the cylinder intersects the cylinder with straight lines. The cutting plane containing a line that intersects a sphere will intersect the sphere with a circle.

SAMPLE PROBLEM 4.12

Find the intersection of plane QRS with tetrahedron $ABCD$ in (*a*). Show the correct visibility of edges of the plane and tetrahedron, but do not show the edges of plane QRS *inside* the tetrahedron.

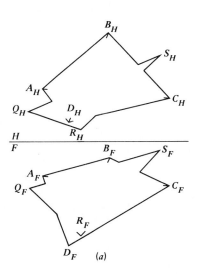

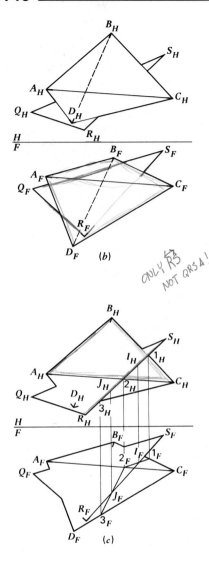

(b)

ONLY RS
NOT QRS4!

(c)

Sample Prob. 4.12

FOR PRISM ON MIDTERM

Solution

The first step in determining the intersection of plane QRS and solid $ABCD$ for this problem would be to draw lines AC and BD lightly, disregarding plane QRS for the moment, as in (b). A simple test for visibility will show that line BD will be hidden in both views.

The second step would be to test for the intersections of lines QS, QR, and RS with any of the faces of the solid. Line QR does not intersect the solid, as can be seen in the horizontal view. Line QS can be found to be above and in front of lines BC, AC, and AD—also a matter of visibility. Line RS is above and in front of lines BC and CD, but it is below and behind line AC. Line $\boxed{RS}$ does intersect the solid; it intersects faces ABC and ADC. Similar tests can be made for the intersection of any of the four lines of solid $ABCD$ with plane QRS. There is no need to be redundant. With the information we already know, line AC is the only line that intersects plane QRS. Our problem of the intersection of a plane and a solid has been resolved into three lesser problems; the intersection of line RS with plane ABC, the intersection of line RS with plane ADC, and the intersection of line AC and plane QRS.

As the next step, in (c) consider only line RS and plane ABC. Ignore all other planes and lines. A cutting plane perpendicular to the horizontal plane could contain line RS and the line of intersection of the cutting plane and plane ABC. All are coincident with the horizontal plane but not in the frontal plane. The point where line $1_F 2_F$ crosses line $R_F S_F$ locates the point of intersection of line RS and plane ABC. Call this point I.

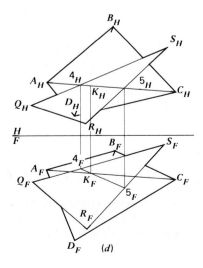

(d)

Next, determine the point of intersection of line RS and plane ADC. Again, using a cutting plane containing line RS and perpendicular to the horizontal plane, the line of intersection of the cutting plane and plane ADC will be line $2_F 3_F$. The point where line $2_F 3_F$ crosses line $R_F S_F$ locates the point of intersection of line RS and plane ABC. Call this point J. The points of intersection (entrance and exit) of line RS and solid $ABCD$ are now shown in (c).

Last, consider line AC and plane QRS. Ignore all other planes and lines. In (d) a cutting plane containing line AC and perpendicular to the horizontal plane would have the line of intersection $4_F 5_F$ seen in the frontal view. Point K is the point of intersection of line AC and plane QRS. Lines IK and JK are the intersection of plane QRS and solid $ABCD$.

The complete visibility of the solid and the plane is shown in (e).

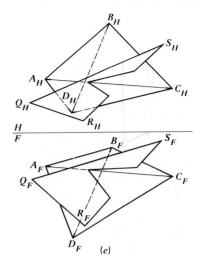

Sample Prob. 4.12

SAMPLE PROBLEM 4.13

Determine the intersection of line m and the cone as given in (a).

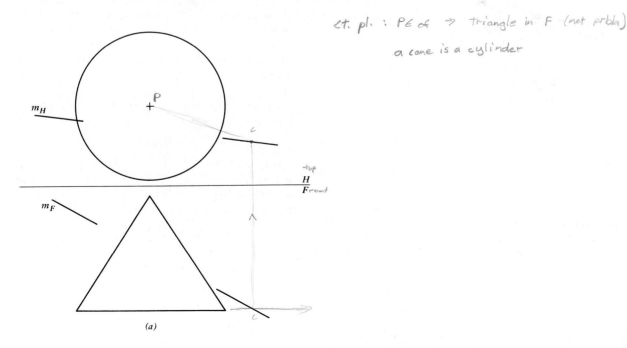

ct. pl. : PE of → triangle in F (not prbh)

a cone is a cylinder

(a)

Solution

Consider the general intersection problem represented in Figure 4.9. The cone is one body, and line m is the second body. We can introduce a plane that intersects both bodies or, stated

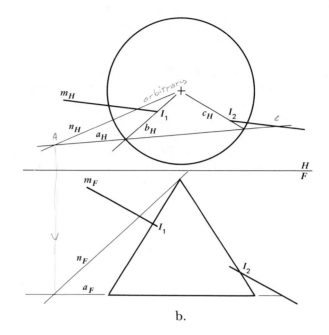

b.

differently, that contains line m and intersects the cone. Note; if we choose a plane that contains the vertex of the cone, this plane will intersect the cone with straight lines. The details of the solution are shown in (b). The cutting plane is described by the original line m and line n. Line n is drawn through the vertex of the cone and must intersect line m, and it can be drawn arbitrarily in either the H or F views. Then the point of intersection of lines m and n is projected to the opposite view. The points where lines m and n pierce the base plane of the cone are seen in the F view and are projected into the H view. Line a is then the intersection of the cutting plane with the base of the cone. The point where line a intersects the base circle of the cone can be seen in projection H; then lines b and c can be constructed in the H view. These two lines are the intersection of the cutting plane and the cone. Where lines b and c intersect line m are the two points where line m pierces the cone. These points are labeled I_1 and I_2.

cutting plane contains
vert. P
ln. M
ln. AC

PROBLEMS

4.14 Determine the following. Show correct visibility in all views.

    a. Where line m intersects plane P.

    b. Where line n intersects plane Q.

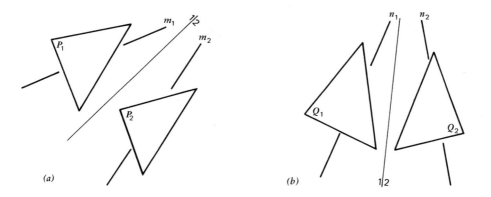

(a)   (b)

4.15 Determine the intersection between planes ABC and XYZ. Show correct visibility in both views.

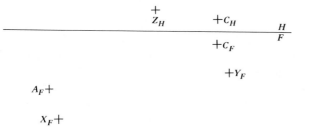

4.16 The planes used in the discussion of the intersection of two planes in Section 4.5 are repeated here. Determine the intersection points of lines QR and SR with plane ABC and the intersection points of lines AC and BC with plane QRS. What does an intersection point outside the bounds of a plane mean? Compare your results with Figure 4.7.

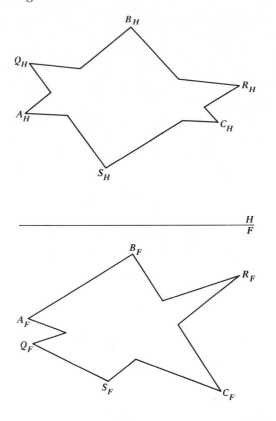

4.17 Determine the line of intersection between the two limited planes.

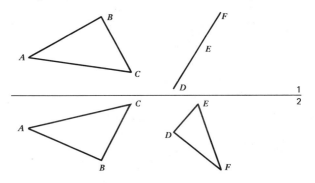

4.18 Plane QRS intersects tetrahedron $ABCD$. Complete the views showing the correct visibility of all edges. Discontinue the lines QR, RS, or SQ within the tetrahedron.

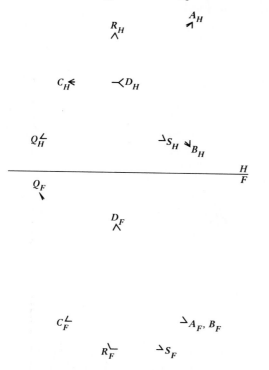

4.19 Find the acute dihedral angle between the two planes. Show correct visibility of edges.

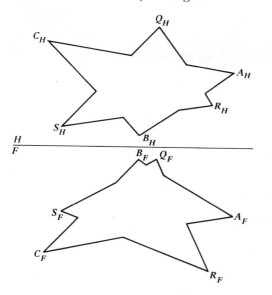

4.20 Complete the views of tetrahedron *ABCD*. Show the correct visibility of edges. Find the intersection of the plane *PRQ* with the tetrahedron. Do not show lines *PR*, *RQ*, or *PQ* within the solid.

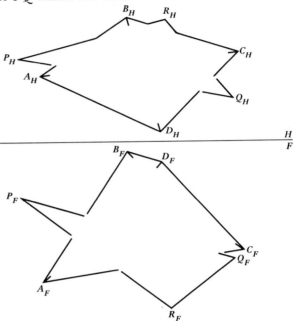

4.21 Establish the intersection lines between the plane and the triangular prism. Use only the two given views and show visibility.

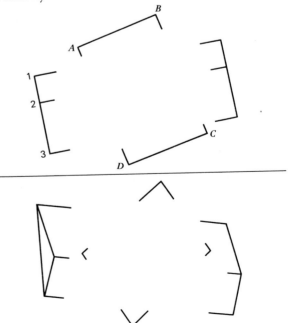

4.22 Determine where line *m* intersects the sphere. Remember that the intersection of the sphere and a cutting plane is a circle.

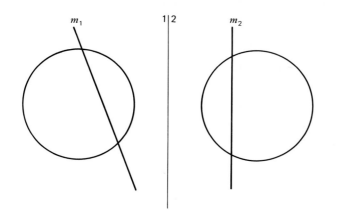

4.23 Find the intersection of the right circular cylindrical leg of the water tower and the spherical portion that forms the bottom of the tank (scale: 1 in. = 8 ft). The centerline of the cylinder and the center of the sphere are in a plane that is parallel to the viewing plane.

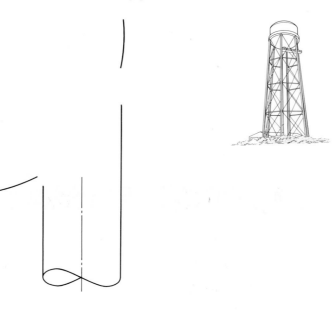

4.24 A truncated, circular, conical, steel pressure vessel is involved in a radiation experiment. An electron beam is to be projected at a target located at point X inside the vessel. The straight beam emerges from an accelerator at point Y. It is necessary to install a "window" that is transparent to the beam where the beam leaves the vessel as well as where it enters. Find the points on the pressure-vessel surface that will be pierced by the beam.

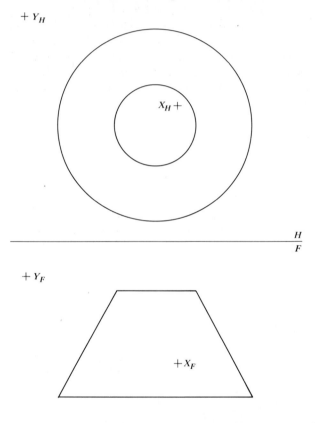

4.8 REAL LINES IN SPACE

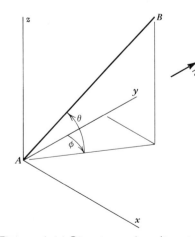

Figure 4.11 Direction of a line in space.

The *direction* of a line in space can be specified by two angles —the angles ϕ and θ shown in Figure 4.11. If the x–y plane is horizontal and the y-axis is oriented toward the north, this same spatial orientation can be represented with two orthogonal views, as shown in Figure 4.12. The two angles ϕ and θ are spherical coordinates. Angle ϕ specifies the direction of the horizontal projection of line AB. Alternatively, we could state the horizontal direction of the line. Angle θ specifies the slope or grade of the line that is the angle line AB makes with a horizontal plane.

On topographical representations, such as maps, a common way to specify the horizontal direction of a line is to state its *bearing*. Bearing is defined as the angle a line makes with a north (N) or south (S) line. It is always an acute angle, which means that bearing must also include a second designation of east or west. North is usually toward the top of a topographical map. It is customary to write

First, N or S.

Then the angle with respect to the N–S line.

Finally, E or W, whichever makes the angle ϕ less than 90°. The bearing of line *AB* in Figure 4.12 is N ϕ° E; the bearing of line *BA* is S ϕ° W. Note that the angle between line *AB* and a north–south line and the direction of line *AB* are both included in the bearing. Figure 4.13 specifies some additional bearing examples.

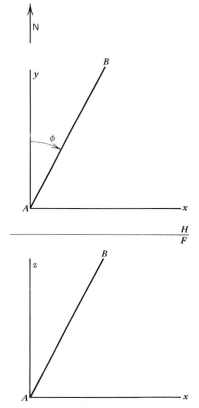

Figure 4.12 Orthogonal views of a line in space.

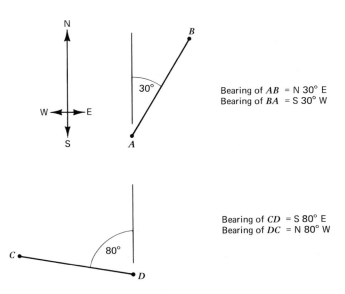

Bearing of *AB* = N 30° E
Bearing of *BA* = S 30° W

Bearing of *CD* = S 80° E
Bearing of *DC* = N 80° W

Figure 4.13 Bearing of a line.

The *slope* of a line is the angle a line makes with a horizontal plane. This angle can be seen in any projection containing both the true length of the line and an edge view of the horizontal plane. If we had the horizontal and frontal views of line *AB* in Figure 4.14, we could project line *AB* into plane 1 to determine the slope of the line. The 1 view contains the true length of line

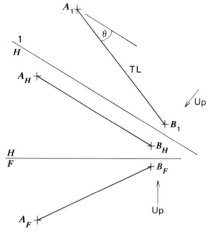

Figure 4.14 Slope of a line.

AB and an edge view of the H plane, which is the H–1 line. Therefore, angle θ is the slope of the line. Note that in this case the slope of line AB is positive, since point B is higher than point A. The slope of line BA is negative. The angle a line makes with any projection plane can be obtained by determining a view that includes the true length of the line and an edge view of the plane.

The *grade* of a line is defined as the ratio of the vertical distance (the rise) to the horizontal distance (the run) between two points on the line. The line and its true length in Figure 4.14 are repeated in Figure 4.15. The vertical distance v between points A and B can be seen in the F view and the 1 view. The horizontal distance h between points A and B shows in the H view and the 1 view. Grade is expressed as percent, so for line AB in Figure 4.15

$$\text{grade} = \frac{v}{h}(100) = \frac{\text{rise}}{\text{run}}(100) \qquad (4.1a)$$

The rise can be positive or negative. Note that the magnitude of the grade of a line can vary from 0 (flat) to ∞ (straight up and down), while the slope of a line can vary from 0 to 90 deg. If we specify the slope as θ and the grade as G, then

$$G = 100 \times \tan \theta \qquad (4.1b)$$

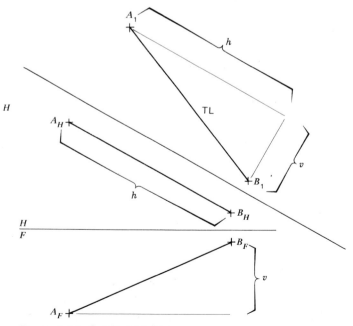

Figure 4.15 Grade of a line.

SAMPLE PROBLEM 4.25

The horizontal and frontal views of line AB and the location of plane 1 are given in (a). Determine slope θ_H of line AB and the angles it makes with the F plane (θ_F) and the 1 plane (θ_1).

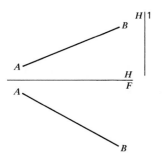

(a)

Solution

To obtain the angle a line makes with a projection plane, we must obtain a view that contains both the true length of the line and the edge view of the projection plane. All the necessary constructions are shown in (b). Projection plane 2 is introduced parallel to the H projection of line AB giving a view showing both the true length of the line and the edge view of projection plane H. Therefore, the angle that the line makes with the H projection plane can be measured in view 2. Projection plane 3 is introduced parallel to the F projection of line AB. As shown, the 3 projection of line AB is in true length and contains an edge view of the F plane. Therefore, angle θ_F can be determined in the 3 view. To get the angle that the line makes with the 1 pro-

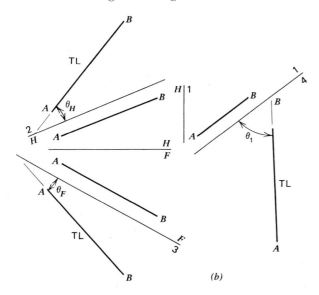

(b)

jection plane, the 1 projection of line *AB* must be completed, and then an adjacent plane 4 is introduced parallel to the 1 projection of *AB*. Projection 4 then gives a true-length view of the line in which the edge view of plane 1 also shows. Therefore, angle θ_1 can be measured in the 4 view.

SAMPLE PROBLEM 4.26

Determine the horizontal and frontal views of line *AB*, if line *AB* has the following characteristics:

1. True length of 50 mm.
2. Grade of +45 percent.
3. Bearing of N 60° E.

Point *A* is given.

Solution

The *H* and *F* views of point *A* and the north direction are shown in (*a*). The bearing of line *AB* can be established as in (*b*). A line making a 60-deg angle with the north direction is added to the H view. Now that we have the direction of the *H* view of the line, we can establish projection plane 1 and locate the 1 projection of point *A*. Since projection plane 1 is parallel to the *H* view of the line, the true length and grade of the line will be

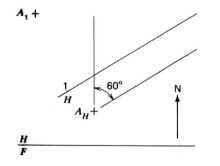

(a)

(b)

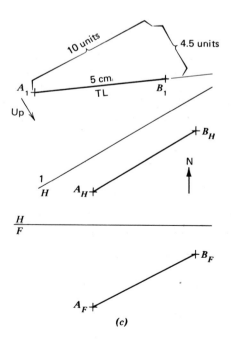

(c)

seen in the 1 projection. Next, the grade of the line can be established, as in (c), by measuring 10 units in the horizontal direction and 4.5 units in the upward direction. The specific number of units is arbitrary, of course, since the ratio of the rise over run defines the grade. Once the direction of the line is established in projection 1, the true length can be laid off, thus establishing the 1 projection of point B. The last step is to project B back into the H and F views.

SAMPLE PROBLEM 4.27

Determine the slope of line AB in Figure 4.14 by rotating the line in the given two views.

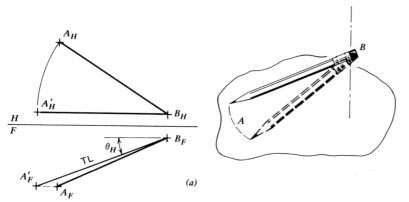

(a)

Solution

Represent a line with your pencil. Place one end of the pencil on a flat horizontal surface and hold the other end above the surface so the pencil makes an acute angle with the surface. Then, if you move the end in contact with the surface while holding the other end fixed, the pencil will sweep out a portion of a cone while maintaining the angle between the line and the

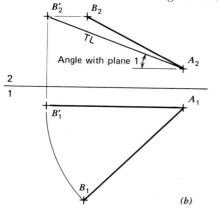

(b)

surface. We can represent this rotation with descriptive geometry. If point B is held fixed and point A moves in a plane parallel to the H plane, the line describes a portion of a cone. The path of A is circular and will therefore show as a circular arc in the H view and a straight line in the F view (a). When the H view is rotated to a position parallel to the H–F line, the line will be in true length in the F view, and the angle θ_H between the line and the H plane (i.e., the slope) can be determined. This *rotation method* can be used to obtain the angle a line makes with any plane. The rotation must save the angle with the plane — that is, move so the end being moved remains in a plane parallel to the plane of interest. Part (b) shows an example.

PROBLEMS

4.28 Find the grade, bearing, and true length of the following.
a. Line AB.

b. Line BA.

4.29 Find the grade, bearing, and true length of the following.
a. Line CD.

b. Line DC.

Answer: (a) −55%, N 70°E, 37 mm

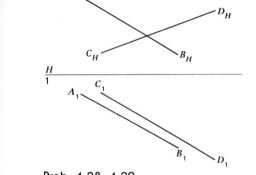

Prob. 4.28, 4.29

4.30 Determine the H and F views of a line AB that has the following characteristics.
a. True length of 25 mm.

b. Slope of 35 deg.

c. Bearing of N 50° W.

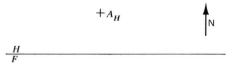

4.31 Determine the H and 1 views of a line AB that has the following characteristics.

 a. True length of $\frac{1}{2}$ in.

 b. Grade of -100 percent.

 c. Bearing of S 45° W.

4.32 Determine the slope of line AB and the angles it makes with planes 1, 2, and 3.

Answer: $-21°,\ 22°,\ 38°,\ 54°$

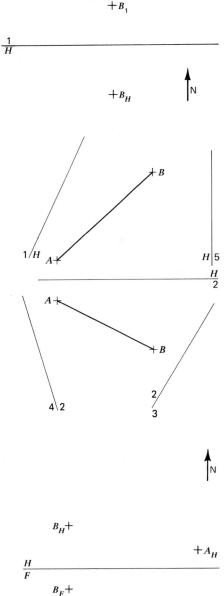

4.33 Determine the slope of line AB in Problem 4.32 and the angles it makes with planes 2, 4, and 5.

4.34 A commercial airliner is at point A traveling on a straight course of bearing N 50° W and is climbing at a 25 percent slope. A small private airplane is at point B traveling on a straight course of bearing N 10° E and is descending at a slope of 25 percent. What is the minimum distance between their flight paths? Which plane would fly above the other at this minimum separation (scale: 1 in. = 4000 ft)?

Answer: 390 ft

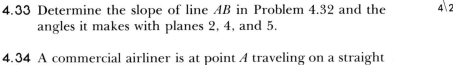

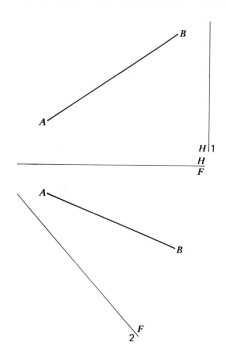

4.35 Using the rotation method, determine the true angle between line AB and projection planes H, F, 1, and 2. Label where you measure angles θ_H, θ_F, θ_1, and θ_2, respectively.

Answer: 18°, 34°, 52°, 22°

4.36 A meteorological research rocket is launched from the ground at point A. The vehicle travels in a straight line and is detected by radar as it goes through point B. Find the angles the trajectory makes with the following.

a. The horizontal (ground) plane.

b. The vertical north-south plane.

c. The vertical east-west plane.

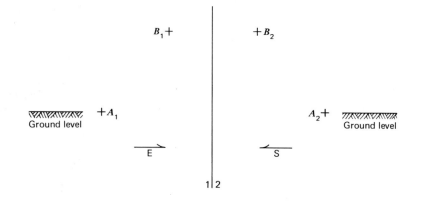

4.37 A light airplane is sighted at point A at an elevation of 1000 ft. It travels horizontally on a bearing of S 30° E for 1 mi to point B. It then turns sharply to a bearing of N 45° E and climbs at a +30 percent slope along a true-length path of 2 mi to point C. At point C it again turns sharply. The new flight path to D is due north on a slope of +20 percent and has a true length of 1.5 mi. Choose a convenient scale.

a. What is the elevation at point C?

b. What is the elevation at point D?

c. How far is point D from point A?

d. What is the slope of the flight path directly from D to A?

4.38 Note that once the true length of the line is known (by the rotation method in this case), the angles with the various projection planes can be obtained simply by using the true length of the line and the fact that when the line is rotated the point (end) being moved takes a path parallel to the plane of interest. For example, angle θ_2 between line AB and projection plane 2 is obtained by finding the intersection of a line through B parallel to the 2 plane (the 1–2 fold line) and the arc centered at A with a radius equal to the true length of the line. Determine the angles line AB makes with the 3, 4, 5, and 6 planes by working only in the 1 and 2 views.

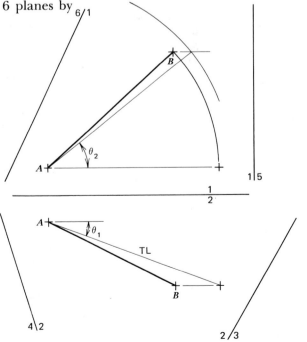

4.9 SHADOWS

Illustrators of shapes and buildings, such as architects, often include the effect of lighting in their work. The representation of shadows and lighted areas can help communicate shape and provide realism to the work.

The determination of the boundaries of light and dark areas caused by shapes and openings is a direct application of the intersection of a line and a surface. Consider the problem of finding the shape of the shadow cast on the horizontal surface by triangle ABC in Figure 4.16a. The direction of the rays of the sun are included in the H and F views. First, line a is constructed parallel to the rays and through corner A; this line represents a ray that passes very close to corner A. Where line a intersects the horizontal surface, it establishes the corner of the shadow corresponding to corner A. This point, A_S, is then located in the H view.

The process can be repeated for the other two corners and, knowing the shadow of a straightedge is straight, the shape of the shadow shown in Figure 4.16b can then be completed.

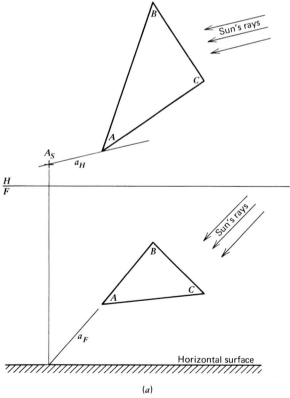

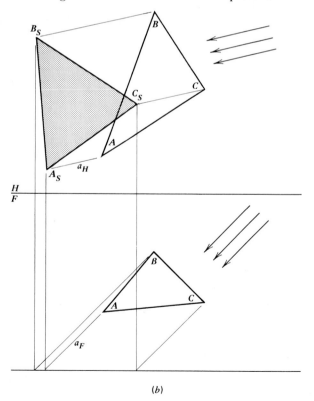

(a)

(b)

Figure 4.16 Shadow cast by triangle.

SAMPLE PROBLEM 4.39

The sun's rays are coming from the direction shown in (*a*). Determine the resulting shadows from the step-shaped solid on itself and on the ground.

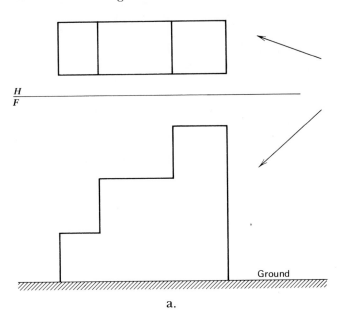

a.

Solution

First, corners on the solid that will influence the shape of the shadow are identified with numbers 1 to 7 (*b*). Rays are then constructed through these points in the front view. Where these various rays intersect the ground or the solid itself shown in the front view, they are identified with the same number and a prime mark. Now, knowing the direction of the rays in the horizontal view or the top view, and where the various rays intersect the ground or the object in the front view, the points of intersection can be projected up into the horizontal view. Note that point 2' is on the second step and all the other points of interest are on the ground. The limit of the shadow in the *H* view from point 1 to 1' corresponds to the vertical corner of the solid. The limit from 1' to 3' corresponds to the line on solid 1–3. The limit of the shadow 4'-5' corresponds to line 4–5 on the object, and so on. We suggest you follow through each limiting line on the shadows to understand their source thoroughly.

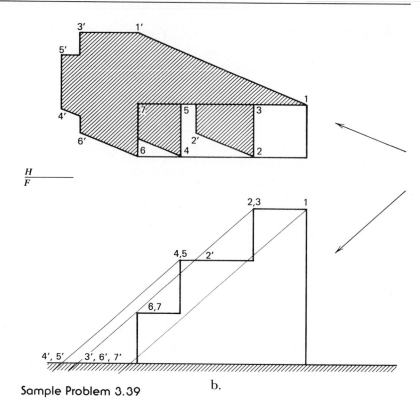

$$\frac{H}{F}$$

Sample Problem 3.39

b.

PROBLEMS

4.40 Determine the true shape of the shadow cast by plane *ABC* on surface *S* shown in edge view in the *F* view.

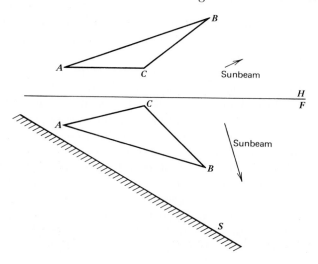

4.41 Rework Sample Problem 4.39 with the direction of the sun's rays as shown.

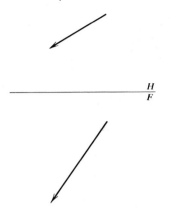

4.42 An open window with a rectangular shade in a "paper-thin" wall is shown with three orthogonal views and an isometric view. The direction of the sun is shown in the H and F views. Determine the resulting shape of the sunspot on the floor.

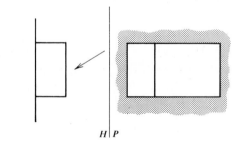

$H \mid P$

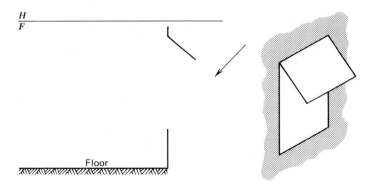

$\dfrac{H}{F}$

Floor

4.43 Determine the true shape of the sunspot cast on the floor when the rays come through the rectangular window in the direction shown. The two views of the window are shown as sections.

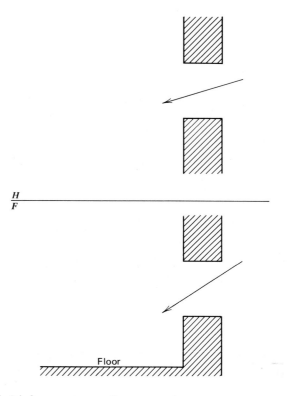

4.44 Light emanates from a point source near a rectangular-shaped opening in a vertical wall as shown. Show the limits of the light spot on the floor.

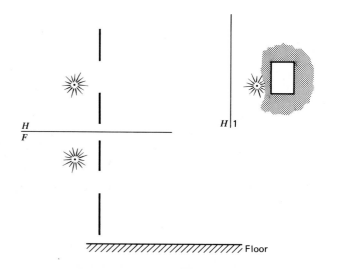

4.45 Find the sun pattern cast on the floor. It is desired to find the effect when the sun's rays are coming from the direction shown.

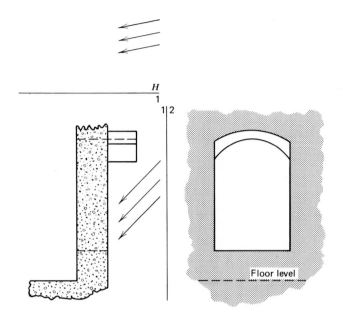

4.10 CONTOURS AND PROFILES

Surfaces can be represented by mathematical models of the general form $z = f(x, y)$, as illustrated in Figure 4.17a. A common use of such mathematical models in engineering analysis

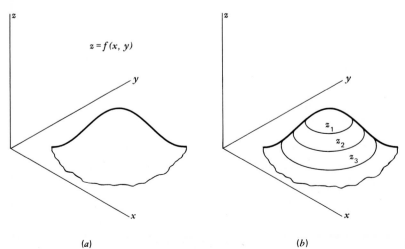

(a)   (b)   Figure 4.17

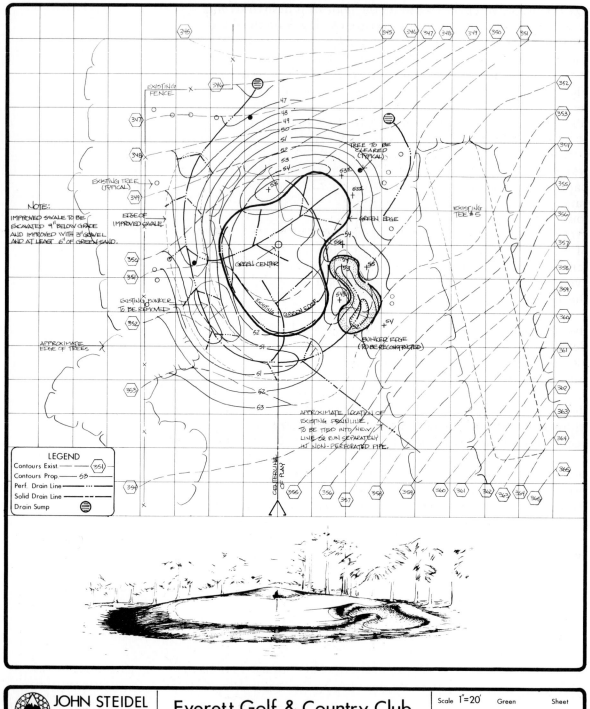

Figure 4.18 A topographical map.

has the x- and y-axes representing distance or position and the z-axis representing such quantities as force, stress, pressure, density, velocity, temperature, or height. These models are contour maps, and lines of a constant z value are contour lines. When curves of constant z values are identified as shown in Figure 4.17b, they usually have specific names. If z represents temperature, the curves are called isothermal lines; if z stands for stress, the curves are isostress lines; if z is height, the curves are isoheight lines or contour lines.

Contour lines are a common method of representing and working with a three-dimensional surface on two-dimensional paper. A map showing contour lines or lines of constant z elevation is called a contour map. A topographical map is a contour map showing lines of constant elevation. A typical topographical map is shown in Figure 4.18; it is a contour drawing of the fourth hole of a golf-course green in Everett, Washington. Elevations of points of interest can be read directly, and average slopes between points can be determined knowing the (horizontal) scale of the map and the contour intervals.

A second and companion way to represent a surface is to see the cross section of the surface. Sectioning will be discussed in detail in Chapter 7, on engineering drawings. A section is a view of either of two parts of a three-dimensional object, where two parts result when a plane intersects the object and divides it. A sectional view is the internal view that results. Consider the intersection plane as a cutting plane and the sectional view as the cutaway view. See the simple example in Figure 4.19.

A profile is a vertical section that results when the intersecting plane is vertical. This plane can be curved. Construction projects such as roadways, pipelines, and canals involve the profile of the ground surface along the path of the construction. An example of a construction profile is shown in Figure 4.20. Note how the exaggeration of the vertical scale helps communicate the shape of the profile, and that the profile represents a curved vertical surface.

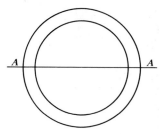

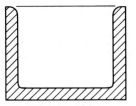

Sectional view of cup resulting from intersection of plane AA with cup; crosshatching is a convention

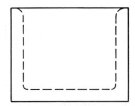

Two orthogonal views of a ceramic cup

Figure 4.19 Section example.

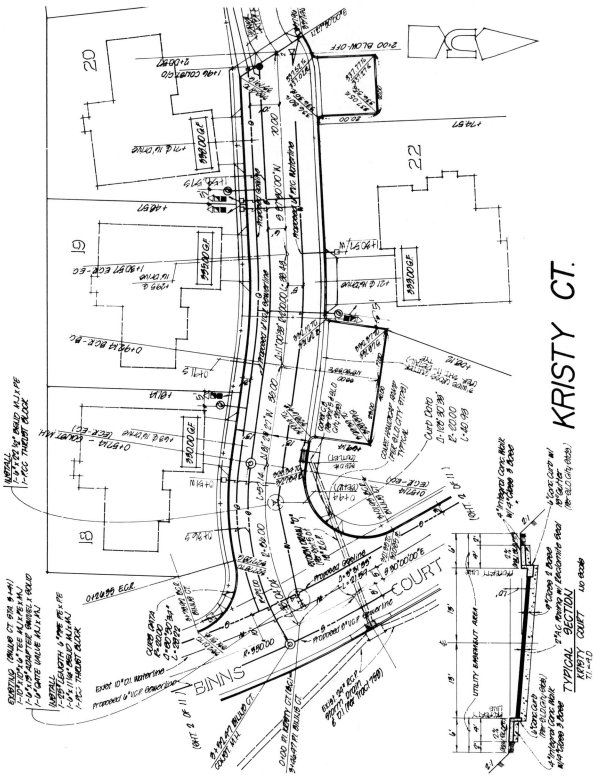

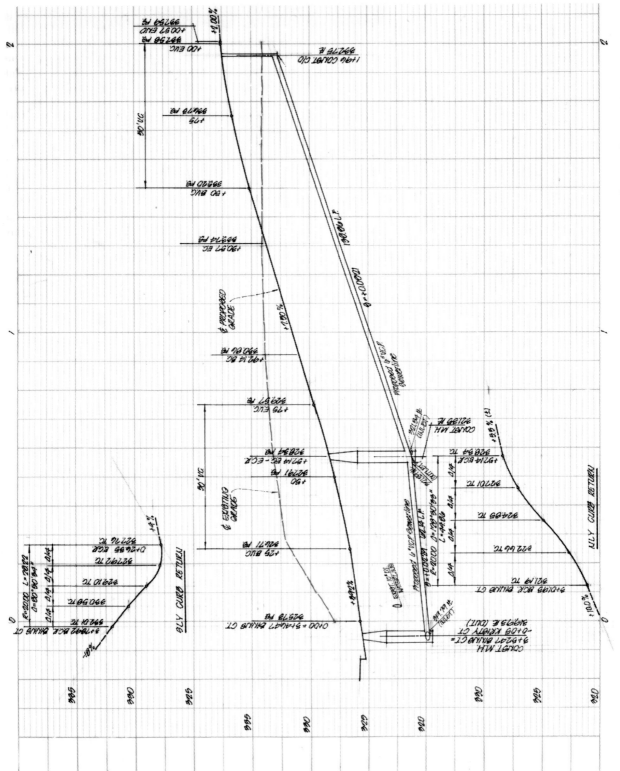

Figure 4.20 Construction profile. (Courtesy Central Coast Engineering and Cuesta Valley Properties, San Luis Obispo, California)

4.11 CUT AND FILL

When a horizontal or gradually sloping path or roadway is constructed, it usually does not match the topography of the area in which it is being built. It must be either built up from the surface or cut into the surface. The determination of the location of the boundaries of these cuts and fills is the subject of this section. The intersections of sloping planes with topographical surfaces is a special kind of intersection problem.

Consider first the situation shown in Figure 4.21. The *H* and *F* views of the cone include contour lines 1 through 5, spaced at equal height intervals on the cone. Sloping plane *ABCD* is shown in both views and is in edge view in the *F* projection. To find the intersection of plane *ABCD* and the cone, we can simply project the intersections of the plane with the contour lines seen in *F* into the *H* view and construct the curve of intersection. You should remember from solid geometry that the intersection of a plane and a cone is an ellipse, which is what we obtain.

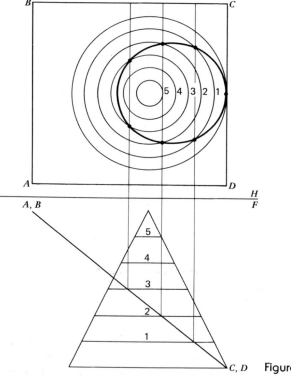

Figure 4.21

We can solve this same problem using only the horizontal view, as shown in Figure 4.22. Contour lines corresponding to heights 1 to 5 have been added to the plane. Recall that contour lines on a plane surface would be straight lines. Now, to find the intersection of the plane and the cone we need only to find the intersection of the various contour lines. For example, the points where straight line 2 on the plane intersect circle 2 on the cone correspond to two intersection points.

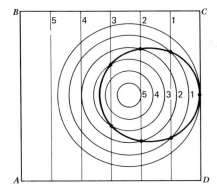

Figure 4.22

You may recall driving on a road through hilly terrain. When the road is above the original surface of the ground, the fill on each side of the road has a plane sloping surface down from the road. When the road is lower than the original ground surface, the cuts made are plane sloping surfaces up from the road.

SAMPLE PROBLEM 4.46

The velocity profile of an open channel flow is shown. Estimate the average velocity of the flow.

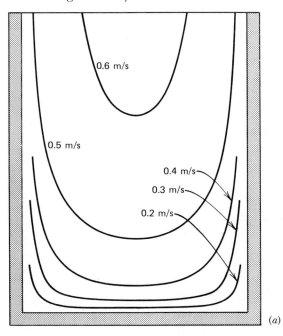

(a)

Solution

The contour lines in (a) represent isovelocity lines or constant-velocity lines. To estimate the average velocity in the channel, we are going to assume that the velocity is constant in areas bounded by solid lines midway between the velocity-contour lines.

In (b) the constant-velocity lines are repeated as dashed

lines and the solid lines are sketched midway between these dashed lines. Next, the area in each velocity range is estimated simply by counting the number of square units in each area. For example, the shaded area corresponds to 0.5 m/s. The product of the velocity and square units gives the flow rate for that section. Then the average flow rate can be obtained by dividing the total of the flow rates by the total cross-sectional area. The accompanying table shows this calculation and the results.

Note that the actual cross-sectional area of the channel is 391 cm² and the estimated area is 388 cm². The counting of square units causes this small error of less than 1%.

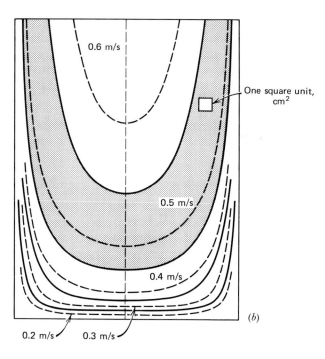

(b)

Velocity, m/s	Square Units, cm²	Product, cm³/s
0.6	122	7320
0.5	150	7500
0.4	56	2240
0.3	32	960
0.2	28	560
	388	18580

$$\frac{18580 \text{ cm}^3/\text{s}}{388 \text{ cm}^2} = 48 \text{ cm/s} = 0.48 \text{ m/s}$$

SAMPLE PROBLEM 4.47

Determine the location of the top of the cut and the bottom (toe) of the fill for the portion of the 5-ft-wide pathway shown. The grade of the cut is to be 1:1 (100 percent), and the grade of the fill is to be 1:2 (50 percent). The pathway is to have a 10 percent slope down to the east, and the elevation of station $1 + 40$ is known to be at ground level (120 ft). Station $1 + 40$ is a conventional way to indicate a point on the path 140 ft horizontally from a reference point. The location of the path and the contour of the immediate area are shown in (*a*).

Solution

This problem is just like the cone and sloping-plane problem we have discussed. The side surfaces of the cut (where the path is below the original ground surface) and the fill (where the path must be built up from the original surface) are sloping planes that intersect the edge of the path (a straight line) and intersect the ground surface (a curve).

At each station the elevation of that point on the path has been added in (*b*). Remember the elevation of station $1 + 40$ is 120 ft and the path has a positive slope to the west (left) of 10 percent. We can now see from the contours that to the left (west) of station $1 + 40$ must be a fill because the path is above the original ground surface, and to the right (east) of station $1 + 40$ a cut must be made.

The final surface shape at station $1 + 20$ is shown as a section in (*c*). In (b) 4-ft intervals have been laid off perpendicular to the pathway at station $1 + 20$. Since the slope is 50 percent, these intervals represent the horizontal distance corresponding to a vertical change of 2 ft. The straight line labeled *a* is nothing more than a 120-ft contour line on the finished fill. Note that when you are at station $1 + 40$ your elevation is 120 ft, but when you are at station $1 + 20$ your elevation is 122 ft and you must move 4 ft horizontally to drop down to a 120-ft elevation. Knowing the slope of the plane surface of the fill to be 50 percent, you can make the additional straight contour lines intersect the corresponding contours on the ground surface locating points on the bottom (toe) of the fill. The work is completed in (*b*).

The cross section of the resulting cut at station $1 + 80$ (180 ft from the base point) is shown in (*d*). The determination of the location of the top of the cut is like the fill problem, with two differences. The slope of the finished surface is 100 percent (1:1 or 45 deg), and the straight contour lines on the finished

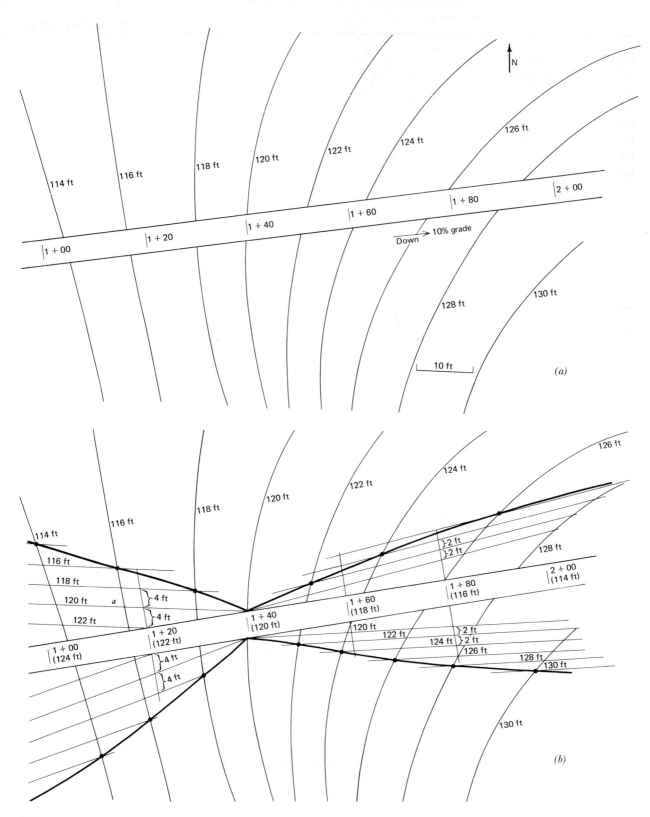

(a)

114 ft   116 ft   118 ft   120 ft   122 ft   124 ft   126 ft

|1 + 00 |1 + 20 |1 + 40 |1 + 60 |1 + 80 |2 + 00

Down → 10% grade

128 ft   130 ft

10 ft

(b)

N

surface move away from the pathway in a different direction when compared to the fill contour lines. Picture yourself walking along the 120-ft contour starting at station $1 + 40$. When you go west [left in (b)] the path is above you; when you go east [right in (b)] you are above the path. The top of the cut is determined by the intersection of the straight contour lines with the corresponding contours on the original ground surface.

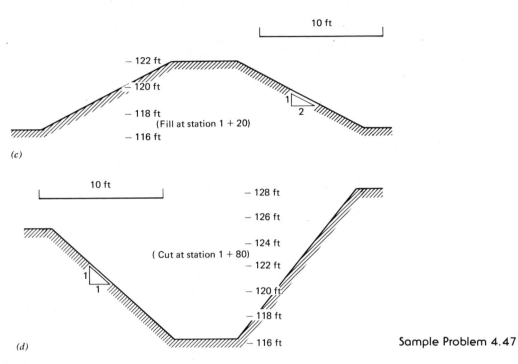

(c)

(d)

Sample Problem 4.47

PROBLEMS

4.48 Make an isometric sketch of the velocity profile in Sample Problem 4.46. Use the z direction to represent velocity with 1 cm = 0.1 m/s and half scale in the x and y directions.

4.49 The velocity (in ft/s) profile in a shallow rectangular channel has been obtained from theory and direct measurement. Determine the average flow rate from the two methods and compare these values.

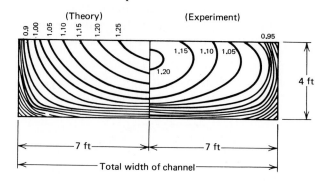

4.50 The following questions refer to the map of Big Bear Creek. Scale: 1 in. = 500 ft

a. If you were to walk from *A* to *B* on a path shown as a straight line on the map, what is the steepest positive or negative slope you would encounter?

b. Can you see point *B* from point *C*?

c. Make an isometric sketch of the top 100 ft of peak *A*.

d. What is the minimum elevation change you must experience to walk from *C* to *D*?

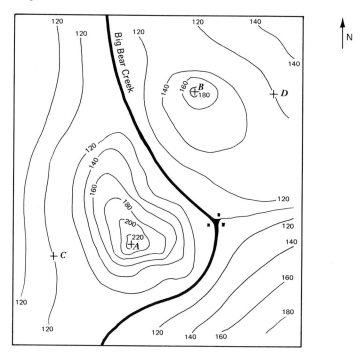

4.51 Points *A* and *B* are two points on two adjacent hills near Big Bear Creek (see Problem 4.50). Find the grade, bearing, and true length of the line of sight from *A* to *B*. Also construct a profile of the ground surface between *A* and *B*.

Answer: −4.6%, N 22°E, 876 ft

4.52 Points *A* and *B* are two points to be connected by a pipeline. Find the grade, bearing, and true length from *A* to *B* (scale 1:1000). Also, draw a profile of the ground surface between *A* and *B*; include the pipeline in your profile.

Answer: 22%, S 41°E, 185 ft

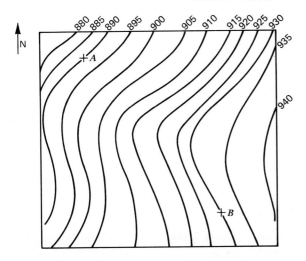

4.53 Referring to Figure 4.18, construct a profile of the land surface along the center of play to the green center.

4.54 A straight 4-ft-wide footpath is to be constructed from point *B* to point *C* over Big Bear Creek (see Problem 4.50). Determine the top-of-the-cuts and the toe-of-the-fills if the slope of the cuts is 40 deg and the slope of the fills is 25 deg.

4.55 Determine the upper limit of the necessary cut (top-of-the-cut) and bottom limit of the required fill (toe-of-the-fill) for the 6 percent grade road shown if the slope of the cut is 1:1 and the slope of the fill is 1:2.

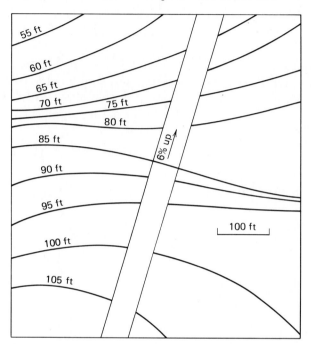

4.12 DEVELOPMENTS

One of the more common applications of descriptive geometry is the "development" of surface areas. Literally, to develop an area means to generate it on a two-dimensional surface such as drawing paper. In its simplest context, a development is a pattern from which flat sheets can be cut, folded, and constructed into geometric shapes. Developments are used to cut ship plate, sheet metal, cloth, glass, plywood, and any other material that is produced in flat sheets. Quite simply, any surface area that can be generated by a straight line can be developed on a two-dimensional surface. Figure 4.23 shows several developments. It is easy to see how a box can be developed from flat pieces, but this statement also extends to the surfaces of cylinders and cones, which can be developed because they can also be generated by a straight line. Conversely, a warped surface cannot be developed.

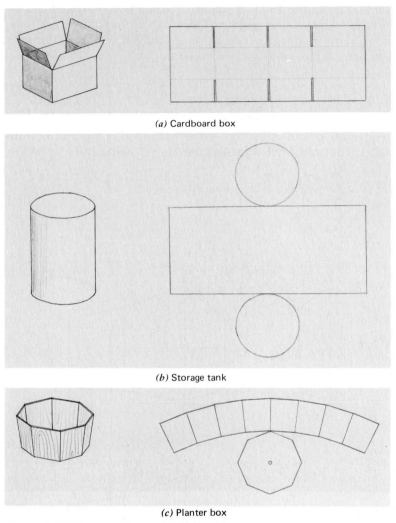

(a) Cardboard box

(b) Storage tank

(c) Planter box

Figure 4.23 Development examples.

The process of development of a surface consists of determining and properly placing the true lengths of elements within that surface. Since these elements always remain the same length and maintain their position with respect to one another, it is easy to see how a plane unwarped surface area can be developed on a two-dimensional surface.

One of the more interesting applications of development is mapping. Surface distances on a solid body can be easily seen and laid out on a development of the surface. A development is a map of a surface area. For example, stairs leading to the top of an oil storage tank could be shown on a development of the tank side. Lines, such as seams or folds on a surface, can be described on a development. This form of mapping is not to be confused with geographic mapping, which is the projection of a warped surface onto a flat surface.

SAMPLE PROBLEM 4.56

The solid shown is a three-sided prism, with top and bottom faces *ABC* and *PQR*. Develop the lateral surface of the prism.

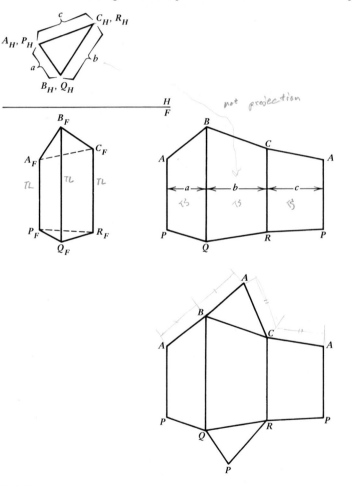

Solution

The prism has three sides, each of which is a plane. Both the top and the bottom faces are also planes. This surface fits our definition of a developable surface, so the entire surface area of the prism can be developed on a two-dimensional surface.

Elements *AP, BQ,* and *CR* are in true length in the frontal view. First, lay out element *AP* to the right of the frontal view. This step is the beginning of your development; it is *not* a profile view. Now, lay out BQ and CR by projecting each from the frontal view. To locate AP, BQ, and CR correctly, you must next space them at the proper distance from each other. As parallel lines, lines AP, BQ, and CR are in point view in the horizontal view. The perpendicular distances between them are a, b, and c, respectively. In the development, these distances are laid out between the elements. Now, draw lines AB, BC, etc. The completed drawing is the development of the lateral surface of the prism. The development is a pattern. The development could be cut out and folded along lines BQ and CR to construct the prism.

The top and bottom faces can be added with a little extra work. Lines AB, BC, and CA are seen in true length in the lateral development. It is a simple matter to construct a top and bottom face that would complete the entire prism. To draw face ABC, locate point A by scribing two arcs, one with a center at B and radius AB and another with center at C and radius AC. Triangle ABC is face ABC, and it is in true shape. Do the same for face PQR.

SAMPLE PROBLEM 4.57

Solid $ABCD$ is a tetrahedron. Develop the lateral surface.

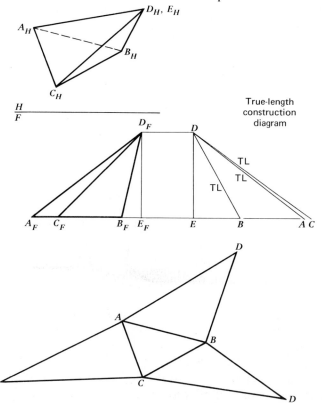

Solution

A tetrahedron has four triangular faces. Only one face, base *ABC*, is shown in true shape in this projection. In it, lines *AB*, *BC*, and *CD* are seen in true length.

Begin the development by redrawing the base of the tetrahedron, triangle *ABC*. Placement of the development is arbitrary. Lines *AD*, *BD*, and *CD* are not in true length, but the true lengths of these lines can be found by a simple construction. Line *DE* is the altitude of the tetrahedron. It is also the vertical rise of each of the lines *AD*, *BD*, and *CD* and it is in true length. The horizontal runs for each line are *AE*, *BE*, and *CE* and are seen in the horizontal view. In the true-length construction diagram, taken from the frontal view, lines *AD*, *BD*, and *CD* are constructed in true length and are now added to the development.

With *A* as a center, scribe an arc *AD*; with *B* as a center, scribe an arc *BD*. These arcs locate point *D*, which completes the development of face *ABD*. Faces *BCD* and *ACD* can be found similarly. Each is in true shape. The completed development could be used as a plane pattern to construct the tetrahedron.

SAMPLE PROBLEM 4.58

Develop the four-sided, symetrical hopper in the figure. The hopper is open at the top and the bottom.

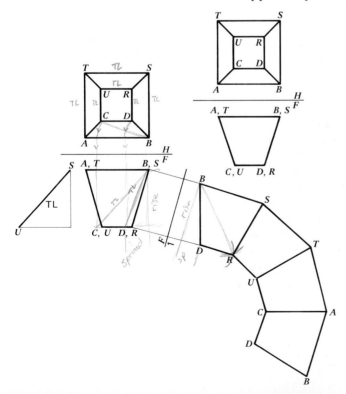

Solution

This problem presents a variation of the two previous problems. Each face is a plane surface bounded by two parallel lines and two nonparallel lines. They are quadrilateral surfaces and not triangles, which makes it a little more difficult to locate points and reconstruct shapes.

Lines *AB* and *CD* are horizontal lines and are in true length in the horizontal view. Plane surface *BDRS* is seen in edge view in the frontal view. Projection plane 1 is now constructed parallel to plane *BDRS*, and plane *BDRS* is projected into plane 1. In that view, plane *BDRS* is in true shape. This projection is the beginning of the development.

Using two additional projection planes, it would be possible to determine the true shape of plane surface *ABCD*. This surface could then be constructed adjacent to the true shape of plane *BDRS*, but there is an easier way to determine the true shape of surface *ABCD*.

Making use of the construction technique of Sample Problem 4.57, find the true length of either diagonal *SU* or diagonal *TR*. Now, locate either point *T* or point *U* on the development. Construct parallel lines *UR* and *ST*. The true lengths of both lines are known. Complete the development.

[handwritten: ← make Δ's out of quad.]

It would also have been possible to use a true-length diagram to find the lengths of *AC* and *BD*, as well as *AD* or *CB*. The approach is a matter of individual preference.

SAMPLE PROBLEM 4.59

Develop the lateral area of the oblique cone shown in orthogonal projection.

Solution

A cone is generated by the circular rotation of a line if one end of the line is fixed. The fixed end is the vertex of the cone.

A development consists of the determination and the proper placement of the true length of elements within that surface. In this case we will choose an element as any line extending from the vertex to the base of the cone, such as a line *OA*. Let us consider 12 elements, each one spaced 30 deg from the next, as measured at the base of the cone. The vertex *O* is common to all 12 elements. Now, forget that the surface is conical. Consider it a 12-sided pyramid. Using only your basic descriptive geometry, it is not difficult to determine the true lengths of all

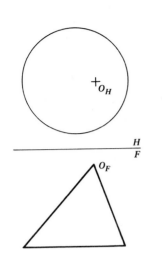

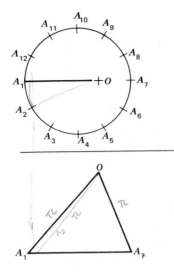

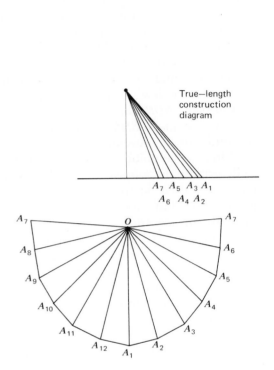

True—length
construction
diagram

Sample Problem 4.59

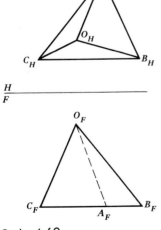

Prob. 4.60

12 elements. The element OA_1 is already in true length in the frontal view. The base is a 12-sided polygon seen in true shape. You have all that is needed to develop the surface.

The only problem with this solution is one of accuracy. We have replaced a cone with a pyramid. Obviously there is an error. The development will not quite fit the surface. The error will be the difference between the chord A_1A_2 and the actual arc distance along the surface. The error could be reduced by taking smaller increments between A_1 and A_2, such as 10, 5, or even 1 deg.

PROBLEMS

4.60 Develop the surface area of the tetrahedron $OABC$ if base ABC is removed. It is to be developed from sheet metal with folds along OA and OB.

4.61 Develop the surface of the right circular cone.

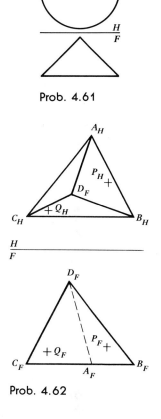

Prob. 4.61

4.62 Points P and Q are on faces ABD and BCD of the tetrahedron $ABCD$. There are four possible straight-line surface paths between P and Q. What is the shortest path and what is the longest path?

Answer: 1.00 in., 1.45 in.

Prob. 4.62

4.63 Points A and B are on the surface of the right circular cone. Determine the shortest surface distance between A and B.

Answer: 22 mm

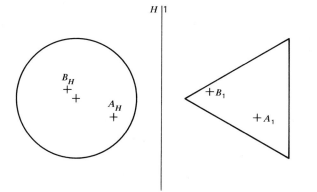

Prob. 4.63

4.64 Develop the surface area of the hopper shown. Note that there must be a seam along lines *FC* and *GD*.

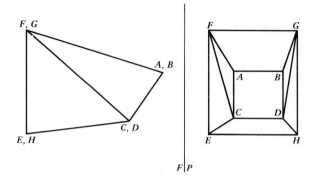

4.65 Develop the Tetrapak container shown. Indicate with crosshatching on the developed surface where two surfaces are glued together.

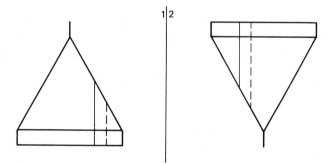

4.66 Develop the lateral surface of the hopper. The *H* and *F* views are drawn to the scale shown. Seam *AB* is horizontal.

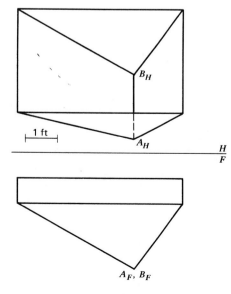

4.67 Develop the circular 2-ft-diameter air duct that provides ventilation between perpendicular walls of two non-adjacent rooms.

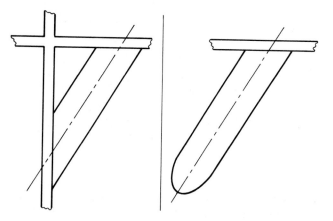

4.68 Develop the conical surface of the modern sheet-metal fireplace shown. Choose a convenient scale and be accurate.

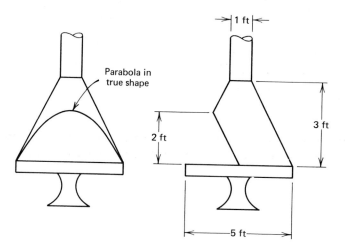

5
GRAPHIC ANALYSIS

5.1 ANALYSIS

Most engineering problems involve the analysis of physical systems and physical quantities. To make these analyses, there must be some representation of these systems and quantities. We call this representation *modeling*, the subject of Chapter 8, and it is very much related to the art of engineering.

Modeling calls for considerable experience and judgment, which you, as a young engineer, do not have. You will learn and use the techniques of engineering analysis in your undergraduate study, and possibly later in your graduate study, long before you are able to model physical systems adequately. Analysis is usually mathematical, but there are some very useful graphic methods of analysis. This chapter is an introduction to several graphic techniques that can be used in engineering analysis.

All of what follows will be covered in much more depth in your subsequent courses. Our goal is to introduce you to a few graphic techniques and demonstrate how you can use your knowledge of graphics in the study and analysis of engineering problems. As you go further into more analysis techniques, we hope you will add to these introductory few. Graphic analyses can serve you well as an aid to description and understanding.

5.2 SCALAR QUANTITIES

Most quantities in engineering have a physical origin or interpretation. *Physical* implies something material, *material* often implies visual, and *visual* does imply graphic. This succession of thought cannot be used for all things, but physical quantities can be graphically presented and quantified, and the techniques of engineering graphics can be used to give them a visual interpretation.

The primary method of quantification is through the use of numbers. A number is the result of a counting (i.e., how much or how many?). How much or how many can be represented by a length. A quantity that can be represented by a length is a *scalar,* and a scalar is completely specified. Time is a scalar quantity. A number and what it represents, such as seconds, minutes, hours, days, weeks, months, or years, are all that is required to specify time. Other examples of scalars are mass, volume, and money. Area and volume are spatial; that is, they are three-dimensional in nature, but their specification is simply a number accompanied by units.

On a straight line, such as in Figure 5.1, an origin 0 and a unit length have been selected. A number of unit lengths can be represented by a distance from the origin. The number 1 is 1 unit length from the origin, the number 2 is 2 unit lengths, and the number 6 is 6 unit lengths from the origin. If the number is an integer or a fraction, it can be represented as a specific point, a specific distance from the origin. This property makes it a *rational number.* Integers and fractions are rational numbers. A fraction is a rational number because it consists of one integer divided by another. Although the work may be tedious, all rational numbers can be located on a straight line with common drawing tools.

Certain other numbers cannot be expressed as rational numbers and are called *irrational numbers.* Irrational numbers cannot be expressed as either integers or fractions. When they are expressed as decimals, they continue on and on. Roots, such as the square root and the cubic root, logarithms, π, and the

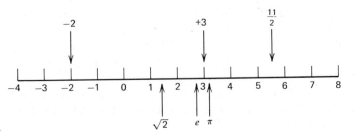

Figure 5.1 Number representation.

base e for Naperian logarithms are often irrational numbers. They can be selected as unit lengths themselves, but there is no mathematically specific point that defines them on any other scale. As examples, π is often used as a unit length to express trigonometric functions, and $\log_{10}$ is often used for functions that vary over a very wide range, such as the Richter number for the magnitude of earthquakes.

5.3 GRAPHIC ADDITION AND MULTIPLICATION

To add and subtract graphically is a simple matter. In Figure 5.2 the number A added to the number B gives

$$A + B = C$$

or

$$A - B = D \qquad (5.1)$$

Subtraction is simply negative addition.

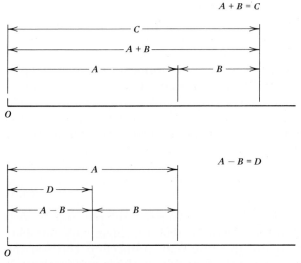

Figure 5.2 Graphic addition and subtraction.

A well-used example of graphic addition and subtraction would be the slide rule. The two scales of a slide rule are *functional scales*, graduated in logarithms. Adding a length on one scale to a length on another is equivalent to adding logarithms, and adding logarithms is multiplication. Subtracting one scale length from another would be division.

Graphic multiplication and division can also be accomplished through similar triangles. Note that for triangles OAB and OCD (Figure 5.3a)

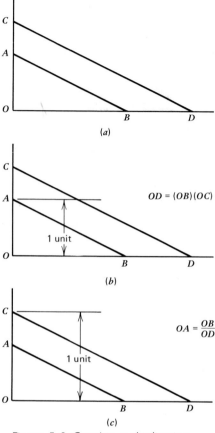

Figure 5.3 Graphic multiplication and division.

$$\frac{OA}{OC} = \frac{OB}{OD}$$

If $OA = 1$ unit, this equation becomes (Figure 5.3b)

$$OD = (OB)\,(OC) \tag{5.2}$$

Thus OD is the product of OB and OC on a scale where $OA = 1$ unit. If $OC = 1$ unit, this equation becomes (Figure 5.3c)

$$OA = \frac{OB}{OD} \tag{5.3}$$

OA is now the quotient of OB divided by OD.

5.4 COMPLEX NUMBERS

The square root of a negative number does not exist, since no real number when squared results in a negative number. But since so much in mathematics involves the square root of a negative number, another unit of length has been invented, the imaginary unit

$$i = \sqrt{-1} \tag{5.4}$$

It is customary to define a complex number z as an ordered pair of real numbers x and y, writing them as

$$z = x + iy \tag{5.5}$$

The real numbers x and y are known as the real and imaginary components of the complex number z.

$$\mathscr{R}(z) = x$$
$$\mathscr{I}(z) = y \tag{5.6}$$

In stating a complex number in two parts, the plus sign is not a sign for scalar addition because real and imaginary numbers cannot be added directly. The only connection between the imaginary unit $\sqrt{-1}$ and the real unit 1 is the definition, $i^2 = -1$.

The graphic representation of a complex number can be shown by plotting the real and imaginary components as the coordinates of a point in the x–y plane. Real numbers exist only as points on the x-axis, and imaginary numbers exist only as points on the y-axis. Complex numbers are numbers anywhere in the x–y plane. When the x–y plane is used to represent complex numbers it is called the *complex plane* or the *Argand* diagram. In Figure 5.4 the number $z = 2 + 3i$ is represented by the point $(2, 3)$ in the x–y plane. The number $z = 0$ is the origin.

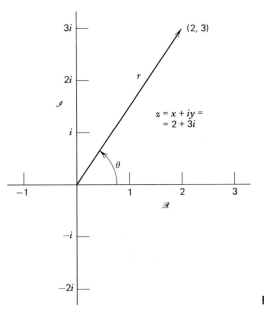

Figure 5.4 Graphic representation of a complex number.

In order to add, subtract, multiply, and divide complex numbers, the graphic form of a complex number is very convenient.

The complex displacement z from the origin can be considered as the distance from the origin to the complex number z. Thus $z = 2 + 3i$ represents a vector from the origin $(0, 0)$ to the point $(2, 3)$. The length of the displacement is the distance

$$r = \sqrt{x^2 + y^2} \tag{5.7}$$

and the direction of the displacement is the angle

$$\theta = \tan^{-1}\frac{y}{x} \tag{5.8}$$

These are also called the *polar coordinates* of the complex number.

The x- and y-coordinates of z are

$$x = r \cos \theta$$
$$y = r \sin \theta$$
$$z = r \, (\cos \theta + i \sin \theta) \tag{5.9}$$

Angle θ is also called the *argument* of z. r is known as the *modulus* of the complex number, and it is the linear distance between the origin and $z(x, y)$.

There are two things to note about the angle θ. First, the complex conjugate $z = x - iy$ would be the mirror image, using the real axis, of $z = x + iy$. Second, angle θ is multivalued; it would have one value between 0 and 2π, another between 2π and 4π, etc. The desired value must be stated.

The sum of two complex numbers is obtained by adding the real and imaginary parts of the complex numbers separately. Thus

$$z_1 = x_1 + iy_1$$
$$z_2 = x_2 + iy_2$$
$$z_1 + z_2 = (x_1 + x_2) + i(y_1 + y_2) \qquad (5.10)$$

Therefore, the vector addition, $z_1 + z_2$, is represented by adding the two complex numbers z_1 and z_2, as in Figure 5.5.

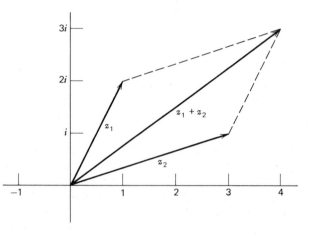

Figure 5.5 Addition of complex numbers.

The product of two complex numbers is also a complex number.

$$z_1 z_2 = r_1 r_2 [\cos (\theta_1 + \theta_2) + i \sin (\theta_1 + \theta_2)] \qquad (5.11)$$

Similarly, dividing one complex number by another gives

$$\frac{z_1}{z_2} = \frac{r_1}{r_2} [\cos (\theta_1 - \theta_2) + i \sin (\theta_1 - \theta_2)] \qquad (5.12)$$

It is not our intent to develop the theory of complex numbers, but one additional thought can be developed to show the graphic usefulness of the concept. If we had a complex number

$$z = r(\cos \theta + i \sin \theta)$$

and z_o were the nth root of z, following the same reasoning used to determine the product of two complex numbers,

$$z_0 = r_0(\cos \theta_0 + i \sin \theta_0)$$

and

$$z^n{}_0 = r^n{}_0(\cos n\theta_0 + i \sin n\theta_0) \qquad (5.13)$$

This is known as *de Moivre's formula*.

$$r_0 = r^{1/n} \qquad (5.14a)$$

and

$$\theta_0 = \frac{\theta}{n} + \frac{2k\pi}{n}, [k = 0, \pm1, \pm2, ..., \pm(n-1)] \qquad (5.14b)$$

Geometrically, n complex numbers would satisfy this equation. They would all lie on a circle with the origin as its center and a radius equal to $r^{1/n}$. The angle of one of the complex roots would be θ/n, and the angle of all the others would be separated from it by multiples of $2\pi/n$. Refer to Sample Problem 5.4 for an example.

SAMPLE PROBLEM 5.1

Line segment AB represents 2 units. Determine the points on the line that represent 1 unit and $1\frac{4}{7}$ units.

Solution
First, the line segment is divided in half by constructing arcs a and b with equal radii and centers at A and B. A straight line through the two points of intersection of the arcs locates the point representing "one," the halfway point, on the line. Then, using some convenient scale, a 7-unit line is drawn at an arbitrary angle with line AB from the midpoint. Finally, similar triangles are used to locate the point representing the value $1\frac{4}{7}$.

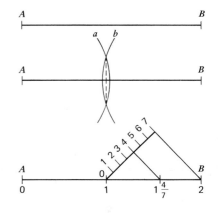

SAMPLE PROBLEM 5.2

Determine $\sqrt{6}$ graphically.

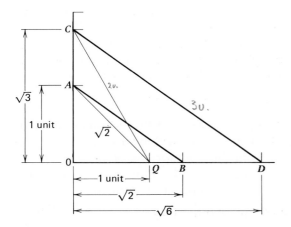

Solution

With a convenient scale, say 1 unit $= 100$ mm, lay off $OQ = 1$ unit horizontally and $OA = 1$ unit vertically. From Q, construct QC at an angle of 60 deg to OQ. This can be done with a triangle, a protractor, or simply by making $QC = 2$ units. Also, lay off $OB = AQ$. Now, construct CD parallel to AB. The following lengths have been found.

$$OA = OQ = 1$$
$$QC = 2$$
$$OC = QC \sin 60 \text{ deg} = \sqrt{3}$$
$$AQ = OB = \sqrt{2}$$
$$OD = (OB)(OC) = \sqrt{2} \times \sqrt{3} = \sqrt{6}$$

In our chosen scale, $OD = 245$ mm or $\sqrt{6} = 2.45$.

SAMPLE PROBLEM 5.3

Given the complex displacements $z_1 = 1 + \sqrt{3}i$ and $z_2 = 1 - i$, determine the vector sum $z = z_1 + z_2$.

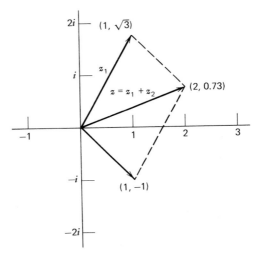

Solution

In the complex plane, $z_1 = 1 + \sqrt{3}i$ is a displacement directed from the origin to the point $(1, \sqrt{3})$. Similarly, $z_2 = 1 - i$ is a displacement directed from the origin to the point $(1, -1)$.

Constructing a parallelogram for the addition of the two displacements, the sum $z = z_1 + z_2$ is the displacement

$$z = 2 + 0.73i$$

This can be checked analytically for the exact answer

$$z = (1 + 1) + (\sqrt{3} - 1)i$$

SAMPLE PROBLEM 5.4

Determine the cubic roots of 1.

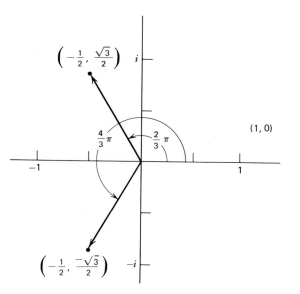

Solution

Obviously, one of the cubic roots of 1 is the number 1 itself, since $1^3 = 1$. But, are there others?

The number 1 in complex form is represented by the point $(1, 0)$ on the real axis.

$$z_1 = 1 + 0i = 1(\cos 0 + i \sin 0)$$

Using Eq. 5.13,

where $r_0 = \sqrt[3]{1} = 1$ and $\theta_0 = \dfrac{0}{3} + \dfrac{2k\pi}{3}$, $[k = 0, 1, 2]$

The solutions all lie on the unit circle where $r = 1$. Note that this circle passes through the point $(1, 0)$, and that at this point $\theta = 0$ $(k = 0)$. Two other points would satisfy Eq. 5.13, one at $\theta = (\frac{2}{3})\pi$ and the other at $(\frac{4}{3})\pi$. These correspond respectively to the conjugate pair

$$z_2 = -\frac{1}{2} + \frac{\sqrt{3}}{2}i$$

and

$$z_3 = -\frac{1}{2} - \frac{\sqrt{3}}{2}i$$

Cubing these complex numbers will verify that they, along with $z_1 = 1 + 0i$, are the cubic roots of 1.

PROBLEMS

5.5 On a linear scale, using the logarithm of 10 as a unit length, graphically find the products of the following numbers by adding their logarithms.

a. 750×35.3

b. 656×297

c. 1450×62.4

d. 1.74×3.21

5.6 Graphically determine the following numbers.

a. 2.3×1.9

b. 5.3×2.9

5.7 Graphically determine the following numbers.

a. 1.3×113

b. 3.1×1.7

On a horizontal axis, with a scale of 1 unit = 10 mm, locate the following numbers.

5.8 $6, 17/4, 7\frac{6}{13}, \sqrt{2}, e$.

5.9 $8, 16/3, 2\frac{6}{7}, \sqrt{5}, \pi$.

5.10 $5, 3/2, 8\frac{7}{11}, \sqrt{3}, \log_{10}550$.

Starting with a unit length, using only a compass and a scale, construct the following numbers.

5.11 a. $\sqrt{8}$

b. $\sqrt{10}$

5.12 a. $\sqrt{20}$

b. $\sqrt{15}$

5.13 Using π as a unit length, plot $y = \tan x$ from $x = 0$ to $x = 4\pi$.

5.14 Using $\log_{10}10$ as a unit length, plot $y = 112.5 + 32.1 \log_{10}R$ from $R = 100$ to $R = 10,000,000$.

5.15 Plot the following complex numbers as distances in the complex plane. State each in polar form.

    a. $-1 + i$

    b. $4 + 3i$

    c. $1 + 4i$

    d. $-5 + 12i$

    e. $-3 + 2i$

5.16 Determine the sum $z = z_1 + z_2$ for the following pairs of complex numbers.

    a. $z_1 = 3 + 2i$; $z_2 = 1 + 4i$.

    b. $z_1 = -3 + i$; $z_2 = -3 - i$.

    c. $z_1 = 2 + 3i$; $z_2 = -2 + 3i$.

    d. $z_1 = -2i$; $z_2 = 2 - 3i$.

5.17 Determine the difference $z = z_1 - z_2$ for the complex numbers given in Problem 5.16.

5.18 Find the sum $z = z_1 + z_2 + z_3$ if $z_1 = 1 + 4i$, $z_2 = -2 + 3i$, $z_3 = 2 - i$.

$$\textit{Answer: } z = 1 + 6i$$

5.19 Find the product $z_1 z_2$, if $z_1 = 3 - 4i$ and $z_2 = 4 + 3i$.

$$\textit{Answer: } z_1 z_2 = 24 - 24 - 7i$$

5.20 Find $z_1 \div z_2$, if $z_2 = \sqrt{2} - \sqrt{2}i$ and $z_1 = -2\sqrt{6} + 2\sqrt{3}i$.

$$\textit{Answer: } z_1 \div z_2 = -2.957 - 0.507i$$

5.21 Determine the two square roots of i.

5.22 Find the cube roots of -1. $\textit{Answer: } z_1 = -1,$

$$z_2 = \tfrac{1}{2} + \frac{\sqrt{3}}{2}i,$$

$$z_3 = \tfrac{1}{2} - \frac{\sqrt{3}}{2}i$$

5.5 VECTOR QUANTITIES

Vectors are very important to engineering analysis. A *vector* quantity requires a direction as well as magnitude specification, and since both magnitude and direction are fundamentally graphic quantities, it is simple to use graphic techniques for

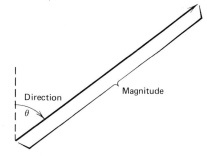

Figure 5.6 Vector quantity.

vector representation. The magnitude of a vector is represented by the scaled length of a line segment (Figure 5.6). The direction of a vector is specified by an arrowhead at one end of the straight-line segment. The representation and manipulation of vector quantities will be discussed in several of the sections to follow, but first we should recognize the more important vector quantities in engineering.

Force, the push or pull of one body on another, is a vector quantity. Consider the 15 mm × 25 mm rectangular bar shown in Figure 5.7. If you place fingers in the holes at each end and pull with a 50-N force, the bar is placed in tension. The forces can be represented as directed line segments, shown as arrows. Here the forces are assumed concentrated; that is, they act along a single line. Your fingers do not pull this way, but, for simplicity, the forces are so represented. Again, this is modeling.

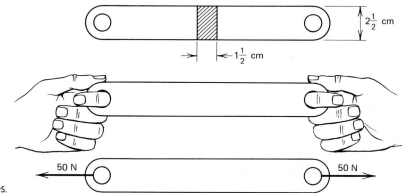

Figure 5.7 Forces.

Moment, the tendency of a force to rotate a body, is a vector quantity. The force **F** in Figure 5.8*a* tends to rotate the body about the axis 0–0. The resulting moment, with units of length times force, is along the axis. The right-hand convention, commonly used for moment vectors, is shown in Figure 5.8*b*. The moment vector is in the direction of the thumb when the fingers are in the rotation direction. Section 5.12 will discuss moments further.

Figure 5.8 Vector representation of moment.

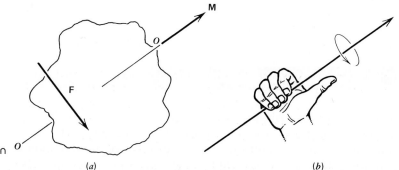

(a)   *(b)*

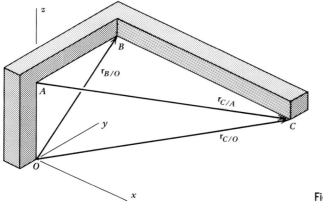

Figure 5.9 Position vectors.

Position of one point with respect to another can be represented with a vector. Position includes a distance and a direction. Figure 5.9 presents some examples. The position vector $\mathbf{r}_{B/O}$ represents the vector displacement of B and O.

You will encounter vector quantities in your study of engineering analysis. Many quantities in mechanics, such as displacement, velocity, acceleration, momentum, and impulse, are vectors. Can you think of others?

5.6 VECTOR ADDITION

Vector quantities can be represented as directed line segments, which are simply scaled lengths of line with an arrowhead to indicate direction. Vectors can be added, subtracted, multiplied, and divided, but these operations must be done carefully, obeying the special properties attributed to vectors.

It is important to become familiar with conventional vector notations. In printed matter, vectors are usually represented with boldfaced type (e.g., **A, B,** and **C**). Representations in regular type (A, B, and C), refer only to the magnitude of the vector. For handwritten work, a number of techniques are used. We suggest using underlining (e.g., $\underline{A}$, $\underline{B}$, and $\underline{C}$), leaving the top of the symbol for other specifications such as a bar ($\bar{A}$) to indicate that A specifies a specific quantity related to the mass center of a body.

Vector addition and subtraction is a geometric process that can be accomplished by graphic or mathematical means. Assume that vectors **A** and **B** in Figure 5.10*a* are two nonparallel vectors that lie in the plane of the paper. Extending the lines of action of **A** and **B** until they intersect (Figure 5.10*b*), and moving the vectors forward or backward along their individual lines of action until the tails meet will form two sides of a parallelogram

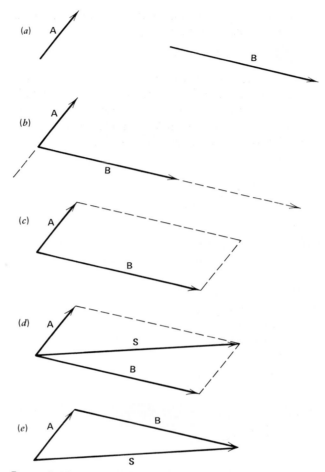

Figure 5.10 Vector addition.

(Figure 5.10c). The diagonal of this parallelogram is vector **S,** which is the sum of **A** and **B** (Figure 5.10d).

$$\mathbf{A} + \mathbf{B} = \mathbf{S} \qquad (5.15)$$

Note that placing the vectors head to tail gives the same results (Figure 5.10e), but it requires *moving* vector **B** from its line of action. Some vectors can not be moved along their line of action or moved from their line of action. These so-called fixed vectors have local effects that can not be ignored. Also, note that for nonparallel vectors, the magnitude of the sum is always smaller than the sum of the magnitudes.

Now, to subtract the two vectors **A** and **B,** as in Figures 5.11a and 5.11b, we write

$$\mathbf{D} = \mathbf{A} - \mathbf{B} \qquad (5.16)$$

which is the vector equation for the difference of **B** from **A.** If we think of $-\mathbf{B}$ as a vector of equal magnitude and opposite sense to $\mathbf{B}$, then we see that (Figure 5.11c)

$$\mathbf{D} = \mathbf{A} + (-\mathbf{B}) \qquad (5.17)$$

and vector subtraction becomes vector addition after the sense change is made. Again, the parallelogram law, or head-to-tail, construction is used (Figure 5.11d). As you become more familiar with the use of vectors in engineering, you will see the need for dealing with vectors in a variety of forms.

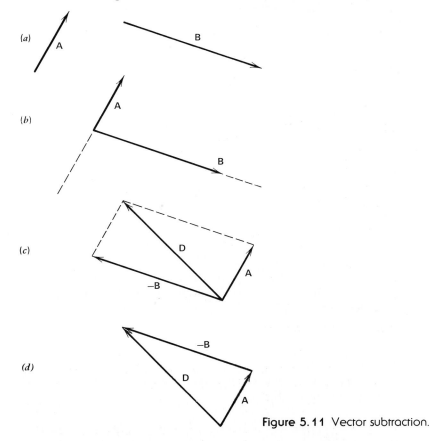

Figure 5.11 Vector subtraction.

5.7 VECTOR COMPONENTS

The *components* of a vector are vectors that, when added, give the original vector. In Figure 5.12, four sets of components are shown for vector $\mathbf{R}$. In Figure 5.12a, $\mathbf{R}_1 + \mathbf{R}_2 = \mathbf{R}$; that is, $\mathbf{R}_1$ and $\mathbf{R}_2$ are the vector components of $\mathbf{R}$. In Figure 5.12b, $\mathbf{R}_x + \mathbf{R}_y = \mathbf{R}$. These are useful because $\mathbf{R}_x$ and $\mathbf{R}_y$ are at right angles to each other or orthogonal.

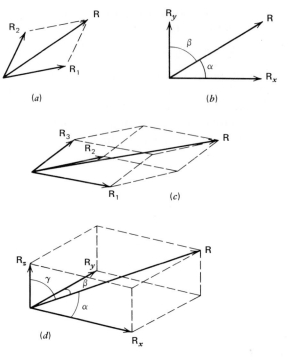

Figure 5.12 Vector components.

In Figure 5.12c, $\mathbf{R}_1 + \mathbf{R}_2 + \mathbf{R}_3 = \mathbf{R}$; in Figure 5.12$d$, $\mathbf{R}_x + \mathbf{R}_y + \mathbf{R}_z = \mathbf{R}$. Both sets $\mathbf{R}_1$, $\mathbf{R}_2$, $\mathbf{R}_3$ and $\mathbf{R}_x$, $\mathbf{R}_y$, $\mathbf{R}_z$ are three-dimensional components of $\mathbf{R}$. There are occasions when it is more convenient to use one or the other. Note that both sets form a parallelepiped within which the vector $\mathbf{R}$ is the diagonal. Using $\mathbf{R}_x$, $\mathbf{R}_y$, and $\mathbf{R}_z$ may be more useful, since we have the geometric relation for the scalar

$$R = \sqrt{R^2{}_x + R^2{}_y + R^2{}_z} \tag{5.18}$$

and the *direction cosines*, $\cos \alpha$, $\cos \beta$, and $\cos \gamma$, of the vector $\mathbf{R}$ can be defined as

$$\cos \alpha = \frac{R_x}{R}, \quad \cos \beta = \frac{R_y}{R}, \quad \cos \gamma = \frac{R_z}{R} \tag{5.19}$$

Figure 5.12d shows the three components of vector $\mathbf{R}$ and the associated angles with the orthogonal x-, y-, and z-axes. For two-dimensional situations, $R_z = 0$, $\cos \gamma = 0$, and $\gamma = 90$ deg.

A vector having unit magnitude is called a *unit vector*. The direction of vector $\mathbf{R}$ can be specified with unit vector $\mathbf{u}_1$ in the same direction as $\mathbf{R}$. The magnitude of vector $\mathbf{R}$ is scalar R. Therefore, as shown in Figure 5.13,

$$\mathbf{R} = R\,\mathbf{u}_1 \quad \text{or} \quad \mathbf{u}_1 = \frac{\mathbf{R}}{R} \tag{5.20}$$

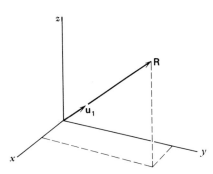

Figure 5.13 The unit vector.

Note that Eq. 5.20 represents two of the same concepts we developed earlier. The first is that a line segment graphically models a quantity; the second is that a unit vector represents a unit length along a directed line. This statement, then, is simply that vector $\mathbf{R}$ is measured by an R number of unit vectors $\mathbf{u}_1$.

This use of unit vectors was developed in some class notes on the elements of vector analysis by Josiah Willard Gibbs (1839–1903), a mathematical physicist who graduated from and taught at Yale University and who developed the application of thermodynamics to chemistry.

The magnitude of unit vector $\mathbf{u}_1$ is unity. Its components along the x-, y-, and z-axes are $\mathbf{u}_{1x}$, $\mathbf{u}_{1y}$, and $\mathbf{u}_{1z}$, and

$$\mathbf{u}_1 = \mathbf{u}_{1x} + \mathbf{u}_{1y} + \mathbf{u}_{1z} \tag{5.21}$$

The coordinate components of the unit vector are the direction cosines of the unit vector (see Figure 5.12d),

$$u_{1x} = u_1 \cos \alpha = \cos \alpha$$
$$u_{1y} = u_1 \cos \beta = \cos \beta$$
$$u_{1z} = u_1 \cos \gamma = \cos \gamma$$

It is easy to verify by geometry that

$$u^2_{1x} + u^2_{1y} + u^2_{1z} = \cos^2 \alpha + \cos^2 \beta + \cos^2 \gamma = 1 \tag{5.22}$$

This is an important concept in vector analysis.

Special symbols are reserved to establish unit vectors along the axes of the basic rectangular coordinate system. These unit vectors have a special designation.

$\mathbf{u}_1$ directed along the x-axis is $\mathbf{i}$.
$\mathbf{u}_1$ directed along the y-axis is $\mathbf{j}$.
$\mathbf{u}_1$ directed along the z-axis is $\mathbf{k}$.

You will note that $\mathbf{i}$, $\mathbf{j}$, and $\mathbf{k}$ are unit vectors, like $\mathbf{u}_1$, but they are *not* components of $\mathbf{u}_1$. It follows that

$$\mathbf{u}_1 = \cos \alpha \, \mathbf{i} + \cos \beta \, \mathbf{j} + \cos \gamma \, \mathbf{k} \tag{5.23}$$

Going back to Eq. 5.20, vector $\mathbf{R}$ can be written as

$$\mathbf{R} = R \, \mathbf{u}_1 \tag{5.20 repeated}$$

or

$$\mathbf{R} = R_x \, \mathbf{i} + R_y \, \mathbf{j} + R_z \, \mathbf{k} \tag{5.24}$$

where R_x, R_y, and R_z are scalar quantities.

Substituting the expressions of direction cosines into Eq. 5.24 will result in

$$\mathbf{R} = R(\cos \alpha \, \mathbf{i} + \cos \beta \, \mathbf{j} + \cos \gamma \, \mathbf{k}) \tag{5.25}$$

When using matrix algebra, the orthogonal components of a vector can be represented as a single row or a column matrix. Therefore, **R** can be represented as

$$[R_x \ R_y \ R_z] \text{ or } \begin{bmatrix} R_x \\ R_y \\ R_z \end{bmatrix}$$

and unit vector $\mathbf{u}_1$ specified as

$$[\cos \alpha \ \cos \beta \ \cos \gamma] \text{ or } \begin{bmatrix} \cos \alpha \\ \cos \beta \\ \cos \gamma \end{bmatrix}$$

5.8 VECTOR MULTIPLICATION

Unlike the case of scalars, there are two methods of multiplying vectors, and the results are different.

The *dot product* is defined as the magnitude of one vector multiplied by the component of the second in the direction of the first or the product of the magnitude of the two vectors and the cosine of the angle between them. The dot product is a scalar quantity.

$$\mathbf{A} \cdot \mathbf{B} = (A)(B) \cos \theta \tag{5.26}$$

In Figure 5.14 it is demonstrated how the dot product can be interpreted as the product of the magnitude of one vector with the projection of the second on the first. It follows that the two vectors **A** and **B** are perpendicular when $\mathbf{A} \cdot \mathbf{B} = 0$, since $\cos \theta = 0$, only at $\theta = \pi/2, 3\pi/2$, etc.

The *cross product* is defined as a vector whose magnitude is the product of the magnitude of the two vectors and the sine of the angle between them.

$$|\mathbf{A} \times \mathbf{B}| = (A)(B) \sin \theta \tag{5.27}$$

The line of action of the vector is perpendicular to both vectors, or, stated another way, it is perpendicular to the plane defined by the two vectors. A vector is simply a directed line segment, and two lines can define a plane. So the line of action is perpendicular to the plane containing vectors **A** and **B.** The direction is defined by the unit vector $\mathbf{u}_1$ in Figure 5.15, and the sense of the cross product is determined by the right-hand rule. Think of the fingers on your right hand as being in the direction of the first vector in the cross product. As your fingers rotate toward the second vector, your right thumb defines the sense of the cross product, as in Figure 5.15. It also follows that when the two vectors are directed along the same line of action, $\mathbf{A} \times \mathbf{B} = 0$, since $\sin \theta = 0$ at $\pi, 2\pi$, etc.

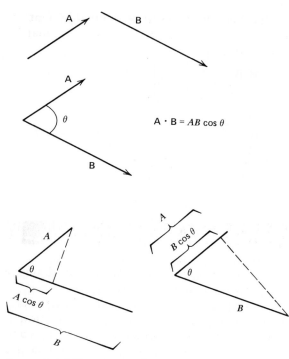

$$\mathbf{A} \cdot \mathbf{B} = AB \cos \theta$$

Figure 5.14 The dot product.

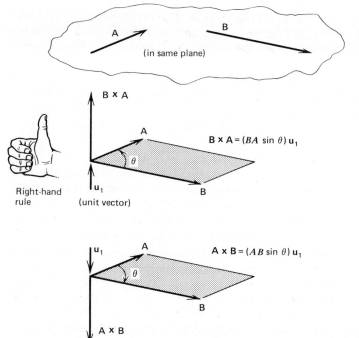

Figure 5.15 The cross product.

In matrix form, the cross product can be defined by a determinant.

$$\mathbf{A} \times \mathbf{B} = \begin{bmatrix} \mathbf{i} & \mathbf{j} & \mathbf{k} \\ A_x & A_y & A_z \\ B_x & B_y & B_z \end{bmatrix} = \mathbf{R} \tag{5.28}$$

Expanding the determinant,

$$\mathbf{R} = (A_y B_z - A_z B_y)\mathbf{i} + (A_z B_x - A_x B_z)\mathbf{j} + (A_x B_y - A_y B_x)\mathbf{k}$$
$$= R_x \mathbf{i} + R_y \mathbf{j} + R_z \mathbf{k}$$

From the expansion, it is obvious that

$$R_x = A_y B_z - A_z B_y$$
$$R_y = A_z B_x - A_x B_z$$
$$R_z = A_x B_y - A_y B_x$$

The product of $AB \sin \theta$ can be represented further as an area, the area of a parallelogram with vectors $\mathbf{A}$ and $\mathbf{B}$ as two of the four sides (see the shaded areas in Figure 5.15). This is an important and useful representation of the vector cross product. Vector $(AB \sin \theta)\, \mathbf{u}_1$ is represented by a line segment perpendicular to the plane containing the two vectors $\mathbf{A}$ and $\mathbf{B}$. Finding this perpendicular is an application of descriptive geometry.

5.9 APPLICATIONS OF VECTOR MULTIPLICATION

Consider two lines OA and OB in Figure 5.16a. The *image* of OB on OA is determined by projecting a ray from B perpendicular to OA at C. OC then becomes the component (image) of OB projected on line OA. Algebraically, OC is obtained by

$$OB \cos \theta = OC \tag{5.29}$$

Using vector notation (Figure 5.16b), and defining the lines as vectors $\mathbf{A}$ and $\mathbf{B}$, the process of projection can be expressed as

$$\mathbf{A} \cdot \mathbf{B} = AB \cos \theta \tag{5.30}$$

The product AB does not have the same unit measure as A or B. It is a new quantity. The projection can be determined using a unit vector in the OA direction.

$$OC = \mathbf{B} \cdot \frac{\mathbf{A}}{A} = \mathbf{B} \cdot \mathbf{u}_A \tag{5.31}$$

Therefore, the mathematical model for the image of one line

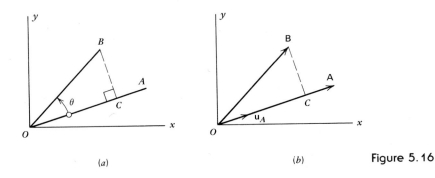

(a)

(b)

Figure 5.16

on another is expressed by the dot product, providing one vector has a unit magnitude.

To plot the projection, the coordinate components of the image must be determined. The terms of Eq. 5.31 for the general three-dimensional case shown in Figure 5.17 may be expanded into their coordinate components.

$$(B_x\mathbf{i} + B_y\mathbf{j} + B_z\mathbf{k}) \cdot (u_x\mathbf{i} + u_y\mathbf{j} + u_z\mathbf{k}) = OC \qquad (5.32)$$

In matrix form, this can be expressed as a simple matrix product with the terms of the first bracketed expression representing a single-row matrix, and those of the second bracketed group expressed as a single-column matrix.

$$[B_x \ B_y \ B_z]\begin{bmatrix} u_x \\ u_y \\ u_z \end{bmatrix} = OC \qquad (5.33)$$

The portion of OA that represents OC can then be expressed as

$$\mathbf{C} = C\,\mathbf{u}_A \qquad (5.34)$$

and can be written in terms of its three coordinate components.

Vector analysis can also be used to generate a line perpendicular to a plane. For example, consider the plane defined by vectors **A** and **B** in Figure 5.18a. The cross product

$$\mathbf{A} \times \mathbf{B} = \mathbf{P} \qquad (5.35)$$

defines vector **P,** which is perpendicular to the plane of **A** and **B.** The direction taken by vector **P** is defined by the right-hand rule. If we look at a true shape view of plane OAB (Figure 5.18b) and rotate the first vector **A** in the cross product counterclockwise into the second vector **B,** the right-hand rule indicates that **P** will be up and out of the paper. Even if the order is reversed so that

$$\mathbf{B} \times \mathbf{A} = \mathbf{Q} \qquad (5.36)$$

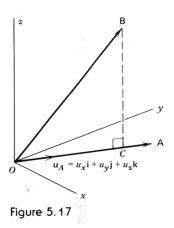

Figure 5.17

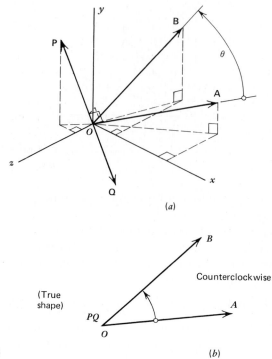

(a)

Figure 5.18

(b)

the perpendicularity is maintained (Figure 5.18a). The magnitude of **P** is

$$P = AB \sin \theta$$

SAMPLE PROBLEM 5.23

Graphically determine sum **S** of vectors **A** and **B** where $\mathbf{A} = 3\mathbf{i} + 2\mathbf{j}$ and $\mathbf{B} = \mathbf{i} + 4\mathbf{j} + 4\mathbf{k}$.

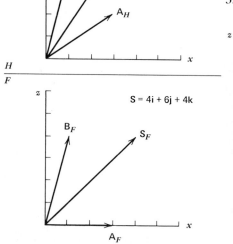

$$S = 4i + 6j + 4k$$

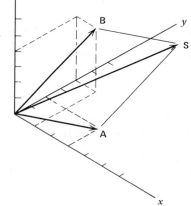

Solution

Orthogonal views or a pictorial view can be used to solve this problem. For illustration purposes both are presented.

Using two orthogonal views, the first step is to establish the views of the two original vectors along with the coordinate system. In the H view, x and y distances are in true length; in the F view, x and z distances are in true length. Using the parallelogram law, the H and F views of sum **S** can be obtained directly from the two views. The components have been scaled from the views.

To utilize an isometric view, first sketch the original vectors. Then form a parallelogram to obtain sum **S.** As we found with the graphic presentation of objects, the orthogonal views give us quantitative information directly, while the isometric view presents a better spatial presentation of the situation.

SAMPLE PROBLEM 5.24

Determine several ways to express vector **V,** which goes from the origin to the point $(20, 12, 9)$ as shown in (a).

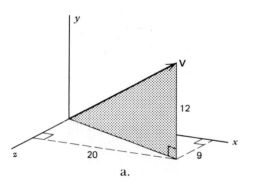

a.

Solution

The magnitude of **V** is

$$V = \sqrt{20^2 + 12^2 + 9^2} = 25$$

and using the unit vector $\mathbf{u}_1$ as in (b),

$$\mathbf{V} = 25\ \mathbf{u}_1$$

Vector **V** can also be written in terms of its components as

$$\mathbf{V} = 20\ \mathbf{i} + 12\ \mathbf{j} + 9\ \mathbf{k}$$

Also,

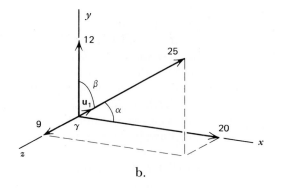

b.

$$\mathbf{u}_1 = \frac{\mathbf{V}}{25} = \frac{20}{25}\,\mathbf{i} + \frac{12}{25}\,\mathbf{j} + \frac{9}{25}\,\mathbf{k}$$

$$\mathbf{u}_1 = 0.8\,\mathbf{i} + 0.48\,\mathbf{j} + 0.36\,\mathbf{k}$$

We can now express the vector using direction cosines

$$\mathbf{V} = 25\,(\cos\alpha\,\mathbf{i} + \cos\beta\,\mathbf{j} + \cos\gamma\,\mathbf{k})$$

where

$$\cos\alpha = 0.8, \quad \cos\beta = 0.48, \quad \cos\gamma = 0.36$$

and

$$0.8^2 + 0.48^2 + 0.36^2 = 1$$

Note that

$$\cos\alpha = \frac{V_x}{V}$$

$$\cos\beta = \frac{V_y}{V}$$

$$\cos\gamma = \frac{V_z}{V}$$

So the components of vector $\mathbf{V}$ can be obtained using the products

$$V_x = V\cos\alpha$$
$$V_y = V\cos\beta$$
$$V_z = V\cos\gamma$$

SAMPLE PROBLEM 5.25

For the two vectors **A** and **B** in Sample Problem 5.23, graphically determine **A** × **B**.

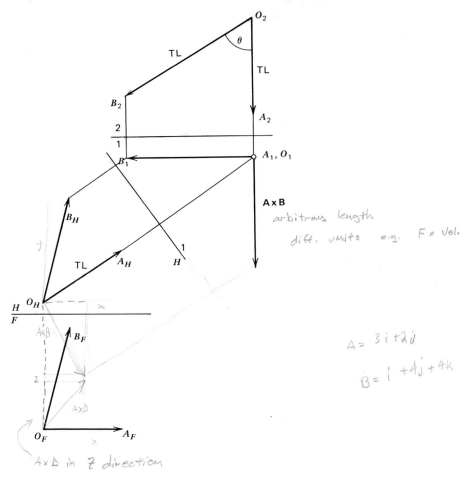

Solution

After the two orthogonal views of the vectors are drawn to scale, plane OAB is defined. Then the true shape of plane OAB is determined in the 2 view. From this true-shape view, the angle between **A** and **B** and the true lengths of **A** and **B** are determined. So

$$|\mathbf{A} \times \mathbf{B}| = AB \sin \theta$$

and **A** × **B** is perpendicular to plane OAB. The cross product **A** × **B** is shown graphically in the 1 view. Note that the right-hand rule specifies the direction of the **A** × **B** vector in the 1 and 2 views.

?. SAMPLE PROBLEM 5.26

Determine the x and y components of the image of line OB on line OA as in (a) ($OB = 2$ in. and $OA = 5$ in.).

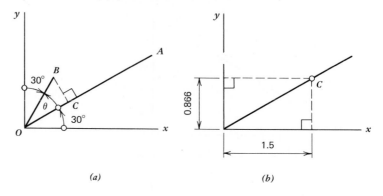

(a)   (b)

Solution

From geometric considerations

$$\theta = 30 \text{ deg.}$$
$$OC = OB \cos 30 \text{ deg.} = 2(0.866) = 1.732 \text{ in.}$$

The mathematical model of this simple example can be expressed by the dot product:

$$\mathbf{B} = B(\cos 60\,\mathbf{i} + \cos 30\,\mathbf{j}) = 2(0.5\,\mathbf{i} + 0.866\,\mathbf{j})$$
$$= 1\,\mathbf{i} + 1.732\,\mathbf{j}$$
$$\mathbf{u}_A = \cos 30\,\mathbf{i} + \cos 60\,\mathbf{j}$$
$$= 0.866\,\mathbf{i} + 0.5\,\mathbf{j}$$
$$C = \mathbf{B} \cdot \mathbf{u}_A = (1 \quad 1.742)\begin{bmatrix} 0.866 \\ 0.5 \end{bmatrix} = 1.732$$

The coordinates of line OC can be determined from the vector representing that line.

$$\mathbf{C} = C\,\mathbf{u}_A = 1.732\ (0.866\,\mathbf{i} + 0.5\,\mathbf{j})$$
$$= 1.5\,\mathbf{i} + 0.866\,\mathbf{j}$$

So that the coordinates of C are then (b)

$$x = 1.5$$
$$y = 0.866$$

SAMPLE PROBLEM 5.27

Consider the physical significance of $\mathbf{A} \times \mathbf{B}$. The two vectors are defined in (a).

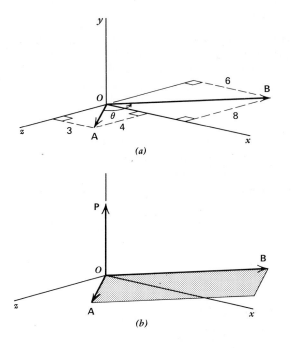

(a)

(b)

Solution

The two vectors can be written

$$\mathbf{A} = 3\,\mathbf{i} + 4\,\mathbf{k}$$
$$\mathbf{B} = 6\,\mathbf{i} - 8\,\mathbf{k}$$

The cross product can be evaluated using the determinate

$$\mathbf{A} \times \mathbf{B} = \begin{bmatrix} \mathbf{i} & \mathbf{j} & \mathbf{k} \\ A_x & A_y & A_z \\ B_x & B_y & B_z \end{bmatrix} = \mathbf{P}$$

For this problem

$$\mathbf{P} = \begin{bmatrix} \mathbf{i} & \mathbf{j} & \mathbf{k} \\ 3 & 0 & 4 \\ 6 & 0 & -8 \end{bmatrix} = (0)\,\mathbf{i} + (24 + 24)\,\mathbf{j} + (0)\,\mathbf{k} = 48\,\mathbf{j}$$

Since the vectors **A** and **B** are in the x–z plane, the perpendicular must be in the y direction, and the right-hand rule establishes the direction to be $+y$. The angle between **A** and **B** can be determined from the dot product

$$\mathbf{u}_A \cdot \mathbf{u}_B = |\mathbf{u}_A||\mathbf{u}_B|\cos\theta = \cos\theta$$

$$\cos\theta = \left| \left(\frac{3}{5}\mathbf{i} + \frac{4}{5}\mathbf{k} \right) \cdot \left(\frac{6}{10}\mathbf{i} - \frac{8}{10}\mathbf{k} \right) \right| = \frac{18}{50} - \frac{32}{50} = -0.28$$

$$\theta = 106 \text{ deg.}$$

As a check

$$\mathbf{A} \times \mathbf{B} = AB \sin \theta = 5(10) \sin 106 \text{ deg} = 48$$

Remember that as shown in (b), the value of the cross product is the area of the parallelogram defined by the two vectors.

SAMPLE PROBLEM 5.28

For the two lines OA and OB, determine the unit vector for a coordinate axis perpendicular to OA and OB.

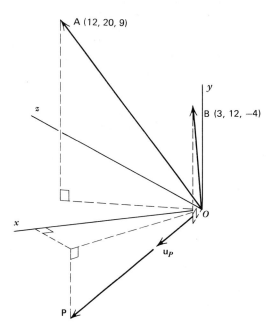

Solution
The perpendicular is established using the cross product

$$\mathbf{B} \times \mathbf{A} = \begin{bmatrix} \mathbf{i} & \mathbf{j} & \mathbf{k} \\ 3 & 12 & -4 \\ 12 & 20 & 9 \end{bmatrix} = 188\,\mathbf{i} - 75\,\mathbf{j} - 84\,\mathbf{k}$$

Unit vector $\mathbf{u}_P$ is therefore

$$\mathbf{u}_P = \frac{(188\,\mathbf{i} - 75\,\mathbf{j} - 84\,\mathbf{k})}{\sqrt{188^2 + 75^2 + 84^2}} = 0.858\,\mathbf{i} - 0.342\,\mathbf{j} - 0.383\,\mathbf{k}$$

SAMPLE PROBLEM 5.29

Determine the coordinates of point Q, which is the end of the image of line OP on the axis defined by $\mathbf{u}_p$.

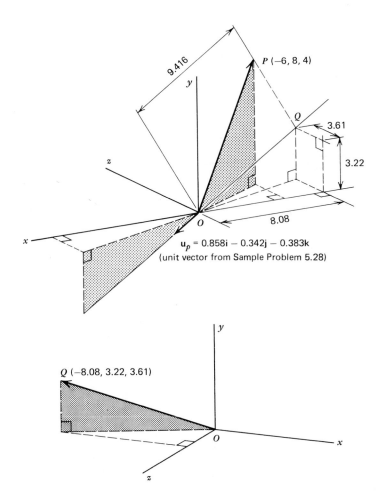

Solution

We can use the dot product to obtain the length of OQ.

$$OQ = \mathbf{P} \cdot \mathbf{u}_p = [-6 \ \ 8 \ \ 4]\begin{bmatrix} 0.858 \\ -0.342 \\ -0.383 \end{bmatrix} = -9.416$$

Then we can write

$$\mathbf{Q} = (OQ)\,\mathbf{u}_p = (-9.416)(0.858\,\mathbf{i} - 0.342\,\mathbf{j} - 0.383\,\mathbf{k})$$
$$= -8.079\,\mathbf{i} + 3.22\,\mathbf{j} + 3.606\,\mathbf{k}$$

and the coordinates of point Q are

$$x_Q = -8.08$$
$$y_Q = +3.22$$
$$z_Q = +3.61$$

The numerical value -9.416 is the coordinate position of Q on the axis defined by $\mathbf{u}_P$. The minus sign shows that it is positioned to the negative side of the origin with regard to that coordinate axis.

PROBLEMS

5.30 Graphically determine the sum of the pairs of vectors specified.

a.

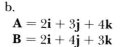

c.

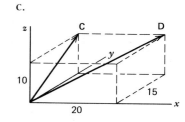

b.

$$\mathbf{A} = 2\mathbf{i} + 3\mathbf{j} + 4\mathbf{k}$$
$$\mathbf{B} = 2\mathbf{i} + 4\mathbf{j} + 3\mathbf{k}$$

Answer: a. $18.3\mathbf{i} + 68.3\mathbf{j}$ N
b. $4\mathbf{i} + 7\mathbf{j} + 7\mathbf{k}$
c. $20\mathbf{i} + 30\mathbf{j} + 20\mathbf{k}$

5.31 Graphically determine the sum of the vectors specified.

a.

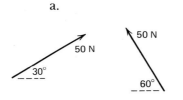

b.

$$\mathbf{E} = 2\mathbf{i} + 6\mathbf{j} + 7\mathbf{k}$$
$$\mathbf{F} = 8\mathbf{i} + 3\mathbf{j} + 9\mathbf{k}$$
$$\mathbf{G} = 5\mathbf{i} \qquad + 4\mathbf{k}$$

c.

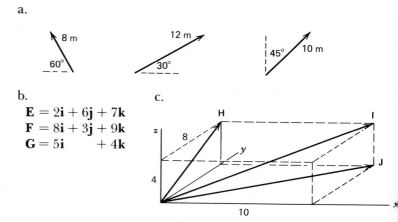

5.32 Determine the sum of the two vectors **A** and **B** by finding the true shape of the plane defined by **A** and **B**.

5.33 The direction cosines for a vector of magnitude 14 units are $\cos \alpha = 0.29$, $\cos \beta = 0.86$, and $\cos \gamma = 0.43$. Determine the x, y, and z components of the vector, and draw an isometric view that shows the vector and its three orthogonal components.

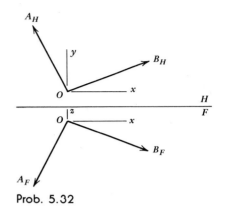

Prob. 5.32

5.34 Two views of vector **B** are shown drawn to scale. Using descriptive geometry, determine the x, y, and z components of **B** and the three direction cosines.

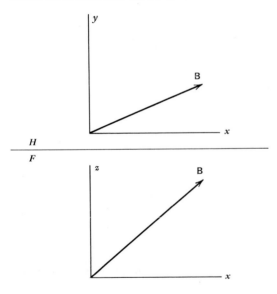

5.35 Determine the dot product of the two vectors **A** and **B** in Problem 5.30.

5.36 The cross product in Sample Problem 5.25 can be determined mathematically as

$$\mathbf{A} \times \mathbf{B} = \begin{bmatrix} \mathbf{i} & \mathbf{j} & \mathbf{k} \\ 3 & 2 & 0 \\ 1 & 4 & 4 \end{bmatrix} = 8\mathbf{i} - 12\mathbf{j} + 10\mathbf{k}$$

By projecting the cross product obtained in Sample Problem 5.25 back into the H and F views, show the result graphically.

5.37 By getting the true shape of the plane that contains **A** and **B,** defined in Problem 5.32, show the magnitude of **A × B.** Note that the area of plane *OAB* is $\frac{1}{2}AB \sin \theta$.

Determine the following graphically.

5.38 E · F

5.39 E × F

5.40 F × E

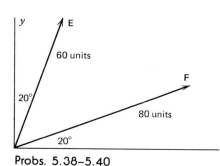

Probs. 5.38–5.40

Determine the following graphically.

5.41 G · H

Answer: 500

5.42 G × H

Answer: $-300\mathbf{i} + 300\mathbf{j} - 500\mathbf{k}$

5.43 H × G

Answer: $+300\mathbf{i} - 300\mathbf{j} + 500\mathbf{k}$

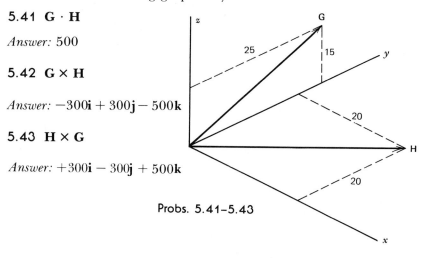

Probs. 5.41–5.43

5.44 Using three nonparallel vectors of your choice, show the following graphically.

    a. Vector addition is commutative; that is, **A + B + C = A + C + B = B + C + A,** etc.

    b. Vector addition is associative; that is, **A + (B + C) = (A + B) + C.**

    c. The scalar product is commutative.

    d. The vector product is not commutative.

5.45 Show the following graphically.

    a. For $\mathbf{A} = 2\mathbf{i} + 3\mathbf{j} + 4\mathbf{k}$

               $\mathbf{B} = 3\mathbf{i} - 2\mathbf{j} + \mathbf{k}$

    that

$$\mathbf{A} + \mathbf{B} = (2 + 3)\mathbf{i} + (3 - 2)\mathbf{j} + (4 + 1)\mathbf{k}$$

    b. $\mathbf{C} \cdot \mathbf{C} = C^2$, $\mathbf{C} \times \mathbf{C} = 0$.

    c. For the unit vectors $\mathbf{i}, \mathbf{j},$ and $\mathbf{k}$ in the $x, y,$ and z directions, respectively,

$$\mathbf{i} \times \mathbf{i} = 0$$
$$\mathbf{i} \times \mathbf{j} = \mathbf{k}$$
$$\mathbf{k} \times \mathbf{j} = -\mathbf{i}$$

Vectors $\mathbf{A}$ and $\mathbf{B}$ are directed from the origin of an x–y–z-axis system to points A and B, respectively. Determine the $x, y,$ and z components of the image of A on B, and draw an isometric view showing vectors $\mathbf{A}$ and $\mathbf{B}$, the image, and the components for the following.

5.46 $\mathbf{A} = 3\mathbf{i} + 4\mathbf{j} + 5\mathbf{k}$

     $\mathbf{B} = 5\mathbf{i} + 4\mathbf{j} + 3\mathbf{k}$

Answer: $4.6\mathbf{i} + 3.7\mathbf{j} + 2.7\mathbf{k}$

5.47 $\mathbf{A} = 4\mathbf{j} + 6\mathbf{k}$

     $\mathbf{B} = 4\mathbf{j} + 7\mathbf{k}$

Vectors $\mathbf{C}$ and $\mathbf{D}$ are directed from the origin of an x–y–z-axis system to points C and D, respectively. Determine a unit vector that is perpendicular to the plane defined by $\mathbf{C}$ and $\mathbf{D}$, and draw an isometric view showing the vectors $\mathbf{C}$ $\mathbf{D}$ and the unit vector.

5.48 $\mathbf{C} = 2\mathbf{i} + 3\mathbf{j}$

     $\mathbf{D} = 3\mathbf{i} + 4\mathbf{k}$

5.49 $\mathbf{C} = 3\mathbf{j} + 4\mathbf{k}$

     $\mathbf{D} = 4\mathbf{i} - 3\mathbf{k}$

Answer: $\mp 0.41\mathbf{i} \pm 0.73\mathbf{j} \mp 0.55\mathbf{k}$

5.10 APPLICATIONS OF GRAPHIC ANALYSIS TO MECHANICS

Mechanics is the physical science that deals with the states of bodies under the action of forces and force systems. The general subject area of *mechanics* is subdivided into solid mechanics, fluid mechanics, and mechanics of materials, and solid mechanics is further subdivided into statics and dynamics.

Bodies at rest are at rest because the force systems acting on them are balanced. This state of force balance is called *equilibrium*, and the study of balanced force systems is called *statics*. When a body is acted on by an unbalanced force system, the body is accelerated. The study of motion and the forces causing motion is called *dynamics*. You will be introduced to the principles of statics and dynamics in your study of physics. Several engineering courses will also cover these areas.

Spatial representation in mechanics is essential. Graphic analysis, representation, and interpretation are central to the study and application of mechanics. The purpose of these sections is to introduce you to several graphic procedures that you will find useful in your study of mechanics.

5.11 CONCURRENT FORCE SYSTEMS

A *force* is the action of one body on another. This action has magnitude, measured in either pounds (lbf) or newtons (N), direction, and point of application. With these three qualities, it is recognized as a vector—or more precisely, a fixed vector (that is, a vector for which the point of application is fixed). In rigid-body mechanics, the point of application is not important, and the force vector may slide along the line of action, which considerably eases the difficulty of our problems. The point of application is important in the mechanics of materials.

The resultant of two or more forces acting on a body is a single force that will replace the effect of the original forces. When the resultant of all forces is zero, the body is in equilibrium. If two forces are in equilibrium, they are *collinear*, since they would have to be equal, opposite, and directed along the same line of action. When a body is acted on by three forces in equilibrium, these three forces must be *concurrent*, which means that all three must pass through one single point. Three concurrent forces in equilibrium must also lie in the same plane, which means that the force diagram exists in two dimensions. Four or more forces in equilibrium may be concurrent but they need not be. Four concurrent forces in equilibrium may all lie in one plane or they may not.

The equilibrium of concurrent force systems is a particularly straightforward graphic analysis, which is why we introduce it here. In Figure 5.19*a*, a mass that weighs 200 N is supported by two rigid members *AB* and *BC* between two rigid walls. Joints *A*, *B*, and *C* can be considered as hinges and are considered frictionless. Members *AB* and *BC* can only sustain axial forces, either tension or compression. There is no way that these members can sustain a transverse force without a transverse force being applied, and none is applied to either member. Moving to Figure 5.19*b*, only three forces act on joint *B*, and these three must be concurrent. The three forces must add vectorially to zero for this system to be in equilibrium. The force polygon in Figure 5.19*c* represents these forces and can be used to determine the magnitudes of the forces in the two members that are both in compression. The force in *BC* will be 400 N, and the force in *AB* will be 346 N. Both are compression. Why? How do we know? Look at Figure 5.19*b*.

In Figure 5.19*b* a portion of the system is separated or isolated from its surroundings. All the forces acting on the isolated body are represented by vectors. This isolation of the body or portion of the body of interest is referred to as a *free-body diagram*, and it is a very important tool in mechanics.

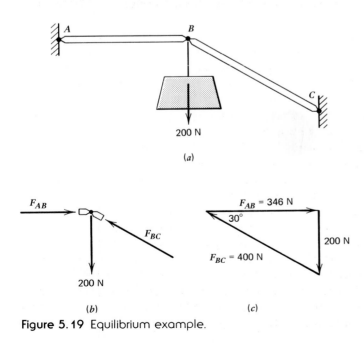

Figure 5.19 Equilibrium example.

The legs of the tripod in Figure 5.20 support compression forces in the directions of the legs. Joint O, where the three legs are attached, must be in equilibrium. This means that the weight force and the three forces in the legs must add vectorially to zero. Again, referring to Figure 5.20, note that if we add the forces in the order of weight, leg A, leg B, and then leg C, force **C** corresponds with the location of leg C, force **B** must be in the plane defined by vector forces **B** and **C,** and vector force **A** must have its head in the plane defined by legs B and C. If the head of vector **A** is not in that plane, it is impossible for the force polygon to close. It will not add to zero. Since the two views of the force polygon are drawn to scale, the forces in the legs can be determined. The representation in Figure 5.20 of the forces acting on the top joint of the tripod is actually a free-body diagram of that joint. It has been *isolated* from its surroundings, and all the forces acting on it are represented.

The free-body diagram consists of the definition of the boundaries of the body or system of interest and the forces acting on that boundary. As you study mechanics and the use of free-body diagrams, you will use principles and techniques of graphics.

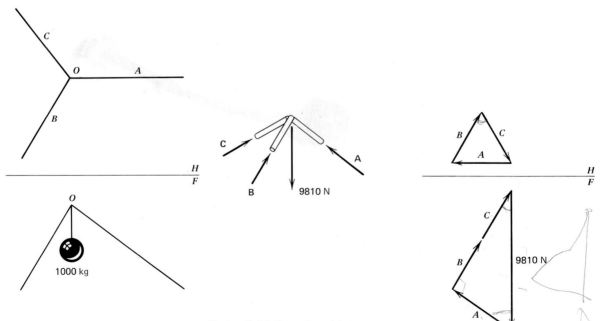

Figure 5.20 Tripod problem.

5.12 MOMENT OF A FORCE

A force has one additional effect on a rigid body: the tendency of a force to rotate a body about any axis that is not parallel to the force vector or does not intersect its line of action. This tendency to cause rotation is called the *moment* of the force.

By definition, the moment of force **F** is written as the vector cross product

$$\mathbf{M}_O = \mathbf{r} \times \mathbf{F} \tag{5.37}$$

See Figure 5.21. The subscript O represents that point about which the moment is taken. The resulting moment vector has a magnitude

$$M_O = rF \sin \theta \tag{5.38}$$

which can be represented by the area of the parallelogram formed by position vector **r** and force vector **F**. Moment vector $\mathbf{M}_O$ is in the direction perpendicular to that plane.

Consider the true shape of the plane defined by the **r** and **F** vectors illustrated in Figure 5.21. The magnitude of the moment or cross product **r** × **F** can be written

$$r(F \sin \theta) \text{ or } (r \sin \theta)F \tag{5.39}$$

which has the area interpretation shown in Figure 5.22.

Let us project this vector diagram of **r** × **F** on both the horizontal and frontal planes. The area the cross product represents is the parallelogram defined by vectors **r** and **F** shown in Figure 5.23. The parallelogram would be distorted in both the horizontal and frontal views. In order to find the measure of this area, we need to find its true shape. This is done with conventional descriptive geometry by finding, first, an edge view of the plane projected in plane 1, and second, the true shape of the plane projected in plane 2. In this view, we can measure graphically the magnitude of the cross product, represented by a directed line segment in projection plane 1. $\mathbf{M}_O$ can then be projected back to the horizontal and frontal planes.

The cross product representing the moment of a force may be expanded into the determinant

$$\mathbf{M}_O = \mathbf{r} \times \mathbf{F} = \begin{bmatrix} \mathbf{i} & \mathbf{j} & \mathbf{k} \\ r_x & r_y & r_z \\ F_x & F_y & F_z \end{bmatrix} \tag{5.40}$$

where **i**, **j**, and **k** are unit vectors in the x, y, and z directions, respectively. Remember that a unit vector is a directed line segment with a magnitude of one unit.

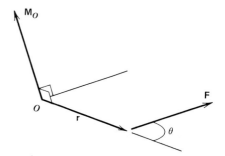

Figure 5.21 Moment about a point.

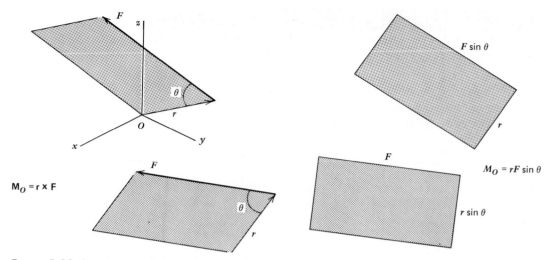

$M_O = r \times F$

Figure 5.22 Area interpretation of moment.

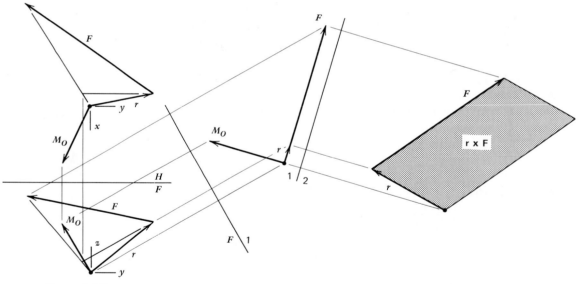

Figure 5.23

If you expand this determinant, you will have **i**, **j**, and **k** components

$$\mathbf{M}_O = (r_y F_z - r_z F_y)\mathbf{i} + (r_z F_x - r_x F_z)\mathbf{j} + (r_x F_y - r_y F_x)\mathbf{k} \quad (5.41)$$

which is understandable. As a vector, the moment of a force can itself be resolved into components

$$\mathbf{M}_O = \mathbf{M}_x + \mathbf{M}_y + \mathbf{M}_z \quad (5.42a)$$

$$= M_x\,\mathbf{i} + M_y\,\mathbf{j} + M_z\,\mathbf{k} \quad (5.42b)$$

Figure 5.24 shows the meaning of a moment about an axis. The x and y components of the force tend to rotate the body around the z-axis. The tendency is quantified by M_z, the z component of $\mathbf{M}_O$.

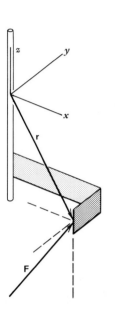

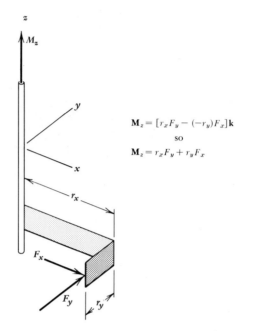

$$\mathbf{M}_z = [r_x F_y - (-r_y) F_x] \mathbf{k}$$

so

$$\mathbf{M}_z = r_x F_y + r_y F_x$$

Figure 5.24 Moment about an axis.

SAMPLE PROBLEM 5.50

What are the forces exerted on the pin connections at A and C if a horizontal 100-N force is applied at B?

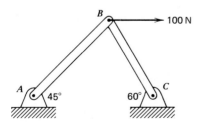

Solution

First, draw the 100-N force to a convenient scale to find the forces at pins A and C. For equilibrium, we know that the members must provide a 100-N force to the left to counteract the

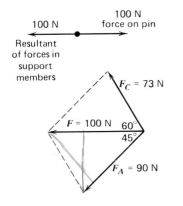

100 N force on pin

100 N

Resultant of forces in support members

$F_C = 73$ N

$F = 100$ N 60°

45°

$F_A = 90$ N

100-N applied force. We also know that the forces in each member are in the same direction as the member; therefore, we can find components of the counteracting 100-N force in the direction of the members. Draw a parallelogram, to scale, showing vectors such that

$$\mathbf{F}_A + \mathbf{F}_C = \mathbf{F}$$

Measure

$$F_A = 90 \text{ N, tension}$$
$$F_C = 73 \text{ N, compression}$$

How do we know one is compression and one is tension and which is which?

SAMPLE PROBLEM 5.51

The lightweight 1-m horizontal column and cord support the 100-N weight as shown. Determine the forces in the column and the cord.

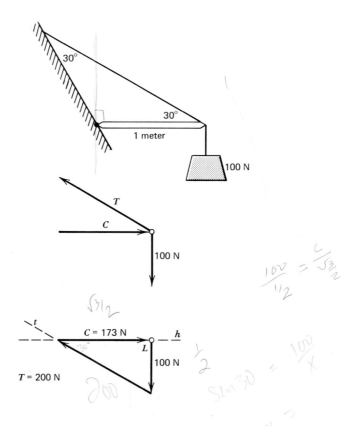

Solution

The three forces acting on the connection must add vectorially to zero. The 100-N force is first drawn to scale. Then the line of action t of the tension force in the cord is drawn, and finally horizontal line h is drawn to close the force polygon. Knowing the 100-N side of the 30 deg right triangle, the other two sides can be determined with trigonometry.

SAMPLE PROBLEM 5.52

A 50-lb force applied to the foot pedal results in a tension T at the connecting link. A free-body diagram of the foot pedal is given. The weight of the pedal is negligible. The drawing is to scale. Find the tension T.

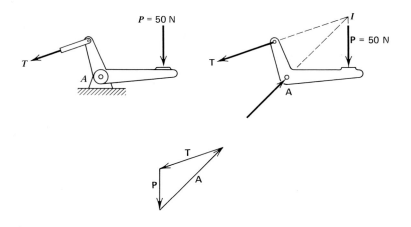

Solution

Three forces act on this foot pedal: the 50-N force, force **T,** and a force at pivot A. Initially, we do not know the magnitude or direction of **A.** The sketch at the right is a free-body diagram of the foot pedal.

On a geometrically true sketch of the foot pedal, draw the lines of action of the 50-N force and force **T.** Extend the lines of action until they intersect at point I.

Since these three forces must be concurrent, force **A** also intersects point I to establish the direction of force **A.**

Now create a vector force diagram for equilibrium (use a convenient scale).

$$\mathbf{P} + \mathbf{A} + \mathbf{T} = 0$$

From the force diagram, scale $T = 95$ N and $A = 118$ N.

SAMPLE PROBLEM 5.53

Determine the forces in all the members of the cantilever truss shown. The truss supports a mass that weighs 1000 lb.

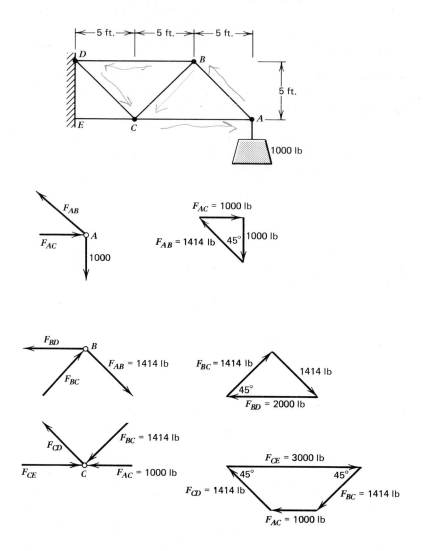

Solution

First, the forces on joint A are represented. Then a force polygon is formed causing the three forces to add to zero. The values are then determined using trigonometry. Member AB is in tension and member AC is in compression. Next, the forces on joint B are represented. Note that the force on B from member AB pulls away from the joint because this member is in tension. The vector polygon and trigonometry give the results. Member BD is in tension, and member BC is in compression. Again, we must remember that members AC and BC are in compression when the forces are represented on joint C. The last force polygon gives us the tension force in member CD and the compressive force in member CE.

SAMPLE PROBLEM 5.54

A 500-lb force is supported by boom AB and two cables as shown (a). Graphically determine the tensions T_1 and T_2 in the cables and the load F supported by the boom.

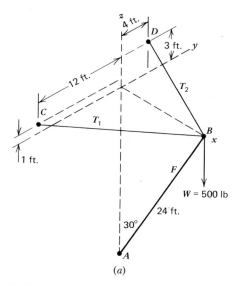

(a)

Solution

Two orthogonal views of the boom and cables are established (b). The 500-lb force is established and represented with a vector drawn to scale; the force vector is in true length in the frontal view and in point view in the horizontal view. The four forces $\mathbf{W}$, $\mathbf{F}$, $\mathbf{T}_1$, and $\mathbf{T}_2$ must form a vector polygon that closes (i.e., adds to zero).

The compression $\mathbf{F}$ in the boom is a force vector in the direction of the boom. A line is constructed from the head of the $\mathbf{W}$ vector parallel to AB in both views. For the force polygon to close $\mathbf{T}_1$ and $\mathbf{T}_2$ must lie in the plane defined by cables BC and BD. Therefore, the head of the $\mathbf{F}$ vector is located where the line drawn parallel to AB intersects plane BCD. This point is labeled P and found by working the intersection of a line and plane problem. For convenience, lines BC and BD were extended to C' and D'. The head end of $\mathbf{T}_1$ must be on line DBD' since $\mathbf{T}_2$ must be coincident with line DBD' for the force polygon to close.

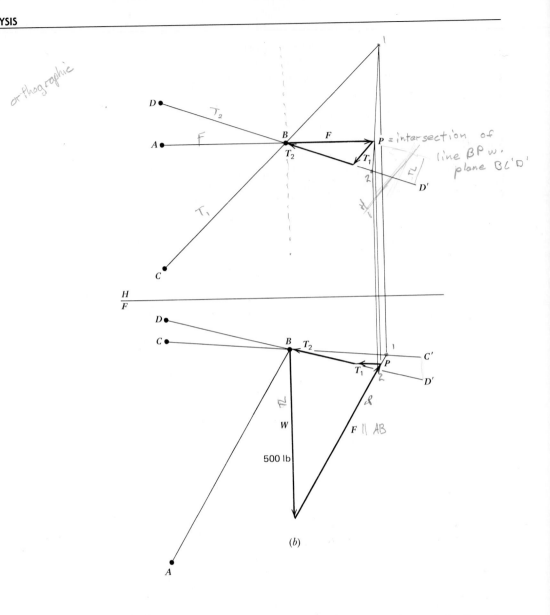

(b)

The magnitudes of the forces can be obtained by finding the true lengths of the vectors. **F** is in true length in the frontal view and is 525 lb. The true lengths of T_1 and T_2 could be found using supplementary views or rotation. These steps are not shown. $T_1 = 100$ lb and $T_2 = 200$ lb.

SAMPLE PROBLEM 5.55

Force vector **F** and its position are shown using horizontal and frontal views (*a*). Lengths are drawn to a convenient scale. Determine the moment of force **F** about a vertical axis through *A*.

Solution

Vector **F** is shown in both horizontal and frontal views (*b*). From point *A*, draw a line to the tail of vector **F**. This is position vector **r** of force vector **F**. The moment of force **F** about point *A* is

$$\mathbf{M}_A = \mathbf{r} \times \mathbf{F}$$

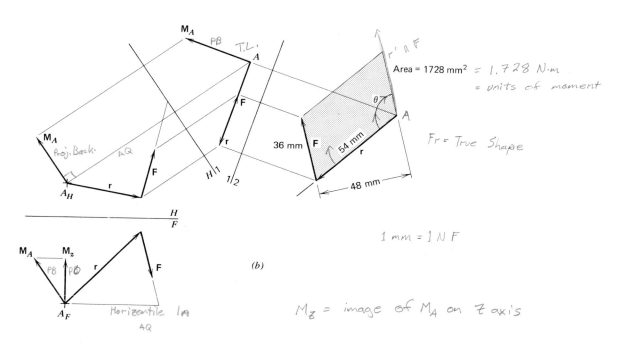

Area = 1728 mm² = 1.728 N·m
= units of moment

F_T = True Shape

1 mm = 1 N F

M_Z = image of M_A on Z axis

(b)

$\mathbf{M}_A$ is represented by a vector perpendicular to the plane containing **r** and **F**.

Draw a horizontal line from point A_F that intersects the line of action of force **F**. Now, find the edge view of the plane containing **r** and **F**. This is shown in view 1. View 2 shows the true shape of the plane containing **r** and **F** and, in particular, the angle θ between the two vectors. The magnitude of **r** is $|\mathbf{r}| = 54$ mm and $|\mathbf{F}|$ is found to be 36 mm or 36 N. The magnitude $|\mathbf{M}_A|$ is

$$|\mathbf{M}_A| = rF \sin \theta = 1.728 \text{ N} \cdot \text{m}$$

The shaded area shown in view 2 also represents the magnitude of $\mathbf{M}_A$.

$$M_A = 36 \times 48 = 1728 \text{ mm}^2 = 1.728 \text{ N} \cdot \text{m}$$

The moment vector is in true length in view 1 and can be projected into views H and F. The vertical component is then obtained by constructing a horizontal line through the head of $\mathbf{M}_A$ in view F; since $\mathbf{M}_z$ shows in true length, from scaling directly we find $|\mathbf{M}_z| = 1.1$ Nm.

PROBLEMS

5.56 Determine the two components of vector $\mathbf{P}$ in the a and b directions.

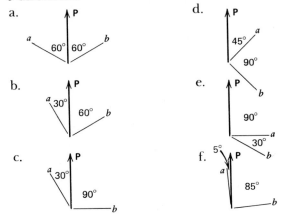

5.57 What are the components of the 80-lb force along lines AB and AC?

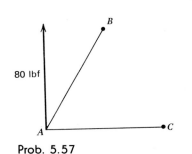

Prob. 5.57

5.58 Determine the magnitudes of the rectangular components of the 80-lb force in the x and y directions.

Answer: $C_x = 64$ lbf
$C_y = 48$ lbf

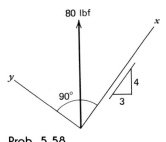

Prob. 5.58

5.59 Using the principles of descriptive geometry, graphically determine the x, y, and z components of the 100-N force. *Hint:* First draw an auxiliary view in which the force will be in true length.

Answer: 25 N, 43.3 N, 86.6 N

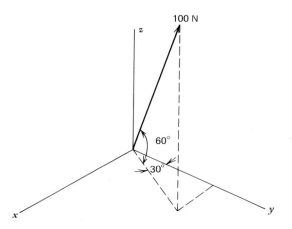

5.60 A cantilever beam is supported by the guy wire AB. The force in the guy wire is 400 lb. Determine the components of the force in the wire parallel to the x-, y- and z-axes.

Answer: 313 lbf, −188 lbf, −164 lbf

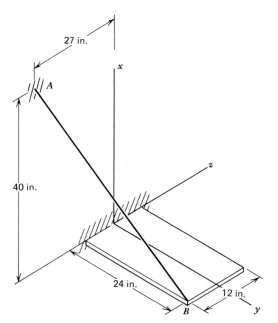

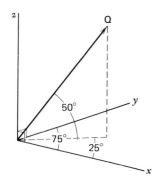

5.61 Determine the x, y, and z components of vector **Q** for the nonorthogonal coordinate system shown. The x- and z-axes are orthogonal, and the y- and z-axes are orthogonal. The angle between the x- and y-axes is 75 deg.

For the following seven problems, determine the forces in all the supporting members (sketches made to scale).

5.62

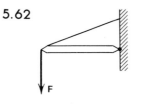

5.63

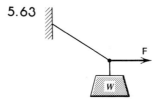

5.64

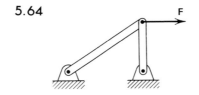

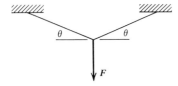

5.65 For the following.
    a. $\theta = 45$ deg.

    b. $\theta = 30$ deg.

    c. $\theta = 15$ deg.

    d. $\theta = 10$ deg.

    e. $\theta = 5$ deg.

5.66

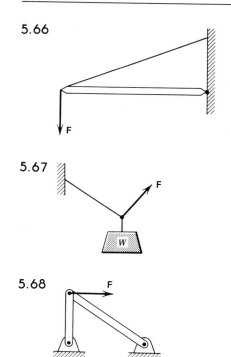

5.67

5.68

5.69 Given D and **W,** find the location of point A relative to the support circle so the forces in all three cords equal **W.**

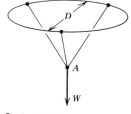

Prob. 5.69

5.70 Determine the forces in the legs of the tripod that supports the 500-kg mass.

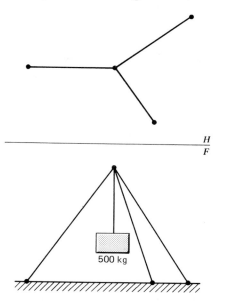

5.71 A 1000-lb load hangs from O, supported by three props. Force P is then applied. What are the resulting forces in props A, B, and C?

Answer: 1350 lb, −105 lb, −165 lb

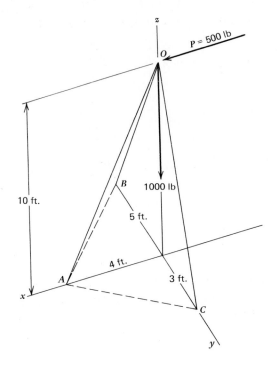

5.72 Determine the forces in the members of the cantilever truss. Neglect the weights of the members.

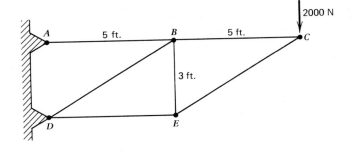

5.73 Determine the forces in the members of the cantilever truss. Neglect the weights of the members.

Answer: $F_{AB} = 4000$ N, Tension $F_{DB} = 5657$ N, Tension
 $F_{BC} = 0$ $\phantom{F_{DB} = 5657 N, T}$ $F_{EB} = 4000$ N, Compression
 $F_{CD} = 0$ $\phantom{F_{DB} = 5657 N, T}$ $F_{EA} = 5657$ N, Tension
 $F_{DE} = 4000$ N, Compression $F_{EF} = 8000$ N, Compression

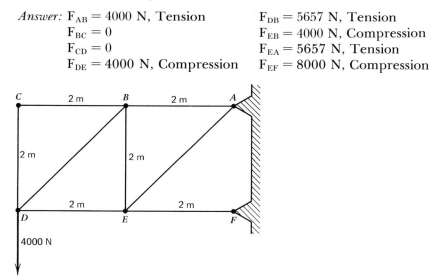

5.74 Determine the moment of the 100-N force about the y-axis and about point O. The point of application of the force is in the x–y plane, and the force vector is parallel to the y–z plane as shown.

Answer: -86.6 N·m; $50\mathbf{i} - 86.6\mathbf{j} + 150\mathbf{k}$, N·m

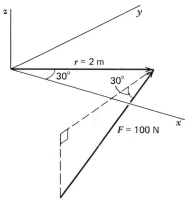

5.75 A steel standpipe is subjected to a 900-lb force by cable AB. Graphically find the moment exerted by the cable about the z-axis of the standpipe.

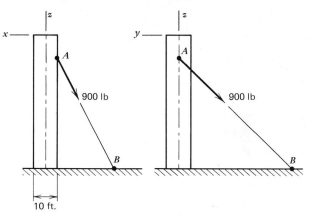

5.76 The steel cleat is held by bolt A whose axis is coincident with the z-axis. Determine the total moment of the 50-N force at bolt A and the component that represents a moment about the bolt axis.

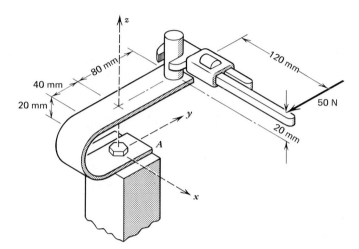

5.77 Determine the resultant moment at point A from the applied load of 50 lbf on the wrench handle. State magnitude and direction.

Answer: $800\mathbf{j} - 600\mathbf{k}$, lb-in.

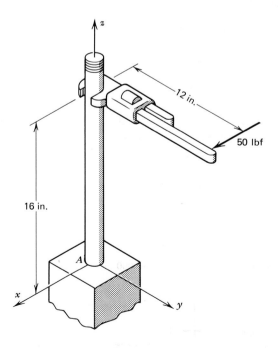

5.78 The 50-lb force acts on the bent bar as shown. The z component of force F is zero. Using Eq. 5.41 directly, determine the tendency of force **F** to rotate the bar about the x-, y-, and z-axes (i.e., determine M_x, M_y, M_z).

Answer: $\mathbf{M}_x = \quad 360$ lb-in.
$\mathbf{M}_y = -480$ lb-in.
$\mathbf{M}_z = \quad -60$ lb-in.

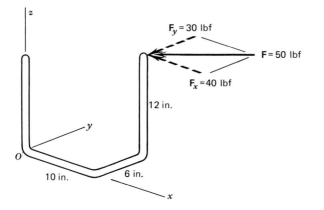

6
COMPUTER GRAPHICS

6.1 INTRODUCTION TO THE GRAPHIC LANGUAGE OF COMPUTERS

Computer graphics is the graphic language by which geometric lines and shapes are generated from a set of mathematical instructions to a computer. These results are usually displayed through some form of computer output, such as the screen of a cathode-ray tube, a printer, or a plotter. Most computers permit the rearrangement of the display, providing the operator understands the basic geometry and the accompanying mathematics by which the original display was generated.

The purpose of this chapter is to acquaint you with the basic geometry and accompanying mathematics necessary to make these analyses. The chapter is not our original work. It is a shortened version of an introduction to computer graphics written by R. Golden and R. Salomon, professors of engineering at the *New Jersey Institute of Technology*. We have made some changes.

No attempt is made to suggest a particular computer language format. It is possible to write a set of computer instructions in any language or format compatible with the computer equipment available to you, once the nature of the required geometry has been analyzed and the appropriate logic developed. It was suggested we give examples of systems using different language formats, such as APL, PASCAL, BASIC, and FORTRAN. As authors, we advocate no special language. We leave the application of the theory suggested in this chapter to you and whatever language is compatible with the computer available to you.

In order to understand computer graphics, you must understand first what a digital computer is and is not. The digital computer is a machine that can do numerical calculations and perform logical operations. It performs these operations extremely rapidly and can do large numbers of them. It cannot perform directly operations of a continuous nature such as integration and differentiation. Therefore, mathematics that effectively accomplish these operations are programmed into a digital computer as repetitive calculations. For instance, instead of integrating, a computer sums. Instead of differentiating, a computer performs ratios. Instead of taking the logarithm of a number, it performs an equivalent series calculation or stores the logarithms in tabular form in its memory. Many of the frequently performed numerical equivalent processes are built into the software (i.e., programming) of the machine as standard routines that can be retrieved when needed.

For computer graphics, it is obvious that lines can be programmed to appear on a cathode-ray tube with the proper interactive electronic circuitry. But how does one translate geometry into electronic circuitry and then retranslate the circuitry back to geometry? The two major features of the modern computer that make computer graphics possible are its ability to store and recall enormous amounts of data and its ability to perform repetitive operations quickly.

As an example, suppose the data for an oblique line in a system of lines were available in terms of one coordinate system. It is important to understand that this line is oblique merely because of its relation to the coordinate system employed to describe it. In another coordinate system, the line might not be oblique; there certainly are coordinate systems in which the line is not. Suppose further that it is desired to describe the position of this line within a system of lines. Problems of this nature are solvable by descriptive geometry, vector analysis, or analytic geometry. Graphic line problems can also be solved with the aid of a computer. All we have to do is to translate the geometry into a numerical language that a computer can use. Whatever it is, that language will involve matrices.

6.2 POSITIONING AND ROTATING A POINT IN SPACE

Since an oblique line has a defined relation to one coordinate system, we can find its position in any other coordinate system. How can the present data be recoordinated into the new system? To do this, we will have to find a method that is compatible with and programmable in a computer. If our view of the data in the initial coordinate system is not convenient, the first question is: In what direction must it be viewed so that it is convenient? The answers to this and to other questions involve positioning and rotating a point in a coordinate system. If we can position and rotate one point, we can position and rotate several. If we can position and rotate several, we can do the same for many. We are only limited by our capacity to handle the necessary numerical calculations.

In the material we will subsequently develop, the viewer line of sight labeled *observer* in the figures refers to that line of sight normal to the plane of the interactive graphic display as shown in Figure 6.1. It is a line that corresponds to the negative direction along the z-axis. When the required view is obtained, it is located in the x–y plane and appears plotted on the output display. In other words, the process of rotation transforms the data into the coordinate system of the front view in orthographic projection, and only the front view.

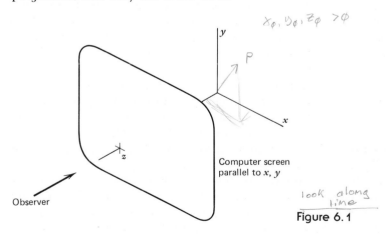

object is moving

#1 y z plane

#2 rot. to z axis

$x_0, y_0, z_0 > 0$

P

y

x

Computer screen
parallel to x, y

z

Observer

look along line

Figure 6.1

When using a computer, it is helpful to turn an object to your own point of view. In effect, you choose a particular direction in which you want to view an object and then turn that viewing arrow until it corresponds to this preferred line of sight. In order to accomplish this, it is merely necessary first to rotate the arrow into the vertical profile (y–z) plane of your sight line,

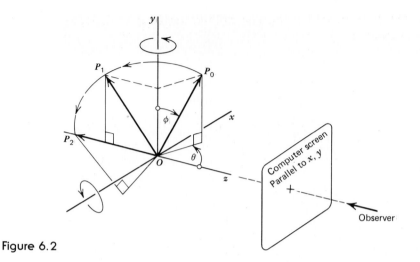

Figure 6.2

and second to lower or raise the arrow within the y–z plane until it assumes the same direction as your view. Figure 6.2 shows this process pictorially.

6.3 MATRIX OPERATIONS

The classic and most natural method of accomplishing the geometry of repositioning is to perform a series of rotations in which a line of sight is rotated until it assumes the desired position with respect to another frame of reference. This frame of reference will be that used by the computer display. The transformation process is mathematically expressed by a series of equations involving the angles of rotation. A separate set of equations is required for each data point and for each rotational step. It does not take a great deal of imagination to visualize that a large number of such equations are necessary to handle even a simple shape.

To illustrate the transformation of the coordinates of a point in one reference frame to coordinates in another, consider

$$[x_0 \ y_0 \ z_0]\begin{bmatrix} R_{11} \ R_{12} \ R_{13} \\ R_{21} \ R_{22} \ R_{23} \\ R_{31} \ R_{32} \ R_{33} \end{bmatrix} = [x_1 \ y_1 \ z_1] \tag{6.1a}$$

$$[P_0][R] = [P_1] \tag{6.1b}$$

where x_0, y_0, and z_0 are the original coordinates of point P and x_1, y_1, and z_1 are the new coordinates. The $[R]$ matrix is called the transformation, or transfer, matrix. The matrix equation

(Eq. 6.1) is nothing more than a shorthand way of specifying the following algebra:

$$x_0 R_{11} + y_0 R_{21} + z_0 R_{31} = x_1$$
$$x_0 R_{12} + y_0 R_{22} + z_0 R_{32} = y_1$$
$$x_0 R_{13} + y_0 R_{23} + z_0 R_{33} = z_1$$

The first matrix $[P_0]$ contains one row and three columns and is properly called a 1×3 matrix. Since the second matrix $[R]$ must have the same number of rows as the first has columns, we know that it must then have three rows. Because the final and desired matrix $[P_1]$ is also a 1×3 matrix, it follows that the multiplier in this instance must also have three columns; as a result the $[R]$ matrix is a 3×3 square matrix. The coordinates of the first array are the coordinates of point P. $[R]$ is a transformation array capable of effecting the rotation of point P about a coordinate axis resulting in a new array $[P_1]$ representing the rotated point P.

If we desire to rotate point P_0 from the initial position shown in Figure 6.3a about the y-axis through 90 deg to a new position, as drawn in Figure 6.3b, we must identify the elements of the transfer matrix $[R]$ in Eq. 6.1b that accomplish this rotation by numerical methods.

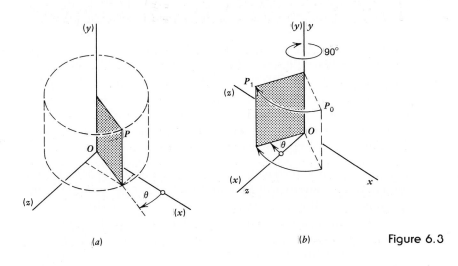

(a) (b) Figure 6.3

When the rotation has taken place, the original x-axis rotates to the position of the z-axis. The original z-axis rotates to its position along the negative x-axis. The y-axis, acting as a fixed pivot line, is unchanged. The question now becomes one of identifying the elements in the 3×3 matrix represented by $[R]$. Let us consider one of the coordinates determined in detail from the following array.

$$[x_0 \; y_0 \; z_0]_P \begin{bmatrix} R_{11} & R_{12} & R_{13} \\ R_{21} & R_{22} & R_{23} \\ R_{31} & R_{32} & R_{33} \end{bmatrix} = [x_1 \; y_1 \; z_1]_P$$

where x_0, y_0, and z_0 are the coordinates of the point P_0 before transformation, and x_1, y_1, and z_1 are the coordinates of P_1 after transformation. Thus, after transformation,

$$x_1 = x_0 \, R_{11} + y_0 \, R_{21} + z_0 \, R_{31}$$

Since z_0 assumes, after rotation, the position of $-x_1$, then R_{31}, the coefficient expressing this transformation, must have a value of -1. In addition, neither x_0 nor y_0 has any projected value on the x-axis after the transition. To reflect this, the two column elements to which they are coupled (R_{11} and R_{21}) must have values of zero. Consequently, the first column of $[R]$ must read

$$\begin{bmatrix} 0 \\ 0 \\ -1 \end{bmatrix}$$

Similar analyses of y_1 and z_1 will result in identifying the total $[R]$ matrix as

$$\begin{bmatrix} 0 & 0 & 0 \\ 0 & 1 & 0 \\ -1 & 0 & 0 \end{bmatrix} \begin{array}{l} = \pm 1 \\ = \pm 1 \end{array}$$

Since the row subscript for all vector matrices is one (1), it is conventionally omitted so that the entire symbolic algebra for this geometric transformation, and hence the numerical model, will read

$$[x_0 \; y_0 \; z_0]_P \begin{bmatrix} 0 & 0 & 1 \\ 0 & 1 & 0 \\ -1 & 0 & 0 \end{bmatrix} = [x_1 \; y_1 \; z_1]_P \qquad (6.2)$$

It may be noted that the sum of the squares of each element in a row or column will be equal to 1. This observation is important, since the property is used to test the validity of the coefficient matrix when it consists of trigonometric functions.

If the view desired by the operator is in the direction of line OP (Figure 6.4a), and this direction is not that of the observer, it will be necessary to transform OP to a position corresponding to that of the observer. In all cases, this can be done by only two separate rotations. The first of these will rotate P about the y-axis into the y–z plane. The arrow end of line OP will be behind the frontal plane (x–y), and the tail of the arrow will be at the origin, as in Figure 6.4b. The second step involves a rotation about the x-axis until point P assumes a position directly

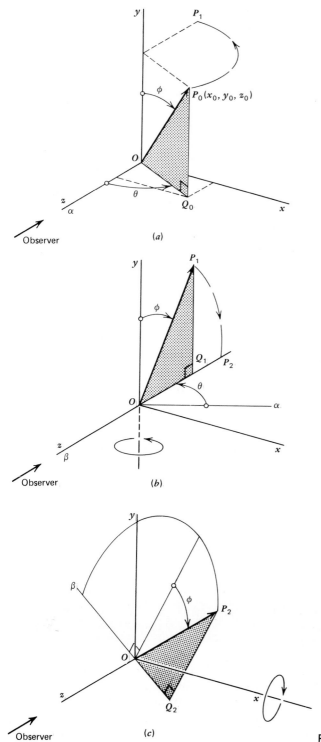

(a)

(b)

(c)

Figure 6.4

behind the origin, and the line of sight OP is that of the observer, as illustrated in Figure 6.4c.

To describe these rotations, line OP must be numerically defined for the computer. These rotations can best be visualized by considering the line of sight to lie in a vertical plane in the form of a right triangle for which the line of sight is the hypotenuse. The coordinates of point P are x_0, y_0, and z_0 (see Figure 6.4a). The plane containing line OP is positioned with respect to the positive direction of the z-axis by a horizontal angle measured in the conventional counterclockwise direction. This angle is θ. Line OP also is positioned with respect to the positive direction of the y-axis by a second angle, ϕ. The matrix transformation for the first rotation in the θ direction is

$$[x_0 \ y_0 \ z_0][R_\theta] = [x_1 \ y_1 \ z_1] \tag{6.3a}$$

and for the second rotation in the ϕ direction, the transformation is

$$[x_1 \ y_1 \ z_1][R_\phi] = [x_2 \ y_2 \ z_2] \tag{6.3b}$$

where the individual transfer matrices are

$$[R_\theta] = \begin{bmatrix} -\cos\theta & 0 & -\sin\theta \\ 0 & 1 & 0 \\ \sin\theta & 0 & -\cos\theta \end{bmatrix} \tag{6.4a}$$

$$[R_\phi] = \begin{bmatrix} 1 & 0 & 0 \\ 0 & \sin\phi & -\cos\phi \\ 0 & \cos\phi & \sin\phi \end{bmatrix} \tag{6.4b}$$

By combining the matrix operations, we can construct a single matrix.

$$[x_0 \ y_0 \ z_0][R_\theta][R_\phi] = [x_0 \ y_0 \ z_0][R] = [x_2 \ y_2 \ z_2] \tag{6.5}$$

$$[R] = \begin{bmatrix} -\cos\theta & -\sin\theta\cos\phi & -\sin\theta\sin\phi \\ 0 & \sin\phi & -\cos\phi \\ \sin\theta & -\cos\theta\cos\phi & -\cos\theta\sin\phi \end{bmatrix} \tag{6.6}$$

6.4 GENERAL ROTATION ABOUT THE COORDINATE AXES

In the material developed with respect to rotation, all rotations are assumed to be taken with the axes passing through the origin. Matrices representing the numerical modeling of such motion can be developed with respect to each axis as an individual transformation. The rotation about the x-axis (Figure 6.5a) is measured counterclockwise from the positive y when

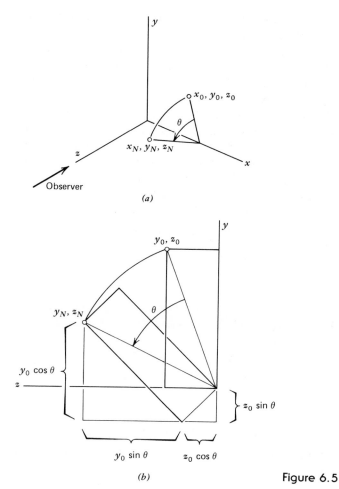

(a)

(b)   Figure 6.5

the motion is viewed from the positive x. This transformation appears as

$$[x_0 \ y_0 \ z_0] \begin{bmatrix} 1 & 0 & 0 \\ 0 & \cos \theta & \sin \theta \\ 0 & -\sin \theta & \cos \theta \end{bmatrix} = [x_N \ y_N \ z_N] \qquad (6.7)$$

The geometric details related to Eq. 6.7 are shown in Figure 6.5b. Rotation about the y-axis is measured counterclockwise from the positive end of the z-axis when viewed from the positive y. The resulting transformation is then

$$[x_0 \ y_0 \ z_0] \begin{bmatrix} \cos \theta & 0 & -\sin \theta \\ 0 & 1 & 0 \\ \sin \theta & 0 & \cos \theta \end{bmatrix} = [x_N \ y_N \ z_N] \qquad (6.8)$$

The corresponding rotation about the z-axis is read counterclockwise from the positive x-axis when viewed from the positive z direction.

$$[x_0 \ y_0 \ z_0] \begin{bmatrix} \cos\theta & \sin\theta & 0 \\ -\sin\theta & \cos\theta & 0 \\ 0 & 0 & 1 \end{bmatrix} = [x_N \ y_N \ z_N] \qquad (6.9)$$

Note that the convention for reading the angle is in agreement with the right-hand rule. That is, the rotational direction, in agreement with the right-hand rule, defines the positive direction of the axis of rotation.

These numerical models may prove useful if the data, instead of being viewed in a particular direction, are being altered by rotation about one of the original coordinate axes and through a specified angular displacement. Note that these are different models from the models generated in Eq. 6.4.

6.5 THE LOGIC OF THE MATRIX

Although the numerical models described in this chapter can be applied "longhand" to produce solutions, these processes are being followed fundamentally to increase your appreciation for the arithmetic generated by the computer. We want to draw attention to the steps outlined in the mathematical development so that a complete awareness can be developed for the kind of information necessary to reduce these techniques to forms acceptable by a computer. An examination of the rotation matrix (Eq. 6.6) may serve to illustrate the process of computer translation.

$$[R] = \begin{bmatrix} -\cos\theta & -\sin\theta\cos\phi & -\sin\theta\sin\phi \\ 0 & \sin\phi & -\cos\phi \\ \sin\theta & -\cos\theta\cos\phi & -\cos\theta\sin\phi \end{bmatrix} \qquad (6.6 \text{ repeated})$$

Once the required line of sight has been established, it is necessary to specify the values of θ and ϕ required to particularize the amount of rotation needed to place that viewing line in the position of the front-view observer. This having been done, the data of the object being viewed can be plotted on the viewing screen and a picture produced to show the detail as it appears when viewed in the desired direction.

If we calculate the values of θ and ϕ with their appropriate trigonometric functions, and only then enter these data in the computer, we defeat the strategy of using a computer. It is far better to allow the computer to calculate all the required variables once the object has been described mathematically, and the coordinates have been determined for the sight direction. This last statement must be interpreted in the light of the particular program being used, since considerations of economy may at times dictate that some small amount of hand calculation be performed.

Let us now examine one avenue of logic to accommodate the rotation matrix to the computer. Again, we emphasize that this logic is not being developed with any special language in mind, but is being presented to make you aware of what factors must be taken into consideration in translating the geometry into the numerical language of the computer.

Figure 6.6 shows a typical position of an oblique line of sight with angles θ and ϕ defined. The coordinates of P are a, b, and c, and they can be used to calculate the absolute values of s and r as

$$r = |\sqrt{a^2 + c^2}|$$

$$s = |\sqrt{a^2 + b^2 + c^2}| = |\sqrt{r^2 + b^2}|$$

The values of θ and ϕ can then be expressed as ratios of the coordinates with r and s. The machine must be given not only these ratios in order to calculate the required angles, but also their quadrant locations. These quadrants are defined by the signs of the coordinates. For example, if a is negative $(-a)$ and c is positive $(+c)$, angle θ is in the fourth quadrant, where the sine of θ is negative. If a is positive and c is negative, angle θ is in the second quadrant and the sine of θ is positive. It is not sufficient simply to identify the ratio of a/c as negative, since this can be applied to either the second or fourth quadrant, and unless the differentiation is made for the computer, it cannot know which option to select.

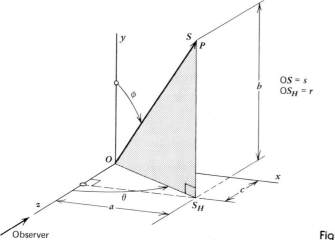

OS = s
OS$_H$ = r

Observer

Figure 6.6

It is necessary to state the particular signs of the coordinate values under consideration and to establish the values of the angles whose functions are used in the matrix.

Most computers distinguish signs by comparing a number to zero so that, for instance, since +3 is greater than zero it is a positive number. On the other hand, −3 is less than zero and is recognized as being negative. This method of comparison, then, is utilized to identify the quadrant locations of angles θ and ϕ. In Figure 6.6, θ is in the first quadrant so both a and c are positive. If this condition is identified as Case 1, it might be expressed as

$$\text{Case 1.} \quad \text{If } a \geq 0$$
$$c \geq 0 \qquad\qquad (6.10a)$$
$$\theta = \frac{\pi}{2} - \sin^{-1}|c/r|$$

In the ratio, c and r have absolute (+) values.

$$\text{Case 2} \quad \text{If } a \geq 0$$
$$c \leq 0 \qquad\qquad (6.10b)$$
$$\theta = \frac{\pi}{2} + \sin^{-1}|c/r|$$

$$\text{Case 3.} \quad \text{If } a \leq 0$$
$$c \leq 0 \qquad\qquad (6.10c)$$
$$\theta = \frac{3\pi}{2} - \sin^{-1}|c/r|$$

$$\text{Case 4.} \quad \text{If } a \leq 0$$
$$c \geq 0 \qquad\qquad (6.10d)$$
$$\theta = \frac{3\pi}{2} + \sin^{-1}|c/r|$$

An examination of the four cases will indicate that for a vertical line of sight

$$a = c = 0$$

the logic breaks down. In this case, the machine is given a special instruction and is told to read

$$\theta = \pi$$

Using the technique of a comparison to zero, the values of ϕ can be covered by two cases.

$$\text{Case 5} \quad b \geq 0$$
$$\phi = \frac{\pi}{2} - \sin^{-1}|b/s| \qquad\qquad (6.10e)$$

In the ratio, b and s have absolute values.

$$\textit{Case 6.} \quad b \leq 0$$

$$\phi = \frac{\pi}{2} + \sin^{-1}|b/s| \qquad\qquad (6.10f)$$

Analysis of this logic will show that no uniqueness of zero values need be considered in this evaluation.

SAMPLE PROBLEM 6.1

Point P is located at $(40, 30, -10)$. Determine the new coordinates for point P after it is rotated 30 deg about the x-axis.

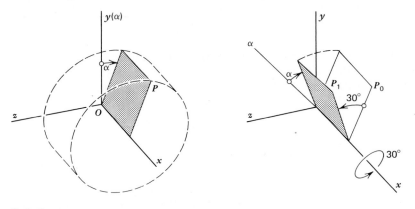

Solution
The rotation shown in the figure can be represented using Eq. 6.7 as

$$[40 \quad 30 \quad -10]\begin{bmatrix} 1 & 0 & 0 \\ 0 & \cos 30° & \sin 30° \\ 0 & -\sin 30° & \cos 30° \end{bmatrix} = [x_1\ y_1\ z_1]$$

Therefore,

$$x_1 = 40$$
$$y_1 = 30(0.866) - 10(-0.5) = 26.0 + 5.0 = 31.0$$
$$z_1 = 30(0.5) - 10(0.866) = 15.0 - 8.7 = 6.3$$

PROBLEMS

6.2 Develop Eq. 6.2 directly from sketches similar to Figure 6.3.

6.3 Develop Eq. 6.4a and 6.4b directly from sketches of the situation.

6.4 Develop Eq. 6.8 and 6.9 directly from sketches similar to Figure 6.5.

6.5 For the specific situation in Sample Problem 6.1, prepare a sketch similar to Figure 6.5 to show graphically what the mathematics does.

6.6 The four corners of a tetrahedron are

$$A(-2, 0, 3), B(2, 2, -1), C(3, 0, 5), D(0, -2, 0).$$

Rotate the tetrahedron through 45 deg about the x-axis and plot the result.

6.7 Rotate the tetrahedron described in Problem 6.6 through 25 deg about the z-axis and plot the result.

6.8 A line of sight slopes down to the right and to the rear at 25 deg to the horizontal (x-z) plane and 40 deg with the profile plane (y-z). Generate the rotational matrix expressing this direction of viewing.

$$Answer: \begin{bmatrix} 0.705 & 0.300 & -0.643 \\ 0 & 0.906 & 0.423 \\ 0.710 & -0.298 & 0.639 \end{bmatrix}$$

6.9 A line of sight **AB** has coordinate data described as $A(0, 0, 0)$, $B(3, -24, 4)$. Determine the matrix that will provide for the proper rotation of this line.

$$Answer: \begin{bmatrix} -0.8 & 0.587 & -0.122 \\ 0 & 0.204 & 0.979 \\ 0.6 & 0.783 & -0.163 \end{bmatrix}$$

6.6 THE OBLIQUE-VIEW MODEL

Once the basic approach employing the symbolism of the matrix has been developed, the first step in generating a view is to establish the direction in which the object is to be viewed. This direction of the line of sight is either given or is selected by the viewer. The selection may be to examine an object for clearance with other items, or it may be to look perpendicular to some auxiliary planes or simply to obtain a better view. In the latter case, it may be necessary to try several views until the best vantage point is obtained. In any event, the first step is to define the direction of sight.

The procedure can be explained best by a simple example, as shown in Figure 6.7. Eight points are needed to define the solid. The solid is 16 units long, 15 units high, and 12 units wide. After an origin is selected to conform to the customary x-, y-, and z-axes, which in this example are selected for corner 1, the corners of the solid are described by evaluating their respective, x-, y-, and z-coordinates with reference to corner 1. The line of sight then is selected. Suppose, for example, it is desired in this particular case to view the body along the diagonal 1–7 (i.e., across opposite corners). It is advisable and also convenient first to select the line of sight, then to place the origin at the tail of the viewing arrow before proceeding to describe the coordinates. This procedure will ensure that the entire object, when viewed along the arrow, will be repositioned behind the viewing plane of the computer screen and thus allow for a con-

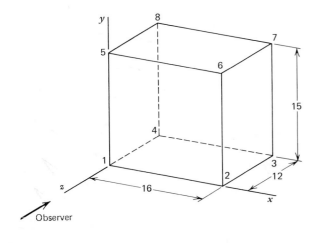

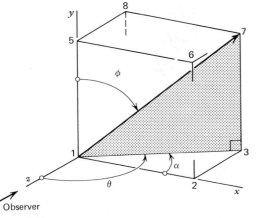

Figure 6.7

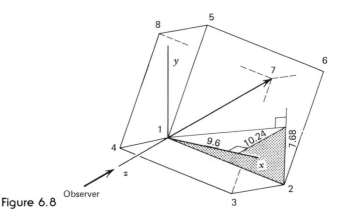

Figure 6.8

dition in which the z-coordinates of the data points, other than zero, will be negative and lie away from the viewer in the front view (Figure 6.8).

At this point, it must be realized that the line of sight, when rotated to its new position, is first turned through an angular displacement associated with the angle designated as θ. Angle θ can be measured by looking along the y-axis. In the horizontal view of the body it measures approximately 127 deg. It must also be noted that the measurement of θ is taken in the conventional counterclockwise direction, measured from the positive z-axis with the final positioning of the arrow corresponding to the negative z direction.

Angle ϕ is the angle measured from the positive y-axis to the line of sight. Remember that this angle can only be viewed in its true measurement while looking perpendicular to the true length of the line of sight along a horizontal direction. This may be determined mathematically or by drawing an auxiliary projection taken perpendicular to the line of sight (1–7) in the horizontal view, as shown in Figure 6.9.

The transformation matrix can now be determined.

$$[R] = \begin{bmatrix} -\cos\theta & -\sin\theta\cos\phi & -\sin\theta\sin\phi \\ 0 & \sin\phi & -\cos\phi \\ \sin\theta & -\cos\theta\cos\phi & -\cos\theta\sin\phi \end{bmatrix}$$

$$= \begin{bmatrix} -(-0.6) & -(0.8)(0.6) & -(0.8)(0.8) \\ 0 & 0.8 & -(0.6) \\ 0.8 & -(-0.6)(0.6) & -(-0.6)(0.8) \end{bmatrix}$$

$$= \begin{bmatrix} 0.6 & -0.48 & -0.64 \\ 0 & 0.8 & -0.6 \\ 0.8 & 0.36 & 0.48 \end{bmatrix}$$

It remains to multiply the original set of data for the corners of the body by this transformation matrix to obtain the new

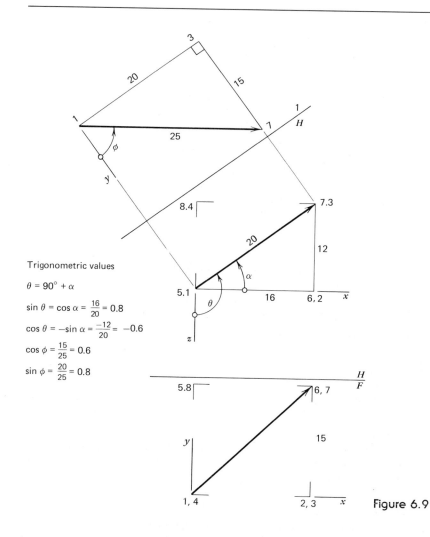

Trigonometric values

$\theta = 90° + \alpha$

$\sin \theta = \cos \alpha = \dfrac{16}{20} = 0.8$

$\cos \theta = -\sin \alpha = \dfrac{-12}{20} = -0.6$

$\cos \phi = \dfrac{15}{25} = 0.6$

$\sin \phi = \dfrac{20}{25} = 0.8$

Figure 6.9

positions (x_N, y_N, and z_N) of the points of the rotated and transformed solid (see Figure 6.8).

Original Data					Transformed Data		
x_0	y_0	z_0			x_N	y_N	z_N
1. 0	0	0	$\begin{bmatrix} 0.6 & -0.48 & -0.64 \\ 0 & 0.8 & -0.6 \\ 0.8 & 0.36 & 0.48 \end{bmatrix}$		0	0	0
2. 16	0	0			9.6	−7.68	−10.24
3. 16	0	−12			0	−12	−16
4. 0	0	−12			−9.6	−4.32	−5.76
5. 0	15	0			0	12	−9
6. 16	15	0			9.6	4.32	−19.24
7. 16	15	−12			0	0	−25
8. 0	15	−12			−9.6	7.68	−14.76

These may be plotted on coordinate paper as in Figure 6.10 and compared to a descriptive geometry solution shown in Figure 6.11 or to the result as viewed on the computer screen (Figure 6.12).

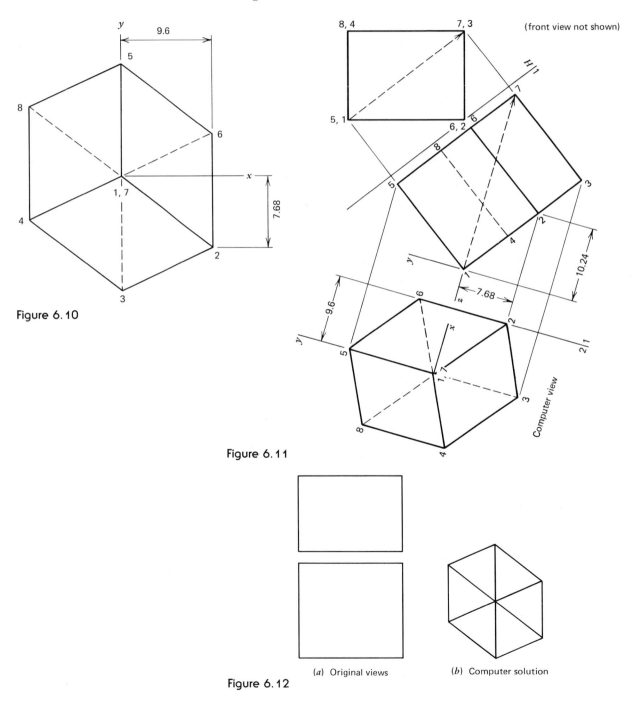

Figure 6.10

Figure 6.11

(a) Original views   (b) Computer solution

Figure 6.12

6.7 DETERMINING THE TRUE SHAPE OF A PLANE

look along C

Finding the true shape of a plane is one of the three basic problems of descriptive geometry. Once you have mastered these three basic problems, there is no limit to the projections you can generate for special purposes. In fact, the problem of the true shape of a plane follows from the other two basic problems. The true length of a line is required before the second basic relationship, the end or point view of a line, can be determined. The edge view of a desired plane is also the projection in which a line in the plane appears in end or point view. Finally, it is possible to view the plane in its true shape by looking perpendicular or normal to the edge view of the plane. This is the process by which the true view of a plane is found using descriptive geometry. It requires three auxiliary views. The process can be somewhat abbreviated in the computer by using the vector cross product introduced in Section 5.8.

The desired line of sight will be one that directly produces the normal direction of sight, which then can be used to generate a set of coordinates representing the figure viewed in that direction. The normal direction can be defined as vector **OC,** which is perpendicular to plane *OAB* in Figure 6.13. The sequence, or order of appearance, of two lines of the required

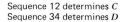

Sequence 12 determines *C*
Sequence 34 determines *D*

Figure 6.13

plane in the vector equation will produce the normal vector as follows.

$$\mathbf{OA} \times \mathbf{OB} = \mathbf{OC} \tag{6.11}$$

These expressions are derived by using the right-hand rule. The determinant representing Eq. 6.11 is

$$\mathbf{OC} = \begin{bmatrix} \mathbf{i} & \mathbf{j} & \mathbf{k} \\ A_x & A_y & A_z \\ B_x & B_y & B_z \end{bmatrix}$$

Expanding the determinant **OC** results in the following expression.

$$\mathbf{OC} = (A_y B_z - A_z B_y)\mathbf{i} + (A_z B_x - A_x B_z)\mathbf{j} + (A_x B_y - A_y B_x)\mathbf{k}$$

$$= C_x \mathbf{i} + C_y \mathbf{j} + C_z \mathbf{k} \tag{6.12}$$

If the origin is placed at O, only the coordinates of A and B need be used. If the origin were not at O, then the terms in the determinant representing lines OA and OB would have to be expressed as differences in coordinates between the ends of the lines.

$$\mathbf{OC} = \begin{bmatrix} \mathbf{i} & \mathbf{j} & \mathbf{k} \\ (x_A - x_O) & (y_A - y_O) & (z_A - z_O) \\ (x_B - x_O) & (y_B - y_O) & (z_B - z_O) \end{bmatrix}$$

6.8 THE ANGLE BETWEEN A LINE AND A PLANE

The direction of sight to produce a set of computer data in a particular coordinate system is determined either by the choice of the analyst or by properties of the geometry under consideration. The angle between a line (ML) and a plane (UVW) is an important measurement in the description of an object (see Figure 6.14). The traditional solution by projections in descriptive geometry involves establishing the true view of a plane (MLN) that contains the line (ML) and is perpendicular to the plane (UVW). The mathematical model to accomplish this solution requires a description of the necessary line of sight and its appropriate transformation matrix.

Consider the pictorial of Figure 6.15, where the angle formed by line DF and the intersection EG of plane DEF with surface ABC is the required angle between plane ABC and line DF. Plane DEF can be established by constructing line DE perpendicular to plane ABC. The direction of this perpendicular

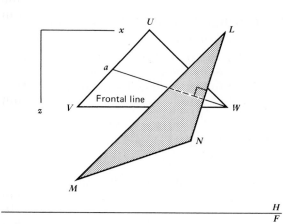

ln ML

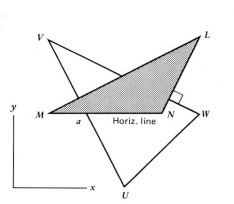

Figure 6.14

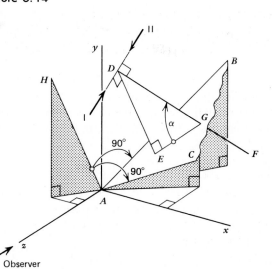

DE || HA

$$I = DF \times HA$$

$$H = HA \times DF$$

$$DF \times (AC \times AB) \text{ shows } \angle$$

Figure 6.15

is the same as that of a parallel line AH, representing the cross product of any two lines considered as vectors (such as AB and AC) in the original plane.

Line AH can then be established as a vector following the convention of the right-hand rule, and the determinant form would read

$$\mathbf{AH} = \mathbf{AC} \times \mathbf{AB} = \begin{bmatrix} \mathbf{i} & \mathbf{j} & \mathbf{k} \\ x_C & y_C & z_C \\ x_B & y_B & z_B \end{bmatrix}$$

In the example shown, using only the coefficients of the unit vectors,

$$x_H = (y_C z_B - y_B z_C)$$
$$y_H = (x_B z_C - x_C z_B)$$
$$z_H = (x_C y_B - y_C x_B)$$

The direction of sight required to evaluate angle α is normal to plane DEF. In the problem, AH can be used in place of DE since it is parallel to DE. Applying the cross-product calculation a second time will produce this view! The line of sight can be written in terms of DF and AH, where the order of the coordinate coefficients (x, y, and z) in the determinant will establish the direction from which the assembly is viewed (view I or view II). This order is established by the application of the right-hand rule.

$$\mathbf{DF} \times \mathbf{HA} \text{ is sight direction I}$$

and is represented by the determinant

$$\begin{vmatrix} \mathbf{i} & \mathbf{j} & \mathbf{k} \\ (x_F - x_D) & (y_F - y_D) & (z_F - z_D) \\ (x_A - x_H) & (y_A - y_H) & (z_A - z_H) \end{vmatrix}$$

Here the origin of coordinates is offset from at least one of the lines so differences in coordinate measurements are required.

$$(y_F - y_D)(z_A - z_H) - (z_F - z_D)(y_A - y_H) = a$$
$$(z_F - z_D)(x_A - x_H) - (x_F - x_D)(z_A - z_H) = b$$
$$(x_F - x_D)(y_A - y_H) - (y_F - y_D)(x_A - x_H) = c$$

Having the x, y, and z coordinates a, b, and c, the transformation matrix may be constructed and applied to data for the assembly in the original coordinate system. Also note that

$$\mathbf{HA} \times \mathbf{DF} \text{ is sight direction II}$$

In vector terms, we have set up an operation called the *triple vector product,* which is the cross product of two vectors where one of them is specified as a cross product of two additional vectors. This product is also a vector.

$$-[\mathbf{DF} \times (\mathbf{AC} \times \mathbf{AB})] \text{ is in sight direction I}$$

$$-[(\mathbf{AC} \times \mathbf{AB}) \times \mathbf{DF}] \text{ is in sight direction II}$$

SAMPLE PROBLEM 6.10

Consider the two views of plane *DEF* shown in (*a*). Display the true shape of the plane as it would appear on a computer screen.

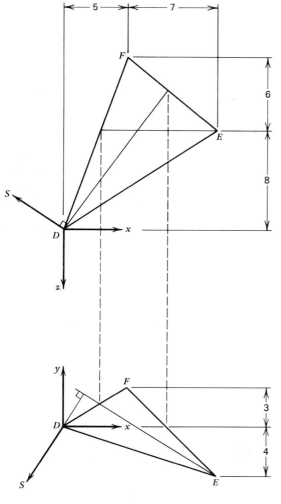

(*a*)

Solution

The normal line DS can be established by conventional methods of descriptive geometry, and to establish the true shape would require two auxiliary projections. This line represents the direction of sight required to establish a true projection of the plane.

To generate the mathematical model of this direction of projection, point D is taken as the origin, and two lines DF and DE are treated as vectors.

The vector cross product is

$$\mathbf{DS} = \mathbf{DF} \times \mathbf{DE}$$

and

$$\mathbf{DS} = \begin{vmatrix} \mathbf{i} & \mathbf{j} & \mathbf{k} \\ F_x & F_y & F_z \\ E_x & E_y & E_z \end{vmatrix} = \begin{vmatrix} \mathbf{i} & \mathbf{j} & \mathbf{k} \\ 5 & 3 & -14 \\ 12 & -4 & -8 \end{vmatrix}$$

From an expansion of this matrix,

$$\mathbf{DS} = -80\mathbf{i} - 128\mathbf{j} - 56\mathbf{k}$$

The direction cosines for vector $\mathbf{DS}$ are

$$\cos \alpha = \frac{-80}{\sqrt{(80)^2 + (128)^2 + (56)^2}} = \frac{-80}{161} = -0.4969$$

$$\cos \beta = \frac{-128}{\sqrt{(80)^2 + (128)^2 + (56)^2}} = \frac{-128}{161} = -0.7950$$

$$\cos \gamma = \frac{-56}{\sqrt{(80)^2 + (128)^2 + (56)^2}} = \frac{-56}{161} = -0.3478$$

All of the direction cosines are negative, which means that the vector is directed as shown in the two views (a). Note that the coordinates of vector $\mathbf{DS}$ are not the same as the coordinates of vectors $\mathbf{DE}$ and $\mathbf{DF}$. As long as we are only interested in the direction of vector $\mathbf{DS}$, this difference is not important.

Using angles θ and ϕ, measured with respect to the positive z- and y-axes, as described in Figure 6.2, their values are

$$\phi = \beta = 142.7 \text{ deg}$$

$$\theta = \sin^{-1} \frac{-80}{\sqrt{(80)^2 + (56)^2}} = 235 \text{ deg}$$

The transformation matrix for this line of sight is

$$R = \begin{bmatrix} -\cos \theta & -\sin \theta \cos \phi & -\sin \theta \sin \phi \\ 0 & \sin \phi & -\cos \phi \\ \sin \theta & -\cos \theta \cos \phi & -\cos \theta \sin \phi \end{bmatrix}$$

$$= \begin{bmatrix} 0.5735 & -0.6513 & 0.4969 \\ 0 & 0.6065 & 0.7950 \\ -0.8192 & -0.4559 & 0.3478 \end{bmatrix}$$

Using the transformation matrix

$$[x_0\ y_0\ z_0][R] = [x_F\ y_F\ z_F]$$

data to produce the coordinate locations of the normal view can be obtained.

Points	Original Data			Final Data		
	x_0	y_0	z_0	x_F	y_F	z_F
D	0	0	0	0	0	0
E	12	−4	−8	13.43	−6.59	0
F	5	3	−14	14.33	4.95	0

These data may be plotted in the conventional *x*- and *y*-coordinates or displayed as shown on the computer terminal (*b*).

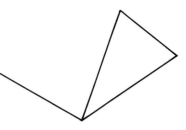

SAMPLE PROBLEM 6.11

The figure (*a*) shows a line *LM* and a plane *UVW*. The plane and the line are described as follows.

$$L(3, 3, -2), M(1, 2, 0)$$
$$U(2, 1, -2), V(1, 3, -1), W(3, 2, -1)$$

Determine the angle between line *LM* and plane *UVW*.

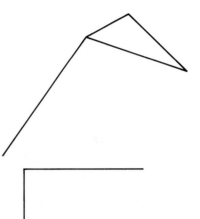

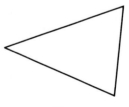

(b)

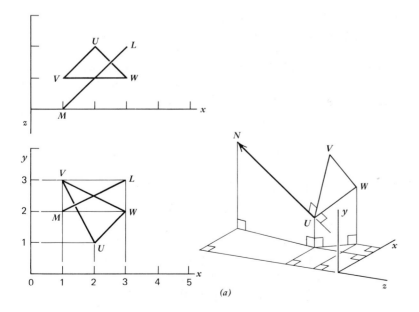

(a)

Solution

The direction of a normal to the plane UVW is first established by the cross product.

$$\text{normal direction} = \mathbf{UV} \times \mathbf{UW} = \mathbf{UN}$$

$$x_N - x_U = (y_V - y_U)(z_w - z_U) - (y_W - y_U)(z_V - z_U)$$
$$= (3 - 1)[-1 - (-2)] - (2 - 1)[-1 - (-2)] = +1$$

Similar calculations will yield

$$y_N - y_U = 2$$
$$z_N - z_U = -3$$

Having established the normal vector, it may now be crossed with line ML.

$$\text{direction of sight} = \mathbf{UN} \times \mathbf{ML}$$

In determinant form, the direction of sight is represented as

$$\begin{vmatrix} \mathbf{i} & \mathbf{j} & \mathbf{k} \\ (x_N - x_U) & (y_N - y_U) & (z_N - z_U) \\ (x_L - x_M) & (y_L - y_M) & (z_L - z_M) \end{vmatrix}$$

The solution of this determinant will lead to

$$a = (y_N - y_U)(z_L - z_M) - (z_N - z_U)(y_L - y_M)$$
$$\text{or, } a = 2(-2 - 0) - (-3)(3 - 2) = -1$$

The other coordinates are evaluated as

$$b = -4$$
$$c = -3$$

Referring to the general definitions shown (b)

$$s = \sqrt{26} \qquad r = \sqrt{10}$$

With these values in hand, the logic of Section 6.5 can be applied. In this instance, Case 3 evaluates θ, while ϕ is expressed by Case 6.

$$\theta = \frac{3\pi}{2} - \sin^{-1}|c/r|$$

$$\phi \quad \frac{\pi}{2} + \sin^{-1}|b/s|$$

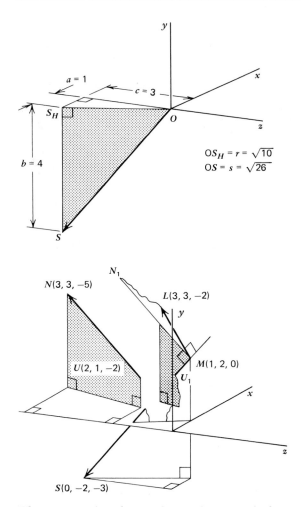

$$OS_H = r = \sqrt{10}$$
$$OS = s = \sqrt{26}$$

The conventional transformation matrix becomes

$$(R) = \begin{bmatrix} 0.947 & -0.248 & 0.196 \\ 0 & 0.620 & 0.784 \\ -0.316 & -0.744 & 0.588 \end{bmatrix}$$

If the original data are specified as x_0, y_0, and z_0, and the new coordinates as x_N, y_N, and z_N, the matrix can be applied as follows:

$$[x_0 \; y_0 \; z_0][R] = [x_N \; y_N \; z_N]$$

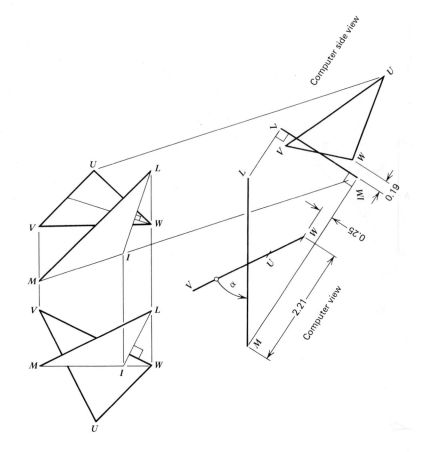

In this problem,

	x_0	y_0	z_0		x_N	y_N	z_N
U	2	1	−2		2.53	1.61	0
V	1	3	−1		1.26	2.36	1.96
W	3	2	−1		3.16	1.24	1.57
L	3	3	−2		3.47	2.60	1.76
M	1	2	0		0.95	0.99	1.76

The traditional solution to this problem appears in (c), while (d) shows the data generated by the matrix plotted on a rectangular grid. The computer solution is in (e).

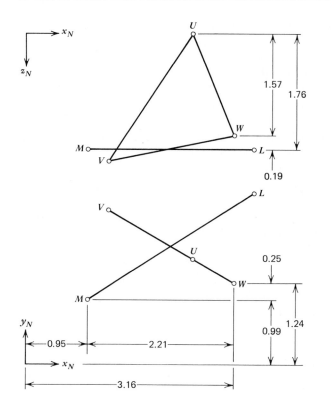

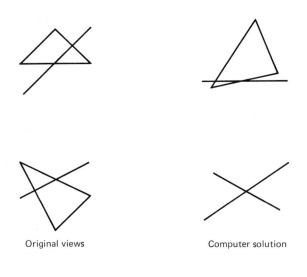

Original views   Computer solution

PROBLEMS

6.12 Generate the data that will enable a computer to "show" a view of the right prism viewed from A to B (dimensions in inches).

6.13 Establish the data of the prism of Problem 6.12 if it is viewed down, forward ($+z$ direction), and to the right at an angle of 25 deg with the horizontal and 15 deg with the frontal. Use point A as the origin of the coordinate system.

Answer:

1	0	0	0
2	−0.958	−0.121	0.259
3	−0.958	−1.571	−0.418
4	0	−1.450	−0.677
5	−0.458	0.648	−1.389
6	−1.416	0.527	−1.130
7	−1.416	−0.923	−1.807
8	−0.458	−0.802	−2.066

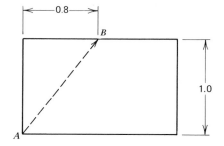

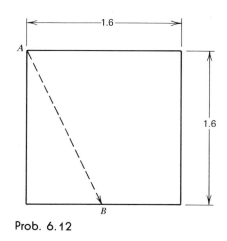

Prob. 6.12

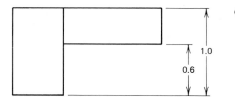

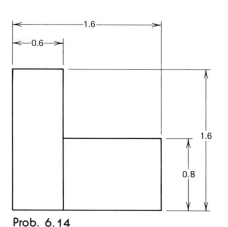

Prob. 6.14

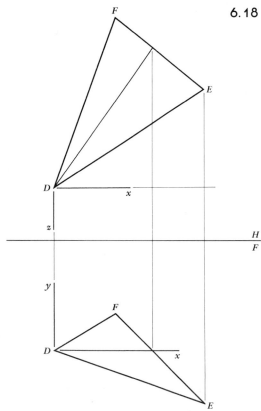

Prob. 6.18

6.14 From the information conveyed by the orthographic views, generate the data for an isometric projection viewed down rearward and to the left (dimensions in inches). Remember that an isometric view is nothing more than an oblique view taken in a specific direction. The first step in this problem is to establish that direction.

6.15 Refer to Problem 6.14. Rotate the block through an angle 150 deg about the y-axis and plot the perspective that results if the observer is placed at $P(0, 2.2, 6)$ (all dimensions in inches).

6.16 Rotate the block of Problem 6.14 15 deg about the x-axis and plot the perspective if the observer is positioned at $P(0.3, 2.5, 4)$.

6.17 A triangle is located at $G(0, 4, 2)$, $E(2, 0, 0)$, $F(4, 3, 3)$. If the triangle is to be viewed from G to E (**GE**), what are the transformed data for F?

Answer: 4.9, 2.3, 2.0

6.18 The two original views of triangle DEF used in Sample Problem 6.10 are repeated here. Using descriptive geometry, determine the true shape of the triangle and compare your result to the computer solution.

6.19 A tetrahedron has its four verticles at $A(2, 4, -1)$, $B(4, 3, 1)$, $C(3, 0, -2)$, $D(0, 1, 2)$. Determine the view of the tetrahedron where line AC is in point view using matrix operations.

6.20 For the tetrahedron in Problem 6.19, establish an axis normal to DBC.

6.21 In Problem 6.19, establish the data for a view of the tetrahedron that will enable the dihedral angle between DAB and CAB to be measured. Therefore, you must view the tetrahedron in the AB direction.

6.22 In Problem 6.19, generate the locational information that will permit the angle between line AB and plane DBC to be measured.

Answer: $\phi = 94.9°$, $\theta = 136.7°$

6.23 A tetrahedron is described as follows: $A(-2, 0, 3)$, $B(2, 2, -1)$, $C(3, 0, 5)$, $D(0, -2, 0)$. Generate the data that would result from viewing ABD in a direction normal to that face.

Answer: A -3.58, $0.09, 0.42$
    B 2.06, $2.14, 0.42$
    C -1.84, $1.16, 5.41$
    D 0 , $-1.95, 0.42$

6.24 Using the tetrahedron of Problem 6.23, develop the data for a view that would provide the angle between AB and ADC.

6.25 With the data given in Problem 6.23, generate a view in which the angle between DC and ABD can be evaluated.

6.26 In Problem 6.23, rotate the tetrahedron through an angle of 30 deg about AC as an axis and plot the results. *Hint.* First view the tetrahedron in the AC direction.

6.27 Generate a view of the tetrahedron in Problem 6.23 by viewing it from C to B.

7
ENGINEERING DRAWINGS

The end product of all engineering is hardware. Engineers make things. This is a cold and simplistic statement, but it is true. Scientists do not make things. Sociologists do not make things. Mathematicians do not make things. And not all engineers make things, but the end result of all engineering is the production or construction of some physical thing that people want. This process of making things to satisfy the needs of society is not simple. It is formal, it is logical, it is carefully done, and engineering drawing is an integral part of the process.

Engineering drawing can be either formal or informal. Formal engineering drawing consists of the working drawings that are actually used to produce an engineering part. It is precise. It has a stylized language and is rigid. Informal engineering drawing consists of any and all drawings and sketches that are used to communicate a concept. Informal engineering drawing is very personal, and in many respects it is the more important of the two.

Most engineering students are quite casual about drawings and drawing conventions. They see them as routine, dull, and subprofessional. Engineering drawings are someone else's responsibility—probably the responsibility of a drafter whom they may never meet. Yet approximately 10 percent of all engineering costs are directly related to the preparation of engineering drawings. Why does this paradox exist?

There are three good reasons for being specific about engineering drawings. Drawings are used (1) to obtain a permanent record of designs, (2) to permit planning and/or to avoid confusion (3) to ensure that the fabrication of that physical thing really is what you wanted.

7.2 ENGINEERING SKETCHES

Engineering sketches are used to communicate thoughts. Although they may be very elaborate and very detailed, they are usually simple line drawings. Sketches exchange ideas before any design commitment is made. They are used in discussions between engineers as a concept grows to a final design. A *working drawing*, however, is a commitment of a concept, and it is a formal document. Engineering sketches precede working drawings, and usually there are many, many sketches before any working drawing is commissioned.

Engineering sketching covers many things, ranging from simple line drawings and free-body diagrams to elaborate renderings. For the most part, engineering sketching is simply one engineer talking to another—or to himself or herself—with the aid of a pencil. Each type of sketch has its special purpose and use. In some cases, standards and symbols have been developed and are accepted as a means to simplify drawing and ensure understanding.

Engineering sketching requires very little artistic ability, but it does require visualization and, what is more important, an understanding of engineering fundamentals. Most engineering sketches will also involve vectors such as forces, fluxes, velocities, stresses, momenta, and accelerations, as well as a visual description. These additions make engineering sketching very different from artistic sketching.

The first four figures in this chapter are engineering sketches. Consider them carefully. Note how much information these drawings convey.

Figure 7.1 is a simple diagram of the loads on a bridge truss. Figure 7.1*b* shows the external loading with the load vectors appearing as arrows. Important dimensions are suggested but the unimportant dimensions are not; *x* and *y* coordinates are fixed. Figures 7.1*c* and 7.1*d* are truss sections. These are called *free-body sections*. Each section is isolated showing all the active forces on that section.

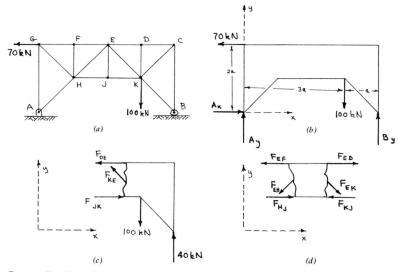

Figure 7.1 Loads on a bridge truss.

Figure 7.2 is a free-body diagram of a latch. Again, active forces are shown as arrows. The V- and H-coordinates are established. Note that this sketch is more formal than Figure 7.1. The engineer has identified the part, dated the sketch for future reference, and identified the sketch as a part of a report. The engineer has also used quadriruled paper for the sketch.

Figure 7.3 is a different sketch showing a scheme for the tension mechanism in a bin-loop tape recorder. Active forces are shown, but this is not a free-body diagram—certainly not a complete free-body diagram. All the active forces are not shown. For example, the pin reaction of the lower bar is not shown. It is not important to this sketch, but it does exist. On the other hand, is there any doubt that the tensioning device is spring loaded? The zigzag line conveys the idea of a spring. What thought does the other detail convey, slightly to the left of the spring? This sketch was taken from a bound notebook, a practical asset if invention or discovery is a factor. The pages in a bound notebook, unlike a loose-leaf notebook, are fixed. In the filing of a patent, records of sketches in a bound notebook are accepted as permanent. Note that this engineer also has used quadriruled paper.

Figure 12.3, in Chapter 12, is an entirely different sketch, showing flow through a nozzle. The sketch shows velocity vectors instead of forces. It is part of an engineering note; it is detailed, fully informative, and is a superb example of engineering sketching. Note the small pictorial sketch that shows a perspective on the flow through the blades.

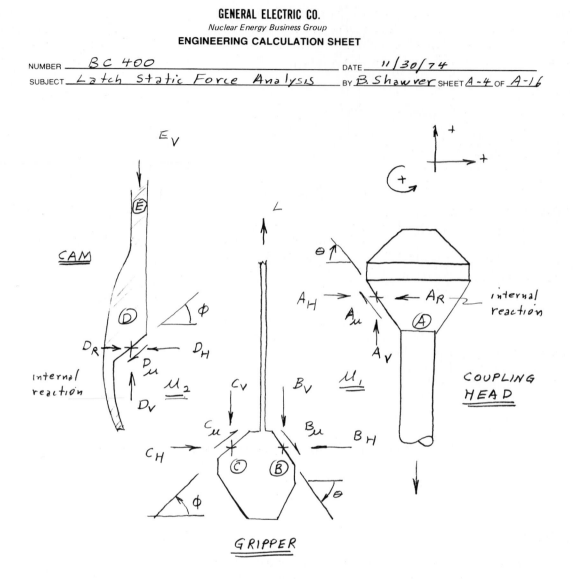

GENERAL ELECTRIC CO.
Nuclear Energy Business Group
ENGINEERING CALCULATION SHEET

NUMBER _____ *BC 400* _____ DATE ___ *11/30/74* ___

SUBJECT ___ *Latch Static Force Analysis* ___ BY *B. Shawver* SHEET *A-4* OF *A-16*

NEBG - 87 (4/80)

Figure 7.2 Engineering sketch in support of application for U.S. Patent 3905634 granted September 16, 1975 to the General Electric Company and assigned to the U.S. Department of Energy. (Courtesy General Electric Company).

Figure 7.4 is an inventor's sketch of fastening schemes for a strike plate for a wood door. It comes from the engineer's notebook; it is signed and dated. It is more artistic than the other sketches, but is there any doubt about how it is put together?

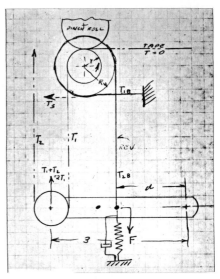

Figure 7.3 Tension mechanism in a bin-loop tape recorder.

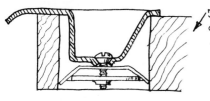

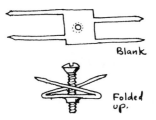

This design, although more difficult to insert into the hole, has the advantage that it draws or pulls the strike down into the hole as the screw is tightened.

E L Schlage Jan 11, 63

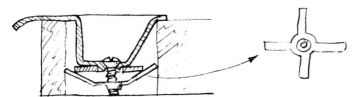

Blank

Folded up.

Figure 7.4 An inventor's sketch. (Courtesy Schlage Lock Company)

PROBLEMS

7.1 The simple line diagram shows the concept of a rope and pulleys. Make an engineering sketch of the upper pulley block.

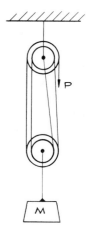

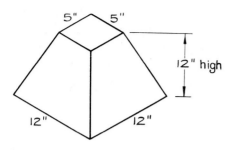

7.2 Sketch a sheet-metal form that can be used to pour a concrete footing.

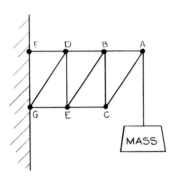

7.3 Prepare sketches that illustrate the forces in members *BD*, *BE*, and *CE*.

7.4 Make an engineering sketch of the bell crank mechanism.

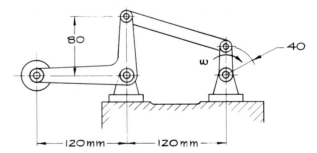

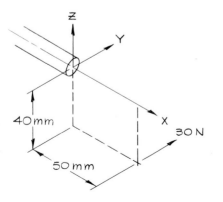

7.5 The 30-N force tends to turn the solid circular shaft. The line of action is located 40 mm below and 50 mm in front of the end of the shaft, and the force is parallel to the *y*-axis. Sketch a mechanical means that would make it possible to transmit this torque to the shaft.

7.6 This problem is the same as Problem 7.5, except that the 30-N force is directed through a point in the *x–y* plane and 60 mm from the end of the shaft and on a line that is at an angle of 45 deg to either axis. The force is parallel to the *z*-axis. Sketch a mechanical means that would make it possible to transmit this torque to the shaft.

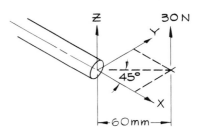

7.7 Sketch a device that can be used to attach and then detach two sections of link chain at a distance of about three links.

7.8 Sketch a stake that will "attach" the tent-support rope to the ground.

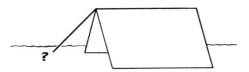

7.3 WORKING DRAWINGS

A working drawing is the formal graphic description of an engineering part. It is also a binding description; it is a commitment of a design to paper. Saying this more specifically, a working drawing is the complete legal description of an engineering part, and in litigation working drawings are *legal documents.* Working drawings are also lasting records of engineering parts. Such records are preserved as long as the part or structure exists.

Contracts are tendered and accepted on the basis of engineering drawings. Contractual performance is measured against working drawings. An engineering part that conforms to the description must be accepted. A part that does not conform need not be accepted. Vast sums of money can change hands. For this reason alone, misinterpreting an engineering drawing could lead directly to lawsuits and judgments.

What is this description? Who describes the engineering part?

Who approves the description? Who interprets the description? If there are questions, who makes the judgments and on what basis? What are the rules of formal engineering drawing? If a controversy does arise about two equally reasonable interpretations of a drawing, one interpretation coming from the originator and the other from the contractor, the legal rule in such cases is "that the interpretation of the party which did not prepare the documents will be considered controlling, since the other party had the opportunity of making any necessary clarification." What this Solomon-like statement means is that the originator of a drawing cannot also interpret that drawing. The person has had the opportunity to originate the drawing and then make it right. That he or she did not change it cannot be used as an excuse if the drawing is later misinterpreted. Think of this the next time you question your instuctor's interpretation of your work.

A *detailed drawing* is the most basic of working drawings. It is a graphic description of a single part, and usually there is only one detailed drawing on any one sheet of paper. The detailed drawing (conventionally shortened to "detail drawing") must be accurate and free of any ambiguity. All the information needed to make the part, including dimensions, tolerances, finishes, and material, is noted on the detail drawing, as well as the names of those responsible for its design.

It is common practice to give the part and the drawing one identifying number. The drawing and the part are therefore identified and inseparable. Figure 7.5 is a detail drawing.

An *assembly drawing* is a drawing of the assembly of parts and is used when the various parts are assembled to make the whole. Each part in the assembly is identified. The identification relates to the accompanying detail drawings. Figure 7.6 is an assembly drawing.

These two types of working drawings constitute the bulk of all working drawings. They both use the principles of orthographic projection laid out in Chapter 3.

Working drawings also use many conventions that are a graphic shorthand that substitutes for rigorous orthographic projection. Both the originator and the reader recognize and understand this graphic shorthand and courts recognize its use. These conventions are few, but the few that do exist should be learned.

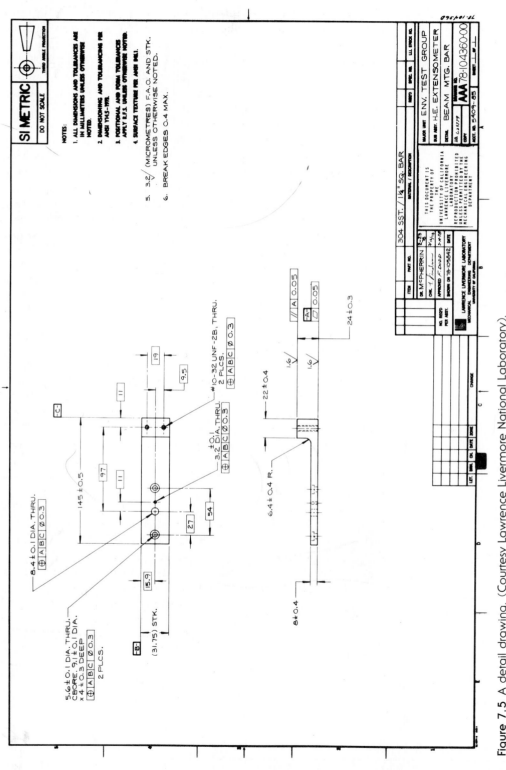

Figure 7.5 A detail drawing. (Courtesy Lawrence Livermore National Laboratory).

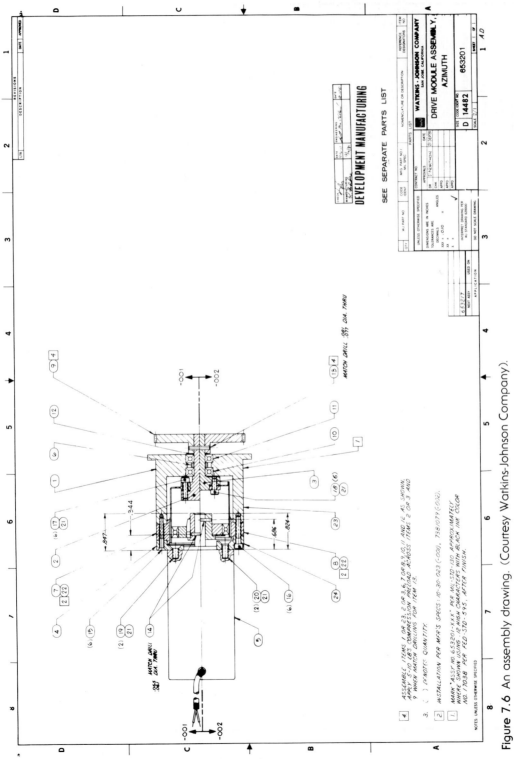

Figure 7.6 An assembly drawing. (Courtesy Watkins-Johnson Company).

7.4 ORTHOGRAPHIC LANGUAGE OF WORKING DRAWINGS

In Chapter 4 we indicated that reading and understanding orthographic drawings is analogous to reading a sentence. The points, lines, letters, and numbers are vocabulary, and orthogonal projection is the grammar.

For detail and assembly drawings, each view is usually an orthogonal projection on the horizontal, frontal, and/or profile plane. In many instances, only two views are given because two views will fully define the position of a point or line in space. For clarity, there may be one or more auxiliary views.

Three orthographic views of objects are shown in Figures 7.7 and 7.8. Note that the horizontal (top) views of both are identical. One object is composed of a cylindrical hole inside a solid cylinder. The other is composed of one solid cylinder on another solid cylinder. How do we discriminate one from the other? Do we need three views? How about Figure 7.9? Do we need three views in Figure 7.9? Can you visualize the solid object?

The orthogonal views in Figures 7.7, 7.8, and 7.9 include some common conventions used in working drawings. Dashed lines represent hidden lines. Centerlines, those with short and long dashed lines, indicate the center or axis of a cylindrical portion of the object. Also note that the intersection lines of the projection planes are omitted, but the orientation of the views follows the rules of orthogonal projection.

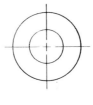

Figure 7.7

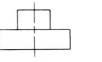

Figure 7.8

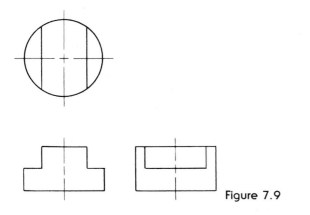

Figure 7.9

7.5 SECTIONS

for internal feature

Decker II? →

The orthographic presentation of a simple object is accomplished by systematically applying the principles of orthographic projection and descriptive geometry. These are precisely followed, and no shortcuts are needed.

When an object is complicated, however, or when several parts are assembled, lines hidden inside the assembly become confusing, even to the most experienced readers of drawings. When this happens, it is usual to *section* one or more views. In a section, some portion of the object is cut away to reveal interior details. A section is imaginary. You do not actually cut anything away, but both the originator of the drawing and the reader recognize what is being done. It is a *convention.* Strict adherence to orthographic projection and the principles of descriptive geometry is abandoned in favor of clarity. This is a good example to show that practical engineering drawing is more a language than a geometry.

As an example of such a situation, consider the simple object shown with three orthogonal projections in Figure 7.10. Descriptively, this would be a rectangular solid with a central hole, two circular bosses, and two counterbored holes within the bosses. An isometric is of some help, but a third view or profile does not help. The profile view adds no information. Our difficulty arises because the interior surface of the counterbored hole is coincident with the top and bottom surfaces of the rectangular solid. A little logic would deduce that these lines would have to be coincident. We cannot leave this to interpretation, however; we must show the interior. We do this by drawing a sectional view.

An imaginary cutting plane slices through the center of the solid. The forward half is then removed, as in Figure 7.11. The view remaining (Figure 7.12) is a cross section of the object with the interior exposed. There is now no question about hidden lines and coincident surfaces. Indeed, hidden lines are rarely shown in a sectional view.

Where the cutting plane slices through a solid part of the object, *crosshatching* is used—an artistic technique that is centuries old and universally recognized. Note that the crosshatching on either side of the centerline is identical, symbolizing that this is one body and that both sides are halves of the same part, but note that crosshatching is a nongeometric technique. Crosshatching lines are parallel lines usually inclined at 45 deg, spaced about 2 mm apart, depending on the area to be crosshatched.

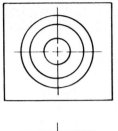

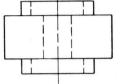

Figure 7.10

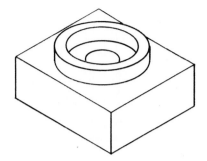

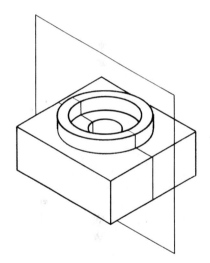

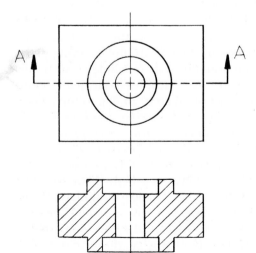

SECTION A-A

Figure 7.12 Cross-sectional view.

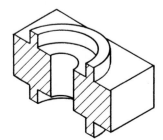

Figure 7.11

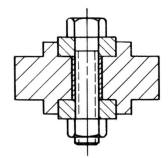

Figure 7.13 Assembly cross section.

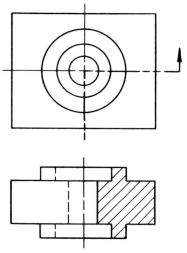

Figure 7.14 Half section.

If we prepare an assembly section, where three mating bushings are shown internal to our piece, crosshatching is varied, either by slanting our lines in a different direction or by using different crosshatching. If there is a central bolt or shaft, as shown in the assembly cross section of Figure 7.13, it is not crosshatched. A bolt has no internal geometry, so crosshatching a bolt would add to the confusion, not detract from it. This is another accepted practice or convention. You would, however, section a shaft if it contained a keyway or internal hole. The shaft would have internal geometry and sectioning, and crosshatching would be needed.

One further convention is the broken line labeled *A–A* (Figure 7.12), which is the edge view of our cutting plane, shown in the horizontal view. The arrowheads show the direction from which the section is viewed. This representation is a convention because the edge view of a plane should be a solid line. Our designation is for a heavy broken line with two arrowheads, labeled *A–A*. It is also quite common for the labeling to be omitted, unless there are two or more sections that can be confused. Then they would be labeled *A–A*, *B–B*, *C–C*, etc.

Figure 7.12 is a full section, meaning that the cutting plane passes through the entire body, and one half is imagined to be removed. Figure 7.14 is a half section. The imaginary cutting plane passes through only one half the body and one quarter is removed. This view takes less time to draw and relies on the inherent symmetry of the part to state that its left and right sections would be identical, which is understood by both the originator and the reader.

In Figure 7.15, we have a different object with two counterbored holes and a different problem. There is still the confusion about hidden lines being coincident with other lines, and the profile view adds enormously to the confusion. A cutting plane parallel to the frontal plane would include one hole or the other, but not both. A cutting plane that includes both holes would have to be on a bias, canted approximately 20 deg to the frontal plane. It is conventional to offset the cutting plane to include both holes, as in Figure 7.16. This is known as an *offset section*.

An example of another conventional practice is shown in Figure 7.17. Here, the cutting plane obviously passed through a web or rib, but that area is not crosshatched. If the web had been crosshatched, it is likely that the reader would have interpreted it as a solid section. Since it is not crosshatched, the reader will question the outside line of the web, look to the horizontal or top view, and recognize what the originator intended. There is much less chance for misinterpretation if the web is not crosshatched than there would be if it had been crosshatched.

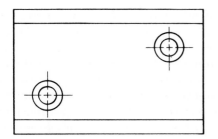

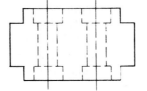

Figure 7.15

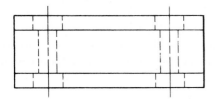

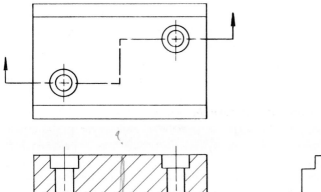

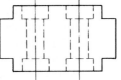

Figure 7.16 Offset section.

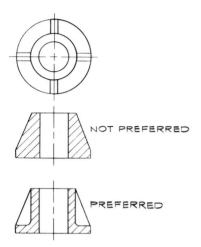

Figure 7.17 Web section.

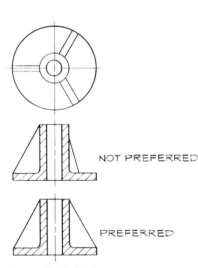

Figure 7.18 Web section.

Still another situation would have arisen if there were an odd number of webs, or ribs, instead of an even number. In Figure 7.18, there are three instead of four, yet the preferred sectional view would be identical. A true projection of the set-back web would be difficult to draw and confusing to read, but it would not be wrong; it is just bad practice. Note that the preferred drawing is really a revolved section, but a cutting-plane line has not been shown; it is not needed. It is understood that one web is rotated into the frontal plane.

Figure 7.19 shows a series of *removed sections* of a propeller blade whose cross section is not constant. The removed sections show what each cross section would be at each particular location called out by the section line. If there is no room, the cross section may be drawn right at the location. It is then called a *rotated section.*

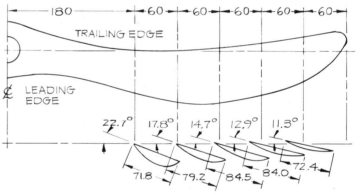

Figure 7.19 Removed sections.

accepted violation

7.6 CONVENTIONS

There are many other conventional practices in addition to sections. Several are shown in Figure 7.20. All are logical.

The first (Figure 7.20*a*) is the convention that uses a line to denote the intersection of two planes, even though the intersection is a curve of small radius. The two planes of the V-groove have a small radius at their intersection, yet their line of intersection is shown in the horizontal (top) view. This is preferred practice. It would be confusing to leave the line of intersection out of the view.

A top view and sectional view of a thick-walled cylinder with a circular hole in the side are shown in Figure 7.20*b*. If the rules of orthogonal projection were strictly adhered to, the lines at

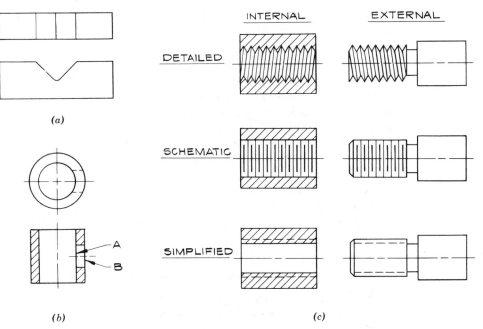

(a)

(b)

(c)

Figure 7.20 Several conventional practices.

A and *B* should be curved. This simpler representation clearly communicates the shape of the object and is much easier to draw. Many simplifications like this can be made in working drawings without reducing the clarity.

Three ways to represent internal and external threads are shown in Figure 7.20c. The detailed representation is sometimes used for large-scale drawings and large threads. Drawing helical threads can be very time consuming and quite unnecessary if there is another simpler way to communicate the existence of threads. Fortunately there is, and it is also shown. When you first looked at the detailed representation, did you even notice that the helical threads were represented by straight lines? Today, the schematic and simplified representations (Figure 7.20c) are used almost exclusively.

Because working drawings represent a major investment of time, effort, and money, any savings in drafting costs can be significant. This is why preferred practices have been introduced and why they are so universally accepted. It is just good practice to eliminate unnecessary drawing. A drafter is an engineering technician who has special training and experience. It is unwise for a drafter to embellish drawings in ink, and that practice has stopped. Most of the drawings in this chapter were made by pencil. Ink is used, however, for drawing on mylar and for computer-made drawings.

Scores of drafting simplifications are now in use. Good drafting practices are constantly being changed and improved. It is very important to understand that communication is the sole reason for engineering drawing. If a drawing is oversimplified and misinterpreted, the added cost of correcting the damage resulting from misinterpretation could be many times the saving.

We suggest that as a part of your study of working drawings you thumb through a drawing manual such as the ANSI standards by the American National Standards Institute or a good engineering drawing textbook. One of the many editions of Thomas French will contain detailed information on conventions and standard practices. French's books (now French and Vierck) have been a standard for engineering drawing for many years. We used them when we were in school, and the father of one of us used French when he was in school.

PROBLEMS

For the next six problems, read the drawing, make an isometric sketch of the object, and decide if all three views are necessary.

7.9

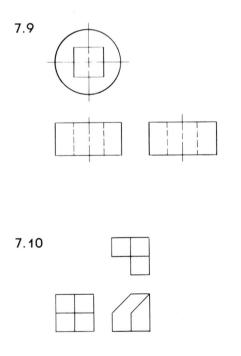

7.10

7.11

7.12

7.13

7.14

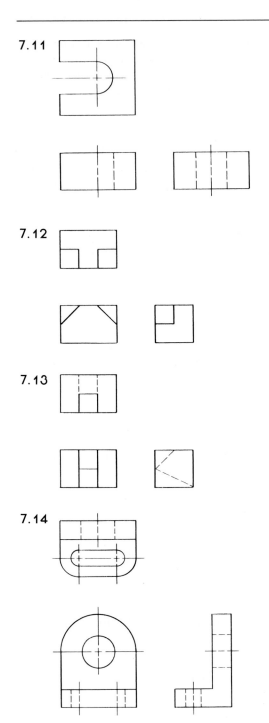

Draw the minimum number of orthogonal views required to represent the following two shapes clearly.

7.15 The concrete footing of Problem 7.2.

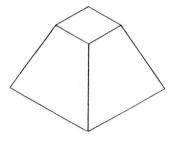

7.16 The shape shown.

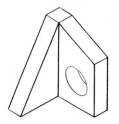

7.17 Represent the flange with one view and a half section.

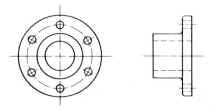

7.18 Represent the flange with one view and a full section.

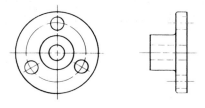

7.19 Represent a coffee cup using one view and a section.

7.20 Make a pictorial drawing that shows the shape of seal 18
shown in the patent assembly section.

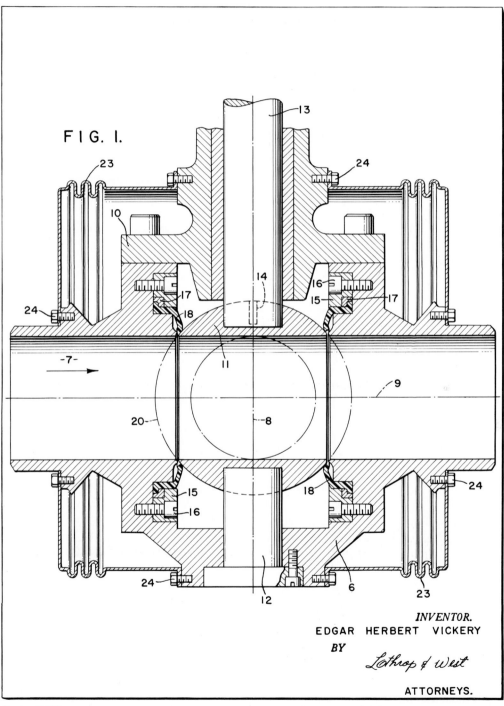

FIG. I.

INVENTOR.
EDGAR HERBERT VICKERY
BY
Lothrop & West
ATTORNEYS.

Prob. 7.20

7.21 Tabulate the conventional ways the plumbing hardware is represented.

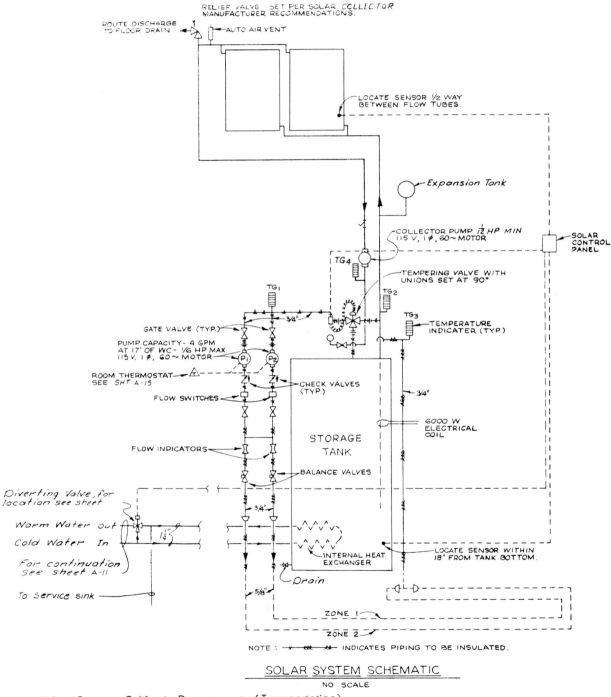

Prob. 7.21 (Courtesy California Department of Transportation).

7.7 DIMENSIONS

A working drawing is normally drawn full size, if at all possible, but it can be drawn to another scale if the size of the object is too large or too small. For better reading, smaller objects are drawn larger than full size, and larger objects are drawn smaller than full size. For convenience, detailed areas and very small parts may be enlarged on a drawing.

Requiring a reader to measure (to scale) a mechanical drawing is against the rules of drafting. Because working drawings are not to be measured, the locations of all points and lines are completely displayed by dimensions. This statement is absolute for mechanical drawing. All needed dimensions are given legibly and conspicuously. For structural drawing, it is occasionally necessary to scale some distances, so the scale of the structural drawing is prominently stated. Note that the word *scale* is used two ways, once as a verb meaning "to measure against some arbitrary standard" and again as a noun that is a ruled standard. Do not confuse the two. They are quite different.

Scales (the noun) are indicated on drawings several ways. If the lengths on a drawing are $\frac{1}{8}$ actual size, the eighth-size scale of the drawing could be indicated as eighth scale, $0.125 = 1.00$, $\frac{1}{8} = 1$, or $1:8$. If the drawing is 10 times the actual size, the scale could be written as ten scale, $10.00 = 1.00$, $10 = 1$, or $10:1$.

Dimensioning is the word used to describe the practice of adding explanatory notes and numbers to an engineering drawing in order to locate points and lines accurately. The proper functioning of a device, product, or system is largely dependent on specifying proper dimensions and tolerances. The authoritative reference on dimensioning is the American National Standard on Engineering Drawing and Related Documentation Practices (document ANSI Y14.5, dated 1981M). Knowledge of the use of machinery, tools, methods, materials, and quality-assurance practices is also needed for proper dimensioning.

Amer. Nat. Standards Inst.

Y 14.5 1982

Dimensions are added to an engineering drawing by placing the required length as a number in an open gap in a line between two arrowheads pointing to the two surfaces being dimensioned. This line is called the *dimension line*. A dimension line, with its arrowheads, shows the extent of a dimension. Any geometric characteristic can be a dimension. Lengths, diameters, angles, etc., are all dimensions.

The two arrowheads and the dimension line can be placed directly between the two surfaces being dimensioned, but the preferred practice is to place dimensions and dimension lines at some distance from the object. To do this, *extension lines* are drawn. These lines are considered as extensions of the surfaces being dimensioned. Normally, extension lines start with a short visible gap from the outline of the part and extend beyond the outermost related dimension line. Extension lines should neither cross one another nor cross dimension lines, but when they do they are drawn unbroken. *exept arrows*

In the case of holes, the location of the center of the hole is necessary—for example, in the drilling of the hole. Thus for drilled holes, *centerlines* are dimensioned. *Leaders* are thin lines drawn for some explanation of design intent. *Notes* are the written explanation. Notes and leaders go together.

Figure 7.21 shows typical notes, dimensions, dimension lines, extension lines, and leaders for one view of a simple part. This variety of lines should be drawn differently. Almost all engineering drawings are drawn in pencil. With modern reproduction practices, ink drawings are not necessary. All drawing lines should be drawn dense, as contrasted to light lines, but leaders, extension lines, and centerlines are drawn as thin lines. The variation in lines is in their thickness, not in density.

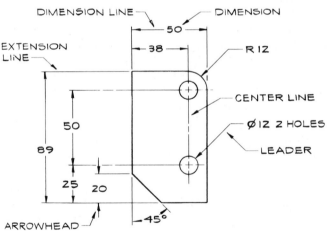

in mm

Figure 7.21 Dimensioning presentation.

There are two kinds of dimensions, *size* and *position*. The size dimension gives the size of a part, such as a hole or a thickness. The position dimension locates the feature that has been sized. Dimensions for size and position must be complete so that the intended features can be produced without any assumptions.

The differences between the two kinds of dimensions are best illustrated by considering a drilled hole. Think of a drilled hole in a solid object. How can you check the position of a hole once the hole is drilled? You can check its size easily, but what about its position? Is it cylindrical? Is the axis oriented correctly? Figure 7.22 shows some typical dimensions of size and position.

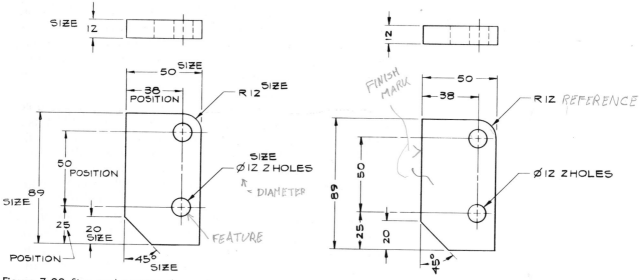

Figure 7.22 Size and position dimensions.

Figure 7.23 Aligned dimensions.

Dimensions are placed in an open space in the dimension line. When the length is small, arrowheads are placed outside the extension line. Note the 12-mm size dimension of the edge view in Figure 7.22. These numbers can be either aligned with the dimension line, which means that you must turn the drawing to read the dimensions, or unidirectional. Unidirectional dimensions are placed to be read from the bottom of the drawing and are preferred. It is much simpler if you do not have to turn the drawing to read dimensions. The vertical dimensions, however, require more space between dimension lines to accommodate the dimension numbers. In Figures 7.21 and 7.22, the dimensions are unidirectional. In Figure 7.23, in order to read the 89-mm and 12-mm dimensions conveniently, the drawing must be rotated 90 deg clockwise.

There are dimensioning conventions. That is, certain accepted practices are used in dimensioning to avoid confusion. They should be followed, but if following the accepted practice leads to more confusion than not following it, the practice should be abandoned. The objective is to present a clean, obvious drawing. The practices of dimensioning include the following.

1. Dimension outside the view.
2. Dimension between views, if possible.
3. Dimension only true lengths.
4. Place dimension lines at least 10 mm away from the view.
5. Place numbers in the middle of the dimension line.

φ.1 UNIDIRECTIONAL
φ.5 USE DECIMALS
φ.9 DIM. ONLY ONE VIEW

PRINCIPLES

6. Space dimension lines uniformly, about 6 mm between lines.

7. Place longer dimension lines outside smaller dimension lines.

8. Don't crowd dimensions.

9. Give no unnecessary dimensions.

1∅ VIOLATE ANYTHING FOR CLARITY

CLARITY

In Figure 7.24, staggered dimensions are used that, in this case, would be acceptable and good practice.

There is one very important practice in dimensioning: to give no unnecessary dimensions. There must only be one method of locating any point or stating any size. No surface, line, or point may be located by more than one dimension. At times, a redundant dimension is given. Such a dimension is called a *reference dimension* and is so labeled — for example, 13 REF.

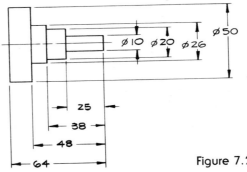

Figure 7.24 Staggered dimensions.

There are a few specific rules for dimensioning parts or features, and all follow common sense. Diameters, symbolized by ϕ, are specified for cylindrical features or holes (Figure 7.25a). Diameters can be measured, but radii cannot. If a radius is to be dimensioned, a small cross is used to show the center of a radius, and the position of that cross is dimensioned (Figure 7.25b). Arcs and angles are dimensioned in Figures 7.25c and 7.25d. Other features are dimensioned in a similar manner.

When several features are dimensioned on the same drawing, they should be positioned either with respect to each other or with respect to a common datum. In Figure 7.26a, the three holes are dimensioned with respect to each other. The center-to-center distance between successive holes can be held with limits. This is known as *chain dimensioning*, and it is a poor practice. Alternately, we could use an end surface as a datum and

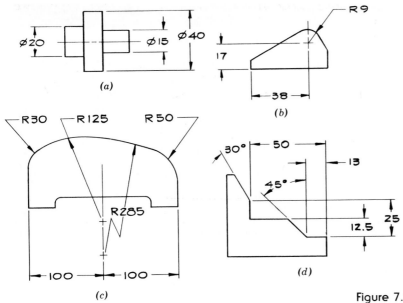

Figure 7.25 Dimensioning diameters and radii.

locate each hole individually from that datum. In Figure 7.26*b*, the bottom and right surfaces are used as datums. This is called *datum dimensioning* and is more accurate in that tolerance errors do not accumulate.

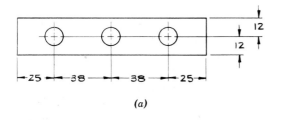

Figure 7.26 Chain *(a)* and datum *(b)* dimensioning.

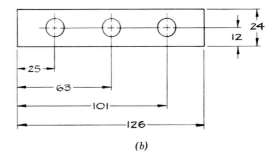

7.8 TOLERANCES

When two parts are to be mated, the fit between them is important. For example, if construction members are to be bolted together, the holes must be located with sufficient accuracy to allow the bolt to slip easily into place. If a sleeve is to fit over a shaft, changing the dimensions of one or both parts will change the kind of fit between the two parts. To control these variables, maximum and minimum dimensions are stated instead of one dimension. The assignment of maximum and minimum dimensions is also called *limit dimensioning*, since it is the assignment of limits on the dimensions. Figure 7.27 shows some examples of limit dimensions. The actual dimension of the finished part must be equal to or within these limiting dimensions or the part is rejected. The difference between the maximum and minimum dimensions is called the *tolerance*.

Obviously, tolerances have much to do with manufacturing and construction costs. A small tolerance means that the part must be made with greater accuracy. This usually means increased cost. Every statement of tolerance means that an inspection may be required. The objective is to use the largest tolerances that can be used successfully. By successfully, we mean that the mating parts can be assembled and perform whatever function they are meant to do.

It is common practice to list a tolerance on every dimension, wherever that dimension is stated on a drawing. If it is not stated, tolerances must be indicated by a general note or included in a reference document. Conventional tolerancing practices are intimately tied to dimensioning practices.

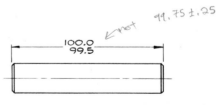

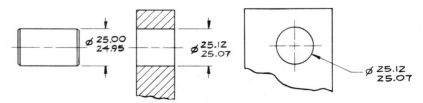

Figure 7.27 Limit dimensions.

The terminology of tolerances is specific, and the terminology adopted by the American National Standard Institute is preferred.

Tolerance. The total amount by which a specific dimension is permitted to vary.

Nominal Size. The approximate dimension or size. It is used only for identification.

Basic Size. The dimension from which the limits or tolerances are derived.

Actual Size. The measured size or dimension.

Allowance. The minimum difference in the dimensions between mating parts. It is the tightest fit—the maximum size shaft fitting in the minimum size hole.

Unilateral Tolerance. In a unilateral tolerance, the allowable tolerance is in one direction. It may be added or subtracted from the basic dimension but not both.

Bilateral Tolerance. For a bilateral tolerance, the allowable tolerance is divided with part being added and the remainder being subtracted from the basic dimension.

Fit. The tightness between two mating parts after they are assembled. There are three classes of fit: clearance, interference, and transition.

There are two systems for stating tolerances on mating parts. Either the inside or the outside part can be used for the basic size. Production economics dictate which this is to be. In most cases, using the hole for the basic dimension is more economical; holes can be made with standard tools. Obviously, these two systems are called the *basic shaft system* or the *basic hole system*. In Figure 7.28, two examples are shown, one using the basic shaft system and the other using the basic hole system. In each case, the tolerances and the allowances are the same and the fit is a clearance fit.

e.g.- ∅.07 = INTERFERANCE FIT
    ∅.07 = CLEARANCE FIT

TOLERANCE ON THE
HOLE = ∅.05

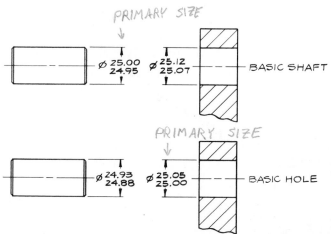

Figure 7.28 Dimensioning mating parts.

In practice, real parts are imperfect in size and shape. Stating a band or zone of tolerance will guarantee that each accepted part will fit as expected, but it will not guarantee that all non-accepted parts will not fit. Obviously, this last part is also important. Economics dictate that usable parts should be accepted and that nonaccepted parts should be unusable. Of course, you must know what you are expecting, and this can be a problem.

In engineering practice, the focus of tolerance dimensioning is in the measurement of the finished piece. What are the actual dimensions? Is perpendicularity true? Are parallel surfaces parallel? Are flat surfaces flat? Are cylindrical surfaces cylindrical? If there is a drilled hole, where is it and how big is it?

As an example of a problem in measurement, consider a flat surface. From where do you begin to measure? The flat surfaces may not be flat, the plane surfaces may not be a plane, and right angles may not be true right angles. For these cases, two measurements of the same dimension can be different! Figure 7.29 exaggerates the problem, but only for effect.

Another problem with dimensioning is tolerance accumulation. If several dimensions are in series, all of them consistently oversized or undersized, the accumulation of these tolerances all in one direction could make the part unusable. As an example, in Figure 7.30, each hole has a nominal diameter of 12 mm, and each is drilled with a spacing tolerance of 0.1 mm in a spacing distance of 38 mm. If the three holes were drilled with an actual spacing of 37.95 mm, the third hole in series could have an accumulated error of 0.15 mm in its position. Other tolerance dimensions could correct this problem if we had reflected our design requirements—what we really want in our dimensioning.

- TOLERANCE BUILD-UP ⇒

PREVENT TOL. BUL.
W. DAT. DIM.

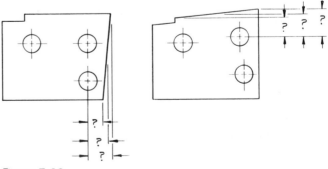

Figure 7.29

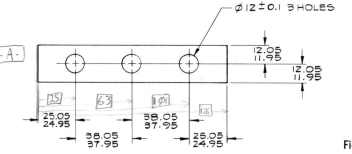

Figure 7.30

7.9 POSITION TOLERANCES

As an example of the problems raised by conventional tolerancing, consider the three-holed part of Figure 7.31. In particular, the tolerance zone for the geometric center of the upper
right hole must lie within a square, 0.1 mm on a side. The
maximum deviation of the true position of the center of the
hole would be 0.07 mm, one half the length of the diagonal. If
a through bolt were placed in this hole and through its mating
hole, the allowance between the bolt and a hole would have to
account for this maximum deviation. That is, if the center of the
hole were at the lower left corner of the tolerance zone, and the
center of the mating hole were at the upper right corner of the
tolerance zone, the bolt would just pass through both holes
without interference. We would allow for a difference in the
true position of the centers of the holes of 0.14 mm.

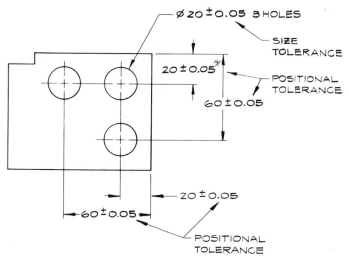

✱ SHOULD = 1.5 × ∅ HOLE

Figure 7.31

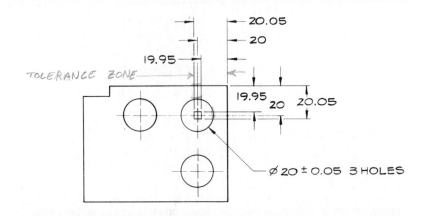

Figure 7.32a SQUARE T. ZONE

Let us now assume that we have allowed for this variation in position and the hole diameters are 20.00. The maximum diameter of the through bolt is held to 19.86 mm. But we only make use of this generous allowance if the centers of the mating holes are located on a diagonal. If the centers of the two mating holes are separated by 0.14 mm but are located on a line other than the diagonal, the parts would be rejected, even though the two holes would mate and receive the bolt. We have used a square tolerance zone, and the center-to-center distance between the two holes would lie outside the allowed tolerance zone except when the two holes line up on a diagonal. Figure 7.32a shows this situation. We are now in the ridiculous situation of rejecting a part that would perform the service for which it was designed simply because the working drawing says that it should be rejected. This is not supposed to happen. Our ability to communicate design intent has been lost! An unacceptable part should not be usable.

This problem is corrected by *position tolerancing*, which locates the theoretically exact position of a feature, as established by basic dimensions. The use of position tolerancing results in a circular tolerance zone, and a circular tolerance zone is 57 percent larger than a square tolerance zone. More parts can be accepted.

In Figure 7.32b, the location tolerance and the size tolerance for the circular hole are separated. Note the boxes drawn around the 20-mm dimensions. They give the symbol for the basic dimension and indicate that the dimension of 20 mm from the given datum is the true position of the center of the hole. Separately, in a note, the position tolerance on the location of the hole is stated as a diameter of 0.14 mm. The symbol ⊕ indicates the true position. This is how we state it. Measuring and inspecting a finished part to check it against the stated dimension in another problem.

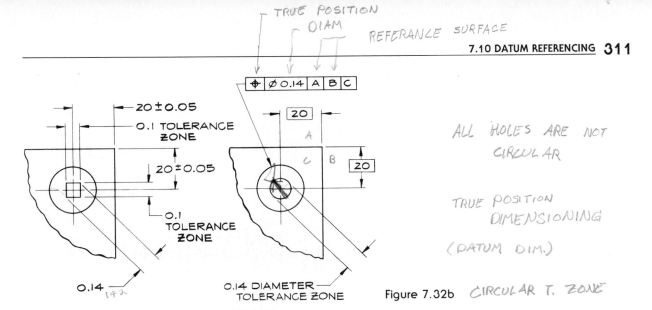

TRUE POSITION
DIAM
REFERANCE SURFACE

$\oplus$ | $\varnothing$ 0.14 | A | B | C

20 ± 0.05

0.1 TOLERANCE ZONE

20 ± 0.05

0.1 TOLERANCE ZONE

0.14

20

A

C

B

20

0.14 DIAMETER TOLERANCE ZONE

Figure 7.32b   CIRCULAR T. ZONE

ALL HOLES ARE NOT CIRCULAR

TRUE POSITION DIMENSIONING

(DATUM DIM.)

Position tolerancing also removes the uncertainty about the origin of measurements. From where are measurements to be made? With conventional tolerancing, the origin is subject to interpretation, and different people interpret differently. Position tolerancing ties down the coordinates for measurement by specifying the datums from which measurements are made.

7.10 DATUM REFERENCING

CIVIL ENG.

Datum referencing is a term used to establish the position of a feature such as a plane surface, a hole, a slot, an angle, or a radius relative to a selected datum. Features selected as datums are called datum features. Usually they are plane surfaces but they need not be. Figure 7.33 shows how both plane surfaces and cylindrical surfaces could be used as datums.

DATUM DIMENSIONING
DATUM = BASE
DATUMS = LATIN NUETER NOUN

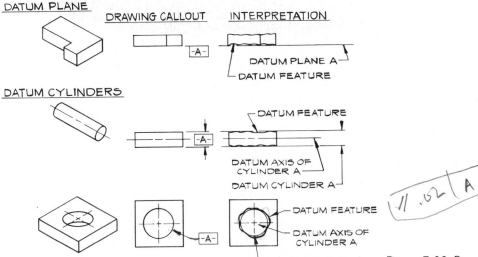

DATUM PLANE

DRAWING CALLOUT   INTERPRETATION

-A-

DATUM PLANE A
DATUM FEATURE

DATUM CYLINDERS

-A-

DATUM FEATURE

DATUM AXIS OF CYLINDER A
DATUM CYLINDER A

DATUM FEATURE

DATUM AXIS OF CYLINDER A

DATUM CYLINDER A   **Figure 7.33** Datum features.

11 .52 | A

The part shown in Figure 7.34 requires three datums, one face and two orthogonal sides. These datums are labeled A, B, and C. They are known as the primary, secondary, and tertiary datums. The order is significant. Datum A is identified by the symbol A, preceded and followed by a dash, all enclosed in a box—that is, $\boxed{-A-}$. The box is connected to surface A by a line. Each datum reference letter is entered in the desired order of precedence, from left to right, in what is called the *feature-control symbol*. Datum reference letter entries need not be in alphabetical order. Note again that A, B, and C are three orthogonal planes. This is fundamental descriptive geometry. The position of any point or line in space can be located with reference to three known orthogonal planes.

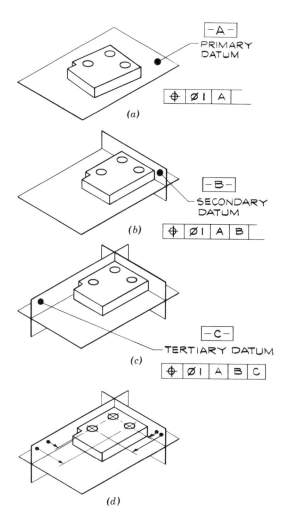

(a)

(b)

(c)

(d)

Figure 7.34 Datum planes.

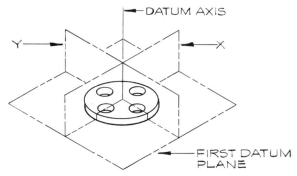

Figure 7.35 Datum axis.

If the part is cylindrical, and no plane surface is available, the cylindrical surface would have to serve as the datum, (Figure 7.33), or we can use the intersection of two of the planes as an axis, placing the line of intersection coincident with the axis of the part (Figure 7.35). Note that this bypasses a crucial step. It is very difficult to locate an axis of a cylinder in space, but locating two orthogonal planes is easy. Thus our engineering drawing statements must be changed. Instead of stating dimensions to a centerline, we must state dimensions to these planes or to another surface that could be used as a substitute.

To establish the datums for positional tolerancing, think ahead to the finished measurement of the completed part. Design intent is the prime consideration. What plane surfaces can be used as datums? If an axis is needed, you will need two intersecting orthogonal planes; remember, you will need three planes overall. Call out these planes on your drawings and state all dimensions and all locations from these three datums. With a little practice, it is simple to use position tolerancing. In Figure 7.36, the same part is drawn as in Figure 7.31 but drawn with position tolerancing. Compare the two. The feature symbol ⓢ will be defined in Section 7.12, and it is correctly used here.

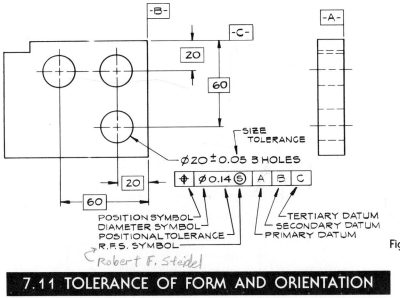

POSITION SYMBOL
DIAMETER SYMBOL
POSITIONAL TOLERANCE
R.F.S. SYMBOL

TERTIARY DATUM
SECONDARY DATUM
PRIMARY DATUM

⌀20 ± 0.05 3 HOLES

SIZE TOLERANCE

Robert F. Steidel

Figure 7.36 Position tolerancing.

7.11 TOLERANCE OF FORM AND ORIENTATION

= SIZE + FEATURE DIM

Straightness, flatness, roundness, and cylindricity are geometric characteristics of form, and perpendicularity, angularity, and parallelism are characteristics of orientation. As in position tolerancing, the characteristics are also stated separately. The benefit of this practice is that by separating the characteristics and measuring and inspecting them separately, the design intent is stated more clearly. This clarity of statement should result in lower costs and improved and faster communication of information. Form and orientation tolerances are specified for features critical to function. They are used where tolerances on size and position do not provide the required control. A larger allowable tolerance could mean a cheaper product. There is also a real possibility that more parts can be accepted. These would be parts that are really usable but rejected under old-style tolerancing procedures. To reject a functional part is as bad as accepting a nonfunctional part.

Form is usually controlled within the tolerance on size. A *form tolerance* is stated by means of a feature-control frame containing the geometric characteristic symbol followed by the allowable tolerance. When the tolerance is on a diameter, the tolerance is preceded by the diameter symbol. *Orientation tolerances* require datums (e.g., one surface must be perpendicular to another). Where a tolerance of position or orientation is related to a datum, the relationship is stated in the feature-control frame by placing the datum reference letter following the tolerance. Vertical lines separate these entries.

The geometric characteristic symbols are shown in Table 7.1. Figure 7.37 shows some of the most common feature controls and what they mean.

TABLE 7.1
Geometric Characteristic Symbols

	Type of Tolerance	Characteristic	Symbol
For individual features	Form	Straightness	—
		Flatness	▱
		Circularity (roundness)	○
		Cylindricity	⌭
For individual or related features	Profile	Profile of a surface	⌒
For related features	Orientation	Angularity	∠
		Perpendicularity	⊥
		Parallelism	//
	Location	Position	⌖
		Concentricity	◎
	Runout	Total runout	⌰

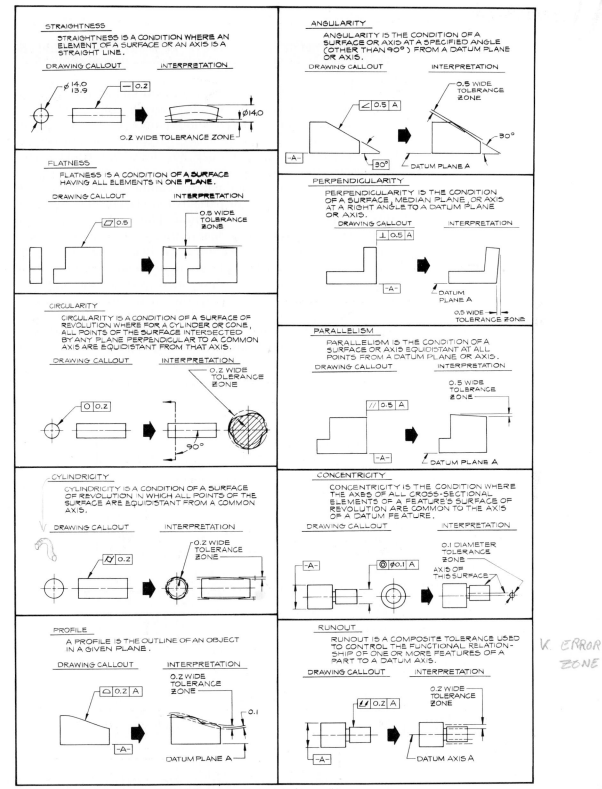

STRAIGHTNESS
STRAIGHTNESS IS A CONDITION WHERE AN ELEMENT OF A SURFACE OR AN AXIS IS A STRAIGHT LINE.

DRAWING CALLOUT   INTERPRETATION

⌀14.0 13.9   — 0.2

⌀14.0

0.2 WIDE TOLERANCE ZONE

FLATNESS
FLATNESS IS A CONDITION OF A SURFACE HAVING ALL ELEMENTS IN ONE PLANE.

DRAWING CALLOUT   INTERPRETATION

▱ 0.5

0.5 WIDE TOLERANCE ZONE

CIRCULARITY
CIRCULARITY IS A CONDITION OF A SURFACE OF REVOLUTION WHERE FOR A CYLINDER OR CONE, ALL POINTS OF THE SURFACE INTERSECTED BY ANY PLANE PERPENDICULAR TO A COMMON AXIS ARE EQUIDISTANT FROM THAT AXIS.

DRAWING CALLOUT   INTERPRETATION

○ 0.2

0.2 WIDE TOLERANCE ZONE

90°

CYLINDRICITY
CYLINDRICITY IS A CONDITION OF A SURFACE OF REVOLUTION IN WHICH ALL POINTS OF THE SURFACE ARE EQUIDISTANT FROM A COMMON AXIS.

DRAWING CALLOUT   INTERPRETATION

⌭ 0.2

0.2 WIDE TOLERANCE ZONE

PROFILE
A PROFILE IS THE OUTLINE OF AN OBJECT IN A GIVEN PLANE.

DRAWING CALLOUT   INTERPRETATION

△ 0.2 A

0.2 WIDE TOLERANCE ZONE

0.1

-A-

DATUM PLANE A

ANGULARITY
ANGULARITY IS THE CONDITION OF A SURFACE OR AXIS AT A SPECIFIED ANGLE (OTHER THAN 90°) FROM A DATUM PLANE OR AXIS.

DRAWING CALLOUT   INTERPRETATION

∠ 0.5 A

0.5 WIDE TOLERANCE ZONE

30°

-A-   30°   DATUM PLANE A

PERPENDICULARITY
PERPENDICULARITY IS THE CONDITION OF A SURFACE, MEDIAN PLANE, OR AXIS AT A RIGHT ANGLE TO A DATUM PLANE OR AXIS.

DRAWING CALLOUT   INTERPRETATION

⊥ 0.5 A

-A-

DATUM PLANE A

0.5 WIDE TOLERANCE ZONE

PARALLELISM
PARALLELISM IS THE CONDITION OF A SURFACE OR AXIS EQUIDISTANT AT ALL POINTS FROM A DATUM PLANE OR AXIS.

DRAWING CALLOUT   INTERPRETATION

∥ 0.5 A

0.5 WIDE TOLERANCE ZONE

-A-   DATUM PLANE A

CONCENTRICITY
CONCENTRICITY IS THE CONDITION WHERE THE AXES OF ALL CROSS-SECTIONAL ELEMENTS OF A FEATURE'S SURFACE OF REVOLUTION ARE COMMON TO THE AXIS OF A DATUM FEATURE.

DRAWING CALLOUT   INTERPRETATION

-A-   ◎ ⌀0.1 A

0.1 DIAMETER TOLERANCE ZONE

AXIS OF THIS SURFACE

RUNOUT
RUNOUT IS A COMPOSITE TOLERANCE USED TO CONTROL THE FUNCTIONAL RELATIONSHIP OF ONE OR MORE FEATURES OF A PART TO A DATUM AXIS.

DRAWING CALLOUT   INTERPRETATION

⟋ 0.2 A

0.2 WIDE TOLERANCE ZONE

-A-   DATUM AXIS A

K. ERROR ZONE

Figure 7.37 Common feature controls.

315

7.12 THE MAXIMUM MATERIAL CONDITION

One other feature control has a decided effect on tolerances: controlling the tolerance by stating it for the condition where a feature of size contains the maximum amount of material. For holes, slots, keyways, and other internal dimensions, this would be the condition of minimum size. For shafts, lugs, and external dimensions, this would be the condition of maximum size. Beginning with the maximum material condition, the tolerance is then allowed to increase for parts that are larger than the minimum or smaller than the maximum. The symbol Ⓜ or the abbreviation MMC is used to denote the *maximum material condition*. If a location tolerance must be held even though the feature is actually smaller or larger than the maximum material condition, that condition is designated as Ⓢ, or the *regardless-of-feature-size condition*. Obviously, one or the other must be selected, but not both. Table 7.2 includes these and other symbols related to the discussion. Figure 7.38 shows how these and other feature control symbols are used. Figure 7.39 shows the maximum material condition for a cylinder, a hole, and a slot.

TABLE 7.2
Other Feature Control Symbols

Term	Abbreviation	Symbol
Maximum material condition	MMC	Ⓜ
Regardless of feature size	RFS	Ⓢ
Diameter	DIA	ϕ
Projected tolerance zero	TOL ZONE PROJ	Ⓟ
Reference	REF	(1.250)
Basic	BAC	3.825

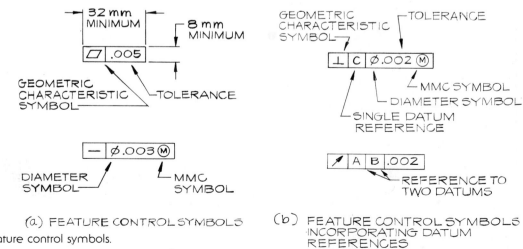

(a) FEATURE CONTROL SYMBOLS

(b) FEATURE CONTROL SYMBOLS INCORPORATING DATUM REFERENCES

Figure 7.38 Feature control symbols.

$\textcircled{M}$= MAX SHAFT IN
MIN HOLE

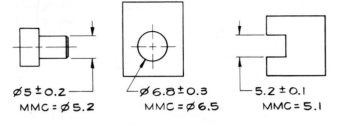

Ø5 ± 0.2
MMC = Ø 5.2

Ø 6.8 ± 0.3
MMC = Ø 6.5

5.2 ± 0.1
MMC = 5.1

MAXIMUM MATERIAL CONDITION = MMC = $\textcircled{M}$ Figure 7.39

As an example of the application of the maximum material condition of a tolerance of form, let us require that a cylindrical pin be straight within a tolerance of 0.04 mm (Figure 7.40). For a cylinder, straightness would be defined as a cylinder, 0.04 mm in diameter, within which the axis of the pin must lie. The condition stated means that under the most extreme conditions, the maximum size cylinder would just fit into the minimum hole. But if the cylinder were smaller, the tolerance zone could be increased. If the cylinder were the minimum size, with a diameter of 4.8 mm, the diameter of the tolerance zone could increase to 0.44 mm!

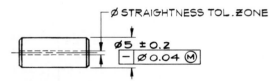

Ø STRAIGHTNESS TOL. ZONE

Ø5 ± 0.2
| — | Ø 0.04 $\textcircled{M}$ |

ACTUAL MEASUREMENT OF DIAMETER	DIAMETER OF TOLERANCE ZONE ALLOWED
5.2	0.04
5.1	0.14
5	0.24
4.9	0.34
4.8	0.44

← STRAIGHTNESS

= K
= K

Figure 7.40

Now let us consider using the maximum material condition for a tolerance on position. A bracket with two holes must fit over two mating cylindrical pins (Figure 7.41*a*). Figure 7.41*b* shows a conventionally toleranced drawing. The maximum material condition would be when the maximum size pins, at the maximum separation distance, must fit within two minimum holes. If the hole sizes are larger, the positional tolerance could be increased. This condition is shown in 7.41*c*. Using maximum

19.98 + 5.06

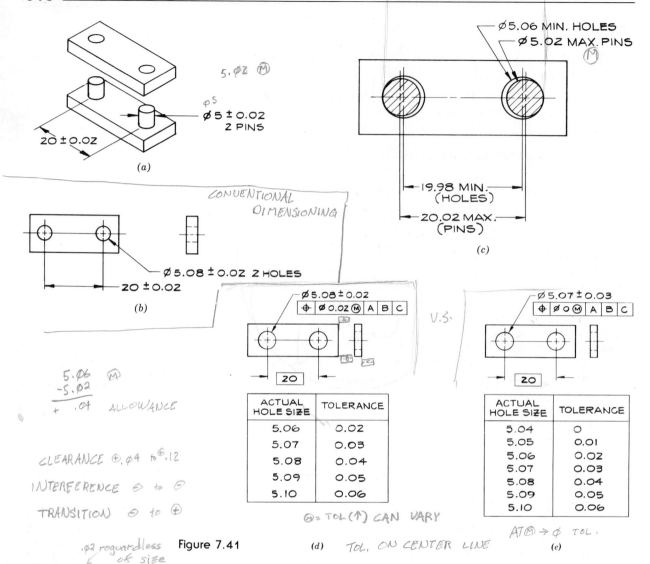

5.02 Ⓜ

φ5

Ø5 ± 0.02
2 PINS

20 ± 0.02

(a)

Ø5.06 MIN. HOLES
Ø5.02 MAX. PINS
Ⓜ

19.98 MIN.
(HOLES)

20.02 MAX.
(PINS)

(c)

CONVENTIONAL
DIMENSIONING

Ø5.08 ± 0.02 2 HOLES

20 ± 0.02

(b)

Ø5.08 ± 0.02

| ⊕ | Ø0.02Ⓜ | A | B | C |

V.S.

20

ACTUAL HOLE SIZE	TOLERANCE
5.06	0.02
5.07	0.03
5.08	0.04
5.09	0.05
5.10	0.06

Ⓜ = TOL (↑) CAN VARY

Ø5.07 ± 0.03

| ⊕ | Ø0Ⓜ | A | B | C |

20

ACTUAL HOLE SIZE	TOLERANCE
5.04	0
5.05	0.01
5.06	0.02
5.07	0.03
5.08	0.04
5.09	0.05
5.10	0.06

ATⓂ → φ TOL.

5.06 Ⓜ
–5.02
+ .04   ALLOWANCE

CLEARANCE ⊕.04 to ⊕.12

INTERFERENCE ⊖ to ⊖

TRANSITION ⊖ to ⊕

.02 regardless
of size

| ⊕ | φ 0.02 Ⓢ | A | B | C |

5.06   .02
5.07   .02
:   .02
.02

Figure 7.41

(d)   TOL. ON CENTER LINE

(e)

material conditions for the hole, the tolerance on diameter could be increased from 0.02 mm to 0.06 mm if the holes were actually 5.10 mm in diameter. What is even more interesting is that we could change the size to 5.07 mm, and the tolerance to 0.03 mm, if zero tolerance were used at the maximum material condition! We have now permitted a larger tolerance and permitted the tolerance to increase with an increase in the diameter of the hole, with no degradation of function (see Figures 7.41*d* and 7.41*e*). Zero tolerance at maximum material conditions permits the acceptance of parts over the widest possible tolerance range. The acceptance of more usable parts means more production at less cost, which is what positional tolerancing is all about.

PROBLEMS

7.22 a. Determine the dimension and tolerance between surface *A* and hole *D*.

b. Determine the dimension and tolerance between hole *B* and hole *C*.

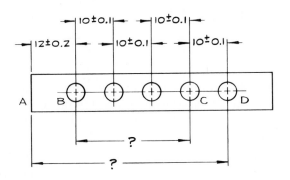

7.23 With dimension lines, arrowheads, and extension lines (no numbers), dimension the object to reflect three design requirements.

a. A relationship between holes *A*, *C*, and *E*.

b. A relationship between holes *B* and *D*.

c. No relationship between *A* and *B*.

7.24 Using equally divided bilateral tolerances, dimension for a clearance fit.

a. Clearance fit to be 0.02.

b. Basic hole size 16 diameter.

c. Total size tolerance 0.16.

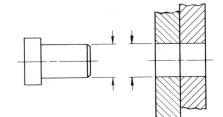

7.25 Dimension and tolerance the pin diameter for an interference fit of 0.01 minimum and a size total tolerance of 0.03.

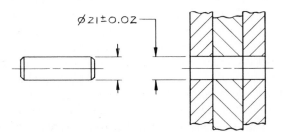

7.26 Dimension and tolerance the pin for a transition fit of 0.01 maximum interference and a size total tolerance of 0.02 applied equally divided bilaterally.

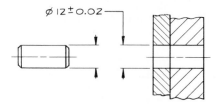

7.27 Dimension and tolerance the tab for a minimum clearance of 0.3 on all sides with a size tolerance of ±0.2. Assume there is no misalignment of grooves to their respective parts.

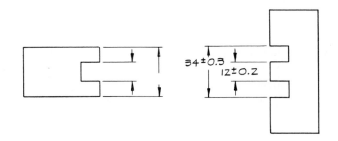

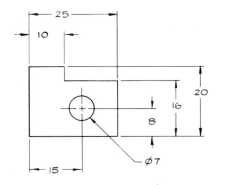

7.28 Specify tolerances for the figure in the following manner.

a. 25 and 20 dimensions—a total tolerance of 0.6 to be applied equally divided bilaterally.

b. 10 and 16 dimensions—a total tolerance of 0.4 to be applied unilaterally with the dimensions shown to be maximum.

c. 8 and 15 dimensions—to be changed to limit dimensioning with the 8 and 15 dimensions specified as the maximum limit and the minimum limit to reflect a 0.1 tolerance.

7.29 a. Limit dimension diameter *A* for an interference fit of 0.05 minimum and a size tolerance of 0.1.

    b. Limit dimension diameter *B* for a clearance fit of 0.1 minimum and a size tolerance of 0.2.

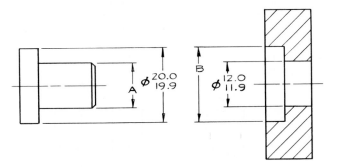

7.30 The gage is to verify a straightness tolerance. Complete the feature control symbol that will specify this requirement.

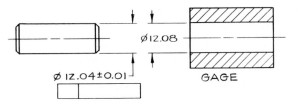

7.31 a. Which case will give the smallest straightness tolerance for the surface element? For the axis?

    b. Which case will give the largest tolerance?

    c. How much straightness tolerance does *A* allow? *B*? *C*? *D*?

    d. If *A* is at MMC, how much straightness tolerance is allowed? *B*? *C*? *D*?

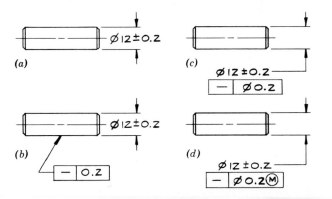

7.32 Compute the allowable straightness tolerance using the following table.

Actual Feature Size	Allowable Straightness Tolerance
11.8	
12.2	
12.05	
12.25	

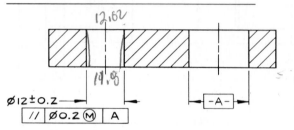

Ø12±0.2

// | Ø0.2 Ⓜ | A

7.33 Compute the allowable perpendicularity tolerance using the following table.

Actual Hole Size	Allowable Perpendicularity Tolerance
11.7	
11.75	
11.9	
12.2	

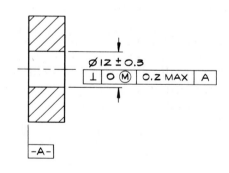

Ø12±0.3

⊥ | O Ⓜ | 0.2 MAX | A

-A-

7.13 PEOPLE ASSOCIATED WITH WORKING DRAWINGS

Four groups of people are intensely interested in the development and uses of working drawings. The first group consists of *engineers*, who have the ultimate responsibility for a working drawing, including the assurance that a drawing reflects the design intent of the engineering part or structure; this includes a statement that the drawing is correct.

The second group are the people who make the working drawing, the *drafters*—professionals trained in engineering drawing. Drafters are completely familiar with conventions and techniques, but they are not responsible for design intent, nor can they be expected to be. It is their responsibility to produce a correct and legal drawing. They are to the engineering profession what nurses are to the medical profession. One cannot exist without the services of the other. It is important for career engineers to recognize this analogy and respect professional engineering drafters for who they are and what they do.

The third group are *machinists or constructors.* This group builds what the engineer has designed and the drafter has committed to paper. They must be able to read the working drawing and make what the drawing represents.

The fourth group are *inspectors,* arbiters who compare part and drawing. It is their responsibility to make a judgment whether or not the part and the drawing are equivalent. We paraphrase a paradox: you cannot make a particular part unless you can measure it, because if you cannot measure it, you would never be able to tell whether you have the part or not! Inspection is intricately involved with testing and measurement. The theory and practice of measurement is a field in itself called *metrology.* = STUDY OF MEASUREMENTS

Obviously, the engineering description must be communicated to all these varied groups, each having an identical understanding of what is being described. The use of conventions, standards, and uniformity is desired and necessary, but there can be only a single interpretation as to what each convention or standard means.

☆ = FORMAL DRW.
I.E. LEAGAL DOCUMENT

☆

A SET OF ENG. DRW.'s
COMPLETELY DESCRIBES A
DESIGN, NOTHING ELSE
NEEDED. 10% to 15% OF TOTAL

1. LEGAL DOCUMENTS
2. NUMBERED (~ page #)
3. KEPT PERMANENTLY

7.14 THE ANATOMY OF A WORKING DRAWING

There are probably as many variations and styles of working drawings as there are people and organizations that create them. But there are also distinct and required similarities between all working drawings. For example, there is a *title block* on every working drawing. This is a space set aside for the identification of the drawing. It is usually placed in the lower right corner of the drawing. The title block contains the name of the company, the date the drawing was produced, the scale of the drawing, the name of the part, an identification number for the drawing, who drew it, who checked it, and who approved it. This is a minimum. There may be much more information.

Refer back to the working drawing in Figure 7.5. It is described as a beam mounting bar for an extensometer. It is to be made from 304 stainless steel. This drawing was made by a drafter by the name of McPherrin on February 23, 1978. It was checked by T. Landram and approved by F. Dodd on March 14, 1978. All this is a part of the permanent record of this particular working drawing and recorded on the drawing.

The identification number of this drawing is AAA 78-104360-00; it is unique. No other drawing has the number at this particular company. The AAA designates the size of the drawing. The number 78 states the year of origin. The numbers 104360 are the unique identification number. The last two numbers, in this case -00, state that this is an original drawing. If there are changes to the drawing, but the drawing remains in use, this number will be changed successively to OA, OB, OC, OD. . . . You know all these things about this particular drawing just from its title block. Other engineering organizations have similar identification systems.

Every statement on a working drawing falls into one of four categories.

> Identification information.
> Shape description.
> Size description.
> Notes and specifications.

The title-block information identifies the drawing and the associated people. The shape of the part being described is presented with what we termed orthographic language in Section 7.4. The rules of descriptive geometry govern shape description. The size of a part and all specific geometric details are defined with dimensions and tolerances. Scaling a drawing should never be required. All other important information not covered in the title block, the shape description, or the size description is included in the form of notes and specifications, which are simply short written descriptions of important information. Figure 7.42 is a working drawing. Note how every aspect of this drawing can be classed in one of the four categories.

7.15 ADDITIONAL TYPES OF WORKING DRAWINGS

There are other types of working drawings in addition to detail and assembly drawings. Some commonly used ones are the following.

- Pictorial drawings.
- Layout drawings.
- Design drawings. *e.g. electrical*
- Notation drawings. *2 words*
- Tabulation drawings.

A *pictorial* drawing is shown in some perspective, the most common being the isometric. The pictorial drawing is used primarily as an added explanation to describe a part or to understand how several parts go together. Figure 7.44 is a pictorial drawing. Note how you picture the part. There is a three-dimensional aspect to pictorial drawings that other engineering drawings do not have.

A *layout* drawing is a two-dimensional diagram of a system. It is more like a pictorial than an assembly drawing. It portrays an assembly, but it is more functional than photographic. Figure 7.45 is a layout of a bag feeding plant to control H_2S emissions in a geothermal power plant.

A *design* drawing is similar to a layout drawing, and the two terms are commonly interchanged. Of all the drawing types we have discussed, the design drawing is the one most often produced by an engineer. The design drawing is typically a full-scale, detailed layout of a part or system being designed. This drawing is an extension of the engineer's mind. It is used to work out details, relationship of parts, fits, etc. Detail and assembly drawings can be made from the design drawings, which are carefully stored as a permanent record. Figure 7.43 shows a design drawing. These sketches were done by the engineer who designed the separator described by the working drawing in Figure 7.42.

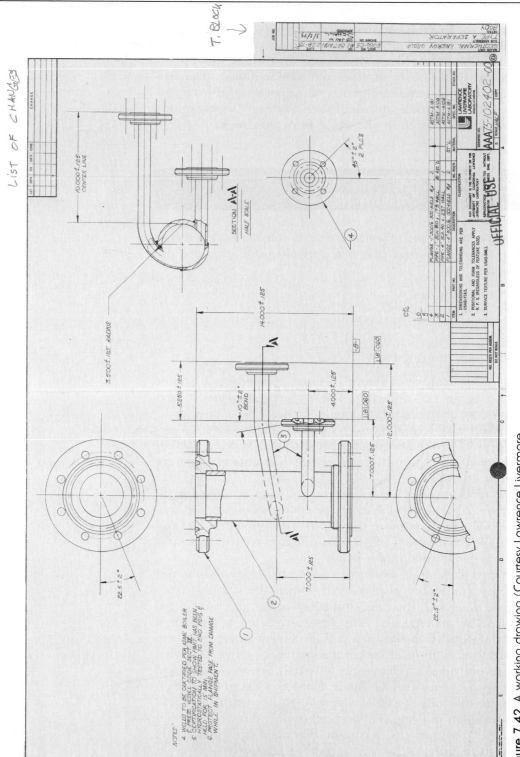

Figure 7.42 A working drawing (Courtesy Lawrence Livermore National Laboratory).

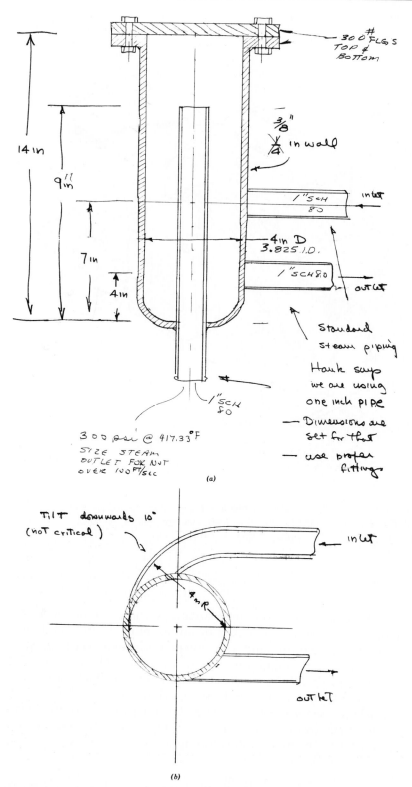

300 FLGS
TOP & BOTTOM

3/8" in wall
4

1" SCH 80   inlet

4in D
3.825 I.D.

1" SCH 80   outlet

Standard
steam piping

Hank says
we are using
one inch pipe

— Dimensions are
set for that

— use proper
fittings

14 in

9 in

7 in

4 in

1" SCH
80

300 psi @ 417.33°F
SIZE STEAM
OUTLET FOR NOT
OVER 100 FT/sec

(a)

Tilt downwards 10°
(not critical)

inlet

4 in R

outlet

(b)

Figure 7.43 A design drawing.

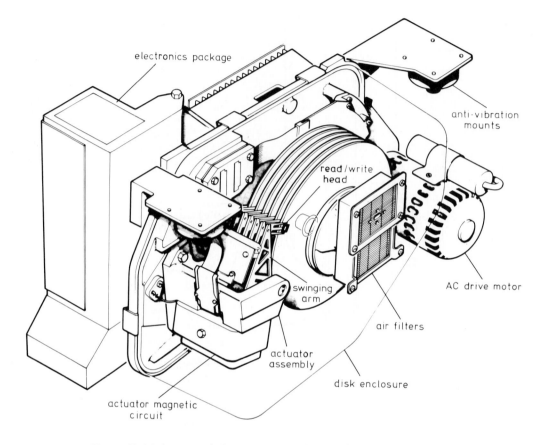

electronics package

anti-vibration mounts

read/write head

AC drive motor

swinging arm

air filters

actuator assembly

disk enclosure

actuator magnetic circuit

Figure 7.44 A pictorial drawing of the IBM 62 PC disk file. (Courtesy of IBM, General Products Division, San Jose, Ca., copyright © "Disk Storage Technology," February 1980, p. 90.)

Many drawings contain tabulations and notations as a part of the description of a part or an assembly. If the tabulations are too numerous and the notation too expansive, separate drawings may contain only one or the other. These are *notation* drawings or *tabulation* drawings (see Figure 7.46 which is a construction schedule and associated notes). It is always surprising to see a sheet of numbers or notes in a set of drawings, but when drawings are viewed as documents, the reasons for painstaking formality and thoroughness are obvious.

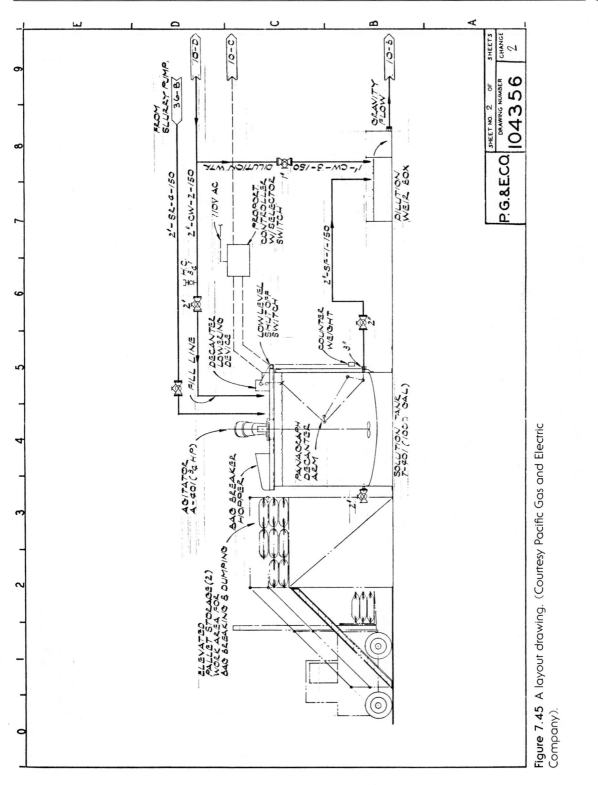

Figure 7.45 A layout drawing. (Courtesy Pacific Gas and Electric Company).

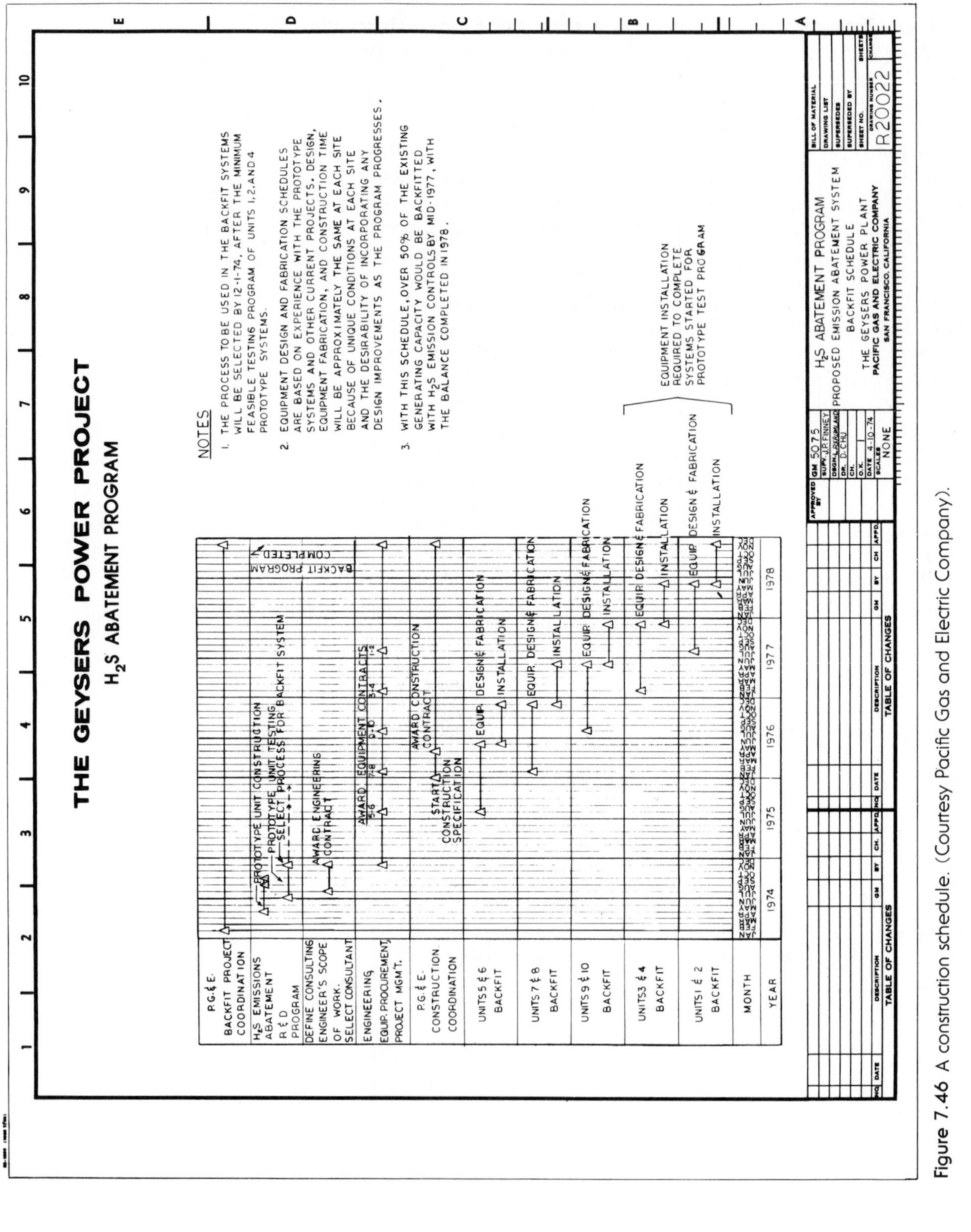

Figure 7.46 A construction schedule. (Courtesy Pacific Gas and Electric Company).

Keep in mind that a set of working drawings is a complete legal description of an engineering device. The set must be complete, and the set must be an entity. There cannot be drawings in one place and notes and descriptions in another.

This chapter begins with the statement, "The end product of all engineering is hardware, . . . and engineering drawing is an integral part of the process." To explain this statement we have focused on the accurate and clear graphic communication of mechanical components and systems, and how orthogonal projections, specific conventions, tolerance specifications, etc., are utilized for this goal. Drawings for buildings and other architectural structures and electrical and electronic drawings are also used to communicate details of the physical situation.

Architectural drawings are based on orthogonal projection, but they have many distinct and different characteristics when compared to working drawings of a machine component. The drawings for a building or structure can be thought of as assembly drawings, since they show to the constructor how the various components such as structural members, fasteners, concrete, sheet material, and roofing are combined. A portion of a drawing from a typical set of building plans is shown in Figure 7.47. Note how sectioning is used to communicate the assembly of the various building materials and components. Note also that dimensions are presented differently in architectural drawings. They are placed above the dimension line, not breaking it, as is done in engineering drawings for machine parts. Further, a short oblique line can be used in place of arrowheads. Some distances are not specified. Most building materials come in standard sizes, and the architect will take this fact into account when drawings are made. The common practice of modular coordination, which uses only dimensions of multiples of 4 in., also simplifies dimensioning and construction.

If any tolerancing information is included with architectural drawings, it will be written in the accompanying specifications, which is usually a sizable document for a building. For example, the architect may want to specify the plumbness of a column or the flatness of a floor. If nothing is said, the constructor will build the building using current accepted standards.

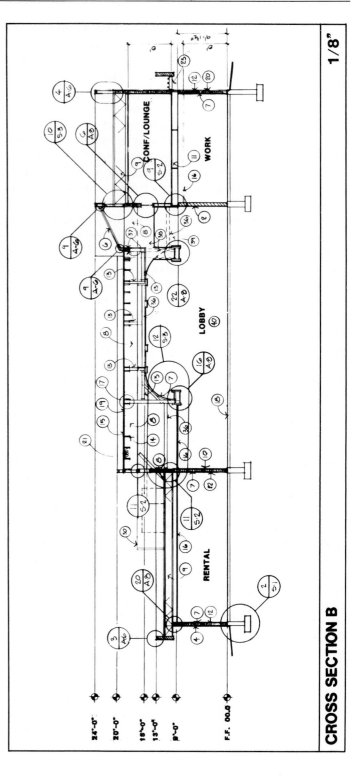

CROSS SECTION B

1/8"

Figure 7.47a A structural drawing. (Courtesy Richmond Rossi Montgomery and Midstate Bank Branch, San Luis Obispo, California).

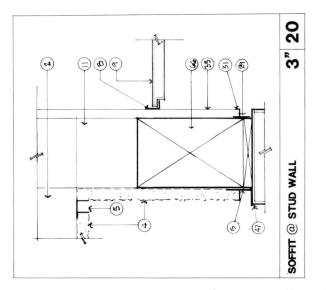

Figure 7.47b

Compare the necessary tolerances for a piston in your automobile engine with the required tolerances for the building in which you live. The piston must be made to thousands of an inch. The specific dimensions and their verification are very important to the function of the piston. Therefore, tolerances must be clearly indicated on the drawing. If buildings are framed so they look square, if the wallboard can be applied without modification, and if the doors and windows fit and function, the building will meet its requirements. Verification of the dimensions of a building is easy. A tape measure that reads to the nearest $\frac{1}{16}$ in. is really all that is required. The relative size and function of a building make its tolerancing requirements much less important than the requirements for an automobile piston.

Electrical and *electronic drawings* are actually charts. They have no physical resemblance to what they are representing. In Figure 7.48, note that relationships of the various components are shown using symbols and notations for the components. This graphic representation of the assembly of the components is sufficient to construct the device. The resulting etched circuit card may look entirely different from this drawing. Note that the drawing is made with a computer. The package (container) in which the device goes might be represented with a more standard working drawing.

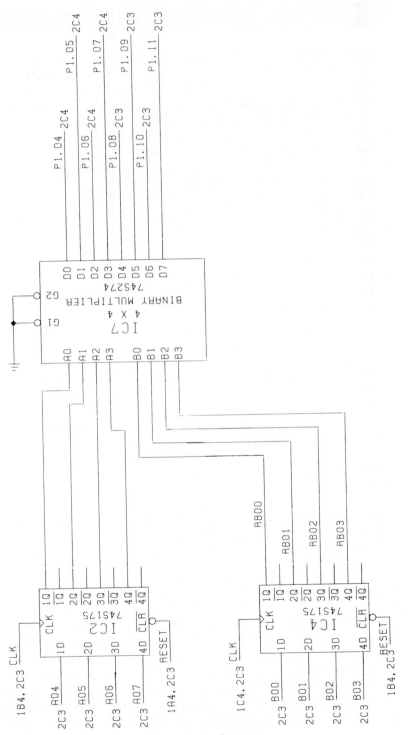

Figure 7.48 A computer-drawn logic diagram.

7.16 CHECKING

There are several levels of responsibility for a working drawing. The drafter is at the first level. It is the drafter's responsibility to create the drawing, to put pencil to paper. While developing a working drawing, the drafter keeps a constant check of the work for accuracy. When the drawing is complete, a copy is made, and it is then reviewed or checked by a responsible *checker,* who corrects drafting errors and omissions. The drawing is then corrected, and the process is repeated until both the drafter and the checker are satisfied.

Checking is a vital part of the drafting process, and it is the second level of responsibility. When it is completed, the drawing is forwarded to the responsible engineer for approval. Approval of the engineering drawing means that the part or assembly could be produced if a decision to make the part or assembly were made. It is the third level of responsibility.

Drawing approval is an engineering responsibility and usually the responsibility of the engineer who commissioned the drawing in the first place. Approval of a working drawing means that the part or assembly represented in that drawing has been accepted functionally. Parts can be produced. Estimates can be made. Contracts can be signed. Other people who need the information contained in this drawing can now have it. Everybody can go to work on production.

There may be more levels of approval, but there are rarely fewer than three. The separation of drawing approval from functional approval is widely accepted practice. Usually the names of the people responsible for all levels are listed on the drawing.

Checking is done in a definite order. The order is up to the checker, but all checking follows a certain pattern and all checking proceeds systematically. As each dimension or feature is verified, a check mark is placed by it—hence the name *checking.* If there is a correction, the error is lined out and the correct comment dimension, note, line, or feature is given.

The following is a suggested sequence.

1. Put yourself in the position of those who are to read the drawing and find out whether it is easy to read and logical. Always do this before checking any individual features. This is important. After you have become accustomed to the drawing you cannot see it as a whole. You will be seeing details. The first step is always to judge whether the drawing stands as a whole.

COMMON SENSE

1. DRW. IS COMMO ⇒ MUST BE ABLE TO READ

OVERALL

↓

DETAIL

2. See that each part is correctly designed and illustrated. See that all necessary views are shown and none is shown that is not necessary. The latter is usually the more difficult of these two tasks.

3. Check overall dimensions, by <u>calculation</u> if possible. Preserve the calculations. Do the dimensions agree?

4. Check position dimensions. Be certain that it is not necessary to do any arithmetic in order to get a needed dimension. *(OR GUESSING)*

5. Check size dimensions.

6. See that the detail drawings are dimensioned to correspond to the assembly dimensions. Will they allow for proper fitting? Watch for interferences. Are there proper clearances for mechanical movement?

7. Check tolerances.

8. Check specifications of finishes.

9. Check to see that all the small details such as screws, bolts, pins, and rivets are standard and that, where possible, stock sizes have been employed.

10. Check notes for completeness and accuracy. Check design list of materials. Are material specifications correct?

11. Check title block.

12. Review the drawing.

7.17 EPILOG

In this chapter we have discussed what an important means of communication working drawings are for engineers, and we have presented what constitutes a proper and complete drawing. Because of the inherent size of the pages of this book and the much larger size of most working drawings, we have not included as many real examples as we would have liked you to see and consider as you study this chapter. If you are not exposed to some full-size real examples of working drawings as a part of the course in which you are using this text, we urge you to seek out and review some examples on your own. Seeing several real examples of how engineers, drafters, machinists, and inspectors communicate with each other should give more meaning to this chapter.

PROBLEMS

7.34 Make detailed working drawings of the parts of the pulley bracket. Do not include the belt.

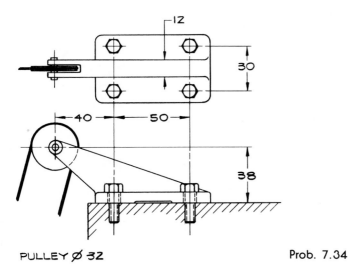

PULLEY Ø 32   Prob. 7.34

7.35 Make an assembly drawing of the pulley-bracket combination.

7.36 The extrusion device consists of a cylindrical chamber, a piston, two end pieces, and tie-down rods. Fluid or gas enters one end, forcing the floating piston down the cylinder and extruding fluid out of the other end plate. Make detail drawings of the extrusion device.

7.37 Make an exploded isometric assembly drawing of the extrusion device.

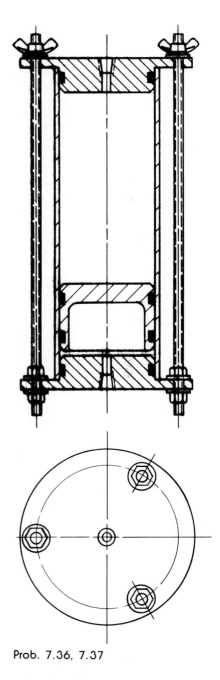

Prob. 7.36, 7.37

7.38 Make detailed working drawings of the bell crank foot pedal.

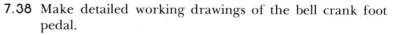

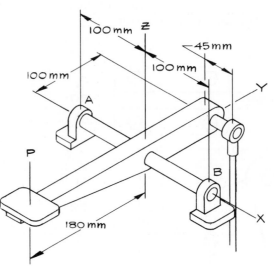

7.39 Make an assembly drawing of the bell crank foot pedal.

7.40 Make a detail drawing of the lower portion of the fixed bearing.

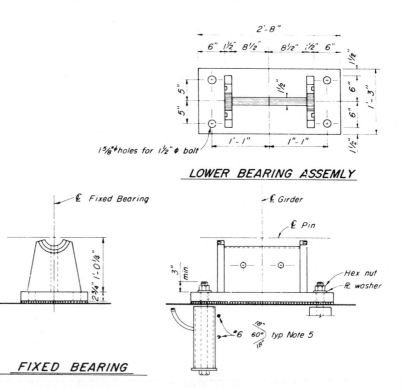

Prob. 7.40 (Courtesy California Department of Transportation).

7.41 Make a detail drawing of the roof-peak connection. Should your drawing include a development?

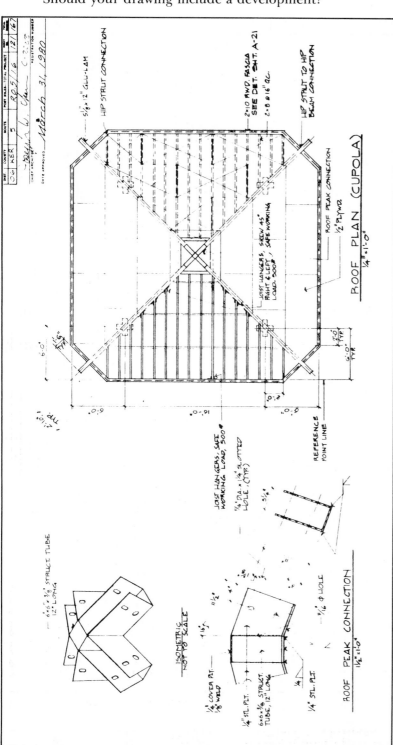

Prob. 7.41 (Courtesy California Department of Transportation).

7.42 Make an isometric drawing of the output shaft. Note that this is part no. 2 in Figure 7.6.

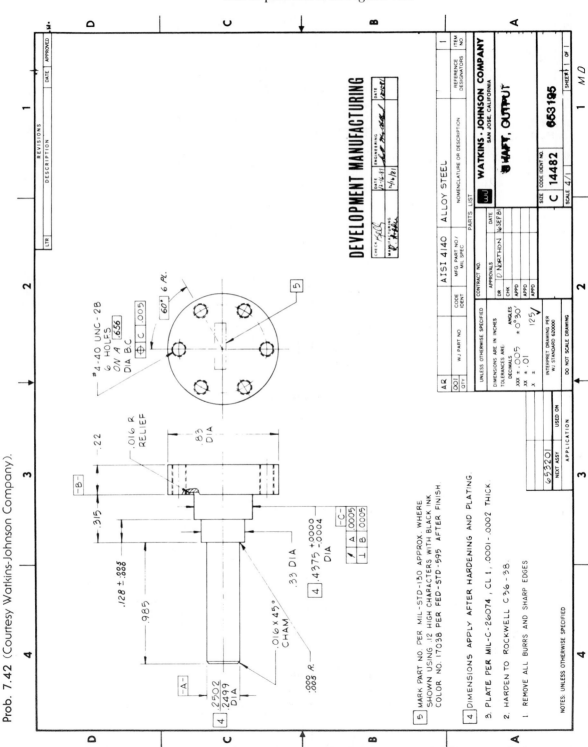

Prob. 7.42 (Courtesy Watkins-Johnson Company).

7.43 The drawing has at least 10 errors. Find them and make the appropriate corrections.

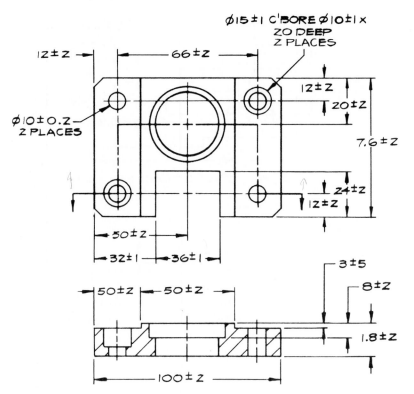

7.44 The drawing has at least 10 errors. Find them and make the appropriate corrections.

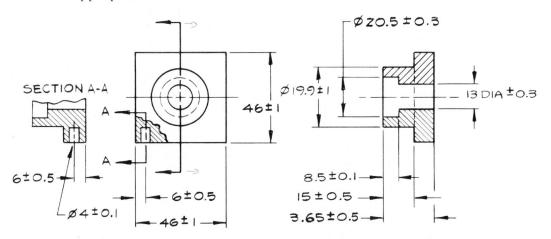

7.45 The drawing has at least 10 errors. Find them and make the appropriate corrections.

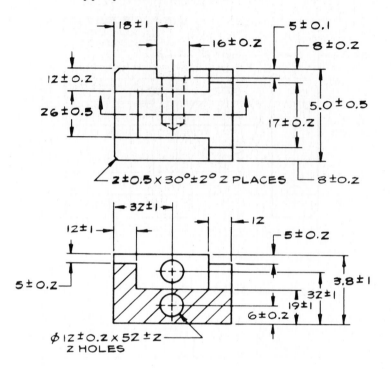

7.46 The drawing has at least 10 errors. Find them and make the appropriate corrections.

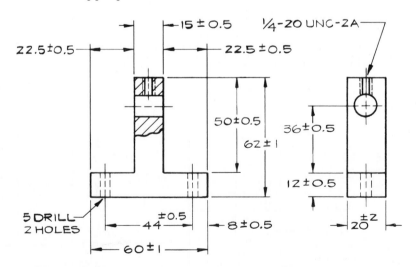

7.47 a. Prepare a simple three-dimensional shape from styrofoam, paper, wood, clay, or some other medium that is easy for you and others to work with.

b. Have a second student prepare an engineering sketch from the shape.

c. Have a third student prepare a detail drawing of the shape from the sketch.

d. Ask a fourth student to construct the object from the detail drawing.

e. Assuming that the original object and the final object differ, make the necessary alterations to the sketch and the detail drawing.

8
MODELING

8.1 WHY MODEL?

Modeling is the most recent innovation of graphic communication in engineering. Modeling has always been an important and fundamental means of communicating designs and concepts, but the recent replacement of assembly drawings by the design model has made model building almost as important to engineering as drafting. Scientists inquire and investigate. Engineers need answers, and when approximate answers will do, one means of finding approximate answers is through modeling.

Engineers use models all the time. *A model is an alternate, and* ← *OF SOMETHING REAL* *usually simplified, representation of something.* Modeling is approximation. We are incapable of accepting and manipulating all the details and complexities of the many components, systems, and factors that concern us with a particular problem. We cannot describe completely what is actually happening. We attack these problems by consciously and subconsciously utilizing models.

Modeling is the means we use to ignore what we cannot understand and to consider what we do understand. The use of models allows us to simulate unfamiliar problems by replacing the unfamiliar with the familiar. The size of a device or object can be scaled up or down; unimportant influences and details can be neglected; systems can be simulated. Symbols are used for clarity and simplification. These scaled and simplified versions, symbols, and simulations replace the original system. The replacement, a model, is imperfect but very useful and very necessary for engineers.

345

8.2 THE USE OF MODELS

A significant portion of your engineering education is spent in learning to model. Applied mathematics, the physical sciences, the engineering sciences, and English composition are all subjects dealing with models and modeling. Engineering graphics can be described as modeling using drawing techniques and geometry.

There are four basic types of models with separate and distinct uses. There are also several subtypes and overlapping types, but four types can be identified.

Symbolic models.
Mathematical models.
Physical models.
Iconic models.

There is no specific order in modeling. One logical argument has physical modeling preceding mathematical modeling, establishing physical principles first. But all engineers are restricted by their own limitations in mathematics. In practice, it is more likely that engineers will fit physical principles to the mathematics that they can handle — usually resulting in an oversimplified linearized model, which may be either good or bad. It is a model, but it may not be a good model. In any event, the order in which the modeling occurs is not important. Both mathematical and physical modeling are used, usually together, and both are subject to the mathematics and physics limitations of the people using the models.

8.3 SYMBOLIC MODELS AND DIAGRAMS

Symbolic models are the most widely used of all models, but symbolic modeling is probably least recognized as a form of modeling. Symbols are simply a shorthand way of representing something. Three academic areas that make considerable use of symbols are chemistry, graphics, and mathematics.

You use dozens and dozens of symbols every day. A *symbol* is a quick means to represent a thought, much quicker than even the shortest of word groupings. This is why symbols are called a shorthand. Because they do represent a thought, they are literally models. Our use of the word *model*, however, is directed to representing a concept by grouping a number of symbols in a *diagram*. The diagram is our model. Diagrams contain sym-

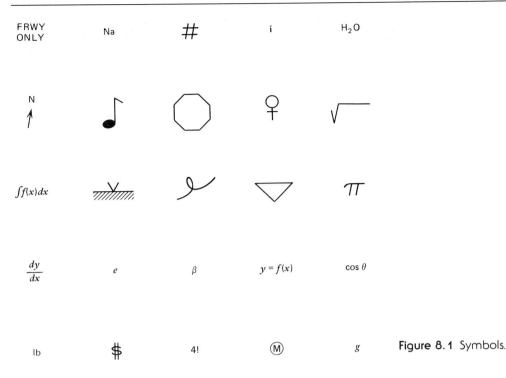

Figure 8.1 Symbols.

bols. For us, one symbol is not enough for a model. We want to consider how several symbols are used together.

In Figure 8.1, a number of symbols are shown, some of which are familiar to you and used by you. Figure 8.2*a* is a diagram of a football play. Athletic coaches use such a diagram, or model, all the time. Note the way people, their movements, and the object of interest, the ball, are geographically represented with symbols.

Diagrams are simplified representations of the structure of a device or system or the functional relationships between the elements in a system. Diagrams may or may not have any resemblance to the real thing, but all diagrams show relationships. The diagrammatic model may represent physical relationships, such as how to put something together; or it may symbolically represent physical relationships using configurations that do not relate to the system's physical makeup, such as a process chart or an electrical circuit; or it may have no physical relationships to the original, simply functioning as a diagram of the procedures to accomplish something.

Figure 8.2*b* models music. This diagram can be described as a language. The symbols are called a score. Recognizing and

flow model w. Titanium CCl_4

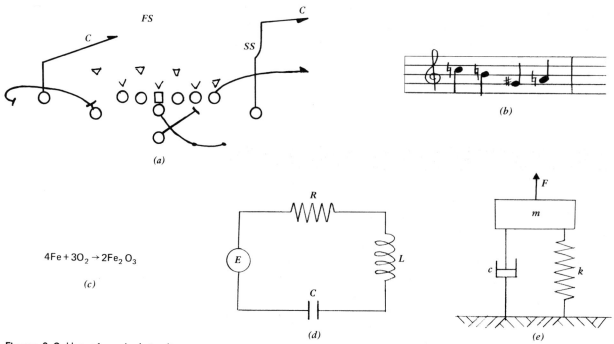

$$4Fe + 3O_2 \rightarrow 2Fe_2O_3$$

(c)

Figure 8.2 Use of symbols in diagrams.

understanding such musical diagrams is called "reading music." That it is regarded as a language is not surprising. We regard all graphics as a language.

The language of chemistry is also a language of symbols. When we wish to describe the corrosion of iron, we can state that the oxygen in the vicinity of the metallic iron is attracted to and reacts with the iron to form iron oxide. This verbal model is lengthy and subject to misinterpretation. Using standard chemical symbols and notations, we can model this reaction as in Figure 8.2c. The chemical symbols for iron (Fe) and oxygen (O), the number of atoms per molecule (subscripts), and the direction of the reaction ($\rightarrow$) are all symbols used in the chemical model of the reaction we often verbally model as "rusting." The symbol for oxygen could be accepted as an abbreviation, but why is Fe the symbol for iron, Pb the symbol for lead, Sn the symbol for tin, or Ag the symbol for silver? These are clearly symbols.

Figure 8.2d is an electrical circuit diagram with resistance R, inductance L, capacitance C, and electromotive potential E. Figure 8.2e is a diagram using symbols to model the dynamics of a mechanical system. Believe it or not, Figure 8.2d is an electrical model of Figure 8.2e, and Figure 8.2e is a mechanical model of Figure 8.2d. The mathematical equations governing the physical principles that each express are identical.

Abbreviations of terms are also symbols. In engineering, most units we use are abbreviations or are represented in symbolic form. Many of these symbols are part of your vocabulary, and you use them as a matter of routine. Such groups of letters as lb, N, in., oz, mm, yd, sec, cm, kg, and h have clear meaning to you.

8.4 DIMENSIONS AND UNITS

To say that the weight of something is 142, the pressure in a vessel is 1020, or the temperature of a solution is 82 is meaningless. To have meaning, there must be an associated *dimension* or *unit*. Both the meaning and the manipulation of numbers and equations require the use of dimensions or units. Do not confuse this use of the word *dimension* with the statement of size, location, and form used in Chapter 7.

Units are the measure of dimensions. The number states a quantity. Thus for a weight to be 142 units, you would also have to know that the unit is one pound force. The weight is therefore 142 pounds force, symbolized as 142 lbf; lbf is the symbol for pounds of force.

Hardly a day will go by without requiring the conversion from a quantity of one type to another: feet to inches, feet to meters or millimeters, minutes to hours, etc. Many of these everyday conversions are second nature, but others require some organization for success. To convert radians to revolutions, do you divide or multiply by 2π, and to convert mass to weight do you divide or multiply by g? Conversion is not always simple.

By structuring your work, conversion of dimensions or units can be easy and accurate. For example, we know there are 12 inches in 1 foot. This relationship can be written as

$$12 \text{ in.} \doteq 1 \text{ ft}$$

where we use the symbol $\doteq$ to indicate dimensional similarity but not numerical similarity. By treating this as an algebraic equation, we can rewrite it

$$\frac{12 \text{ in.}}{1 \text{ ft}} \doteq 1$$

Now to convert 6 ft to inches we can simply multiply by the preceding equation, which is dimensionally equivalent to one. Therefore

$$6 \text{ ft} \times \frac{12 \text{ in.}}{1 \text{ ft}} = 72 \text{ in.}$$

$\omega = \angle v = rev./sec$

$r = 14"$, $v_p = 60 \text{ m/hr}$

$v_p = \omega r$

$\omega = \dfrac{v_\phi}{r}$

$\dfrac{60 \text{ m}}{\text{hr}} \times \dfrac{5280}{\text{m}} \times \dfrac{\text{hr}}{3600 \text{ sec}} \times \dfrac{12"}{14"} =$

$75 \text{ rev/sec} = 75 \dfrac{2\pi}{\text{rad}}/\text{sec}$

$RPM = 720?$

To convert 72 in. to feet we form

$$72 \text{ in.} \times \frac{1 \text{ ft}}{12 \text{ in.}} = 6 \text{ ft}$$

Again, we treat the units algebraically.

We, the authors of this text, have used conventional units (lb, ft, sec) most of our professional lives. It is probable that you will use the SI system of units for your technical engineering work, although there is considerable likelihood that public acceptance of SI units in the United States will be slow. For this reason, you may have to use two systems, one at work and one at home, converting from one to the other. The SI system makes clear and logical use of symbols and is preferred. There is only one accepted symbol for each physical quantity, and there are specific conventions concerning its presentation and combination. Use SI as your standard. In this book, we have used both conventional and SI units.

8.5 MATHEMATICAL MODELS

Mathematics includes the subjects of arithmetic, algebra, geometry, trigonometry, analytical geometry, differential calculus, and integral calculus—all of which you will study and utilize as an engineer.

The abstract science of mathematics is one of the most utilized modeling tools engineers or scientists have at their disposal. It can be regarded as a game, which like any game has a set of governing rules. On the other hand, mathematics is a science that helps satisfy human needs and helps us understand nature. The development of mathematical theory usually precedes its actual use in satisfying human needs by years. Mathematics helps us put word problems into an algebraic form, which we can use more easily to deal with the problems. This process helps us obtain the necessary solutions through reasoning, producing mathematical models.

Mathematical models are in various forms. Equations utilizing and relating a variety of symbols are the most common form. Graphs are an alternate way to present mathematical models and empirical data. They present a graphic interpretation of a mathematical relationship. In Chapters 9, 10, and 11, we will discuss graphic mathematics and the presentation of data.

Perhaps the most profound example of mathematical modeling is the mathematical statements of what we call Newton's laws. Galileo made his original observations in dynamics some 100 years earlier than Newton, but his findings were philosophical. Sir Isaac Newton (1642–1727) formulated the basic laws of kinetics in mathematical form in his monumental work, *Philosophiae Naturalis Principia Mathematics*, published in 1687.

Newton's second law is uniformly accepted as the fundamental principle of classic dynamics. If $\mathbf{R}_1$ is the resultant of a vector-force system on a physical body, its acceleration will be vector $\mathbf{a}_1$. If vector $\mathbf{R}_2$ is the resultant, then the acceleration of the body will be vector $\mathbf{a}_2$, and

$$\frac{R_1}{a_1} = \frac{R_2}{a_2} = \frac{R_n}{a_n} = \text{constant}$$

This constant of proportionality is called mass. It is a quantitative measure of inertia. Newton's second law can be stated *mathematically* as

$$\mathbf{F} = m\mathbf{a}$$

The resultant of the unbalanced force system is equal to the product of mass and the acceleration of the mass $\mathbf{a}$. This is a mathematical statement of a physical observation, using symbols $\mathbf{R}_1$, $\mathbf{R}_2$, $\mathbf{a}_1$, $\mathbf{a}_2$, $\mathbf{a}$, and $\mathbf{F}$. No amount of mathematics however can establish the statement without physical observations. It is a true *mathematical model*, and much of engineering is based on this model.

[handwritten margin notes:]
$F, + a$ are vector quantity

$$F = \frac{d}{dt} (mv)$$

$$v \frac{dm}{dt} + m \cdot \frac{dv}{dt} \quad \text{ONLY FOR}$$

$$\text{RELATIVE} (v)$$

SAMPLE PROBLEM 8.1

An automobile is traveling at a speed of 55 mi/h along a highway. The tires have a diameter of 28 in. Determine the rotating speed of the wheels in revolutions per second.

Solution
We know that

$$\frac{5280 \text{ ft}}{1 \text{ mi}} \doteq 1$$

$$\frac{60 \text{ sec}}{1 \text{ min}} \doteq 1$$

$$\frac{60 \text{ min}}{1 \text{ h}} \doteq 1$$

By multiplying 55 mi/h by 1 and causing the units to cancel, we form

$$\frac{55 \text{ mi}}{\text{h}} \times \frac{1 \text{ h}}{60 \text{ min}} \times \frac{1 \text{ min}}{60 \text{ sec}} \times \frac{5280 \text{ ft}}{1 \text{ mi}} = 80.7 \text{ ft/sec}$$

Now, the circumference of each wheel is π diameters or

$$\pi(28 \text{ in.}) \frac{1 \text{ ft}}{12 \text{ in.}} = 7.33 \text{ ft}$$

This means that the car will move forward 7.33 ft for every revolution of the wheels. Thus the rotating speed of the wheels is

$$80.7 \frac{\text{ft}}{\text{sec}} \times \frac{1 \text{ rev}}{7.33 \text{ ft}} = 11 \frac{\text{rev}}{\text{sec}}$$

SAMPLE PROBLEM 8.2

A stone is dropped from a high bridge over a gorge. It is seen to enter the stream 3 sec after release. How far is it from the bridge deck to the stream?

Solution

The distance traveled by a free-falling body at the surface of the earth is expressed by the algebraic equation

$$s = \frac{1}{2} g t^2$$

This is a mathematical model of a physical law, and it is known to be a true expression. s is a symbol for the distance traveled, t is the time of the fall, g is the acceleration of gravity, and $g = 32.17$ ft/sec^2 in this latitude. Substituting $t = 3$ sec

$$s = \frac{1}{2} (32.17)(3)^2 = 144.8 \text{ ft}$$

The equation does neglect the time it takes for light to travel from the water surface to the bridge and is a simplification, but the error is very small and can be neglected. If we were to measure the time from the release of the stone to the time we heard the sound of the stone's entry into the water, the error would be much, much larger. Can it be neglected?

PROBLEMS

8.3 Symbols usually have some real-world basis. Several standard symbols from three areas of interest to engineers are shown with a list of their meanings. Can you match the meaning with the symbol?

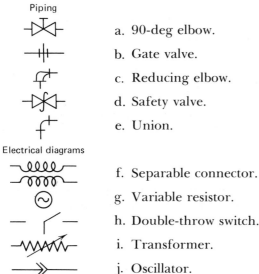

Piping

1.   a. 90-deg elbow.

2.   b. Gate valve.

3.   c. Reducing elbow.

4.   d. Safety valve.

5.   e. Union.

Electrical diagrams

6.   f. Separable connector.

7.   g. Variable resistor.

8.   h. Double-throw switch.

9.   i. Transformer.

10.   j. Oscillator.

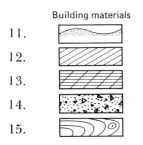

Building materials

11.   k. Wood.

12.   i. Electrical insulation.

13.   m. Concrete.

14.   n. Sand.

15.   o. Cast iron

8.4 Make the following conversions.

a. 17 years to seconds.

b. 12 m/sec to kilometers/hour.

c. 55 mi/h to light years/century.

d. 40 m/sec^2 to g's.

e. 2000 Btu to foot-pounds.

8.5 Make the following conversions.

    a. 2000 Btu to joules.

    b. 0.1 hp (an average person) to watts.

    c. 20 rev/min to radians/second.

    d. Your height in inches, feet, yards, meters, millimeters, kilometers, and miles.

    e. 10.8 light years to kilometers.

8.6 List at least 10 symbols that have been used in the written portions of this chapter.

8.7 Develop a set of symbols that could be used in one of the following locations.

    a. Swimming pool.   b. Classroom.   c. Machine shop.

8.8 State the meaning of each symbol used.

    a. $z = a + bi$ d. $\ln z$

    b. $\mathbf{A} \times \mathbf{B}$

                                 e. $\displaystyle\sum_{n=1}^{\infty} x_n$

    c. $\cosh z$

8.9 State the meaning of each symbol used. Parts of this problem may not look familiar unless you have had integral calculus.

    a. $\displaystyle\int_{x_1}^{x_2} f(x)\ dx$ d. $\dfrac{dy}{dx}$

    b. $\dfrac{\Delta y}{\Delta x}$ e. $\dfrac{\partial y}{\partial x}$

    c. $5!$

8.10 A pipeline 8 in. in diameter can deliver 500 barrels of oil per hour. What is the average velocity of the oil in the pipe (1 barrel = 42 gallons)?

$$Q = A \times v$$

Answer: 2.23 ft/sec

8.11 An automobile was purchased 5 years ago for $4800. Assume uniform depreciation to its present value of $800. Make a mathematical expression or symbolic model for the cost per mile for N miles traveled per year if the costs are as follows. (Assume a mileage of 20 mi/gal.)

License	$50/year
Insurance	$200/6 months
Repairs	$300/year
Service	$80/4000 mi
Tires	$50/5000 mi
Fuel	$1.40/gal

Answer: $0.10 + \dfrac{1550}{N}$, $/mi

8.12 A circular steel bar, 1 m long and 50 mm in diameter, is elongated 0.2 mm by a uniform tension. What is the tensile force in the bar? Strain $\epsilon = \sigma/E$, where σ is the stress and $E = 30 \times 10^6$ lbf/in.^{2}.

Answer: 81200 N

8.13 A small reservoir is 100 acres in area. The power that can be generated by releasing water from the reservoir is $P = Q\,h\,\gamma$, where P is power, Q is quantity flow, h is the head difference from the reservoir to the hydraulic turbine, which happens to be 90 ft, and γ is the weight density of water. What average power can be generated if the reservoir is allowed to draw down 2 ft in 4 h?

Answer: 4556 kW

8.14 Point A is on the periphery of a wheel that starts from rest and accelerates with a constant acceleration of 100 rev/min/sec. The linear speed of a point on the periphery of the wheel is the product of the angular velocity of the wheel in radians per second and the radius of the wheel, which is 750 mm. After 5 sec, what is the velocity of point A ($v = \omega r$, $\omega = \alpha t$)?

Answer: 39.27 m/sec

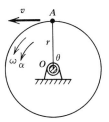

8.6 PHYSICAL MODELS

A *physical model* represents the physical function of a component or system. This function is usually expressed in symbols. A physical model simplifies the original system, but all the important functional correspondence remains. This simplification is a deliberate step that is influenced by knowledge of the subsequent analysis and knowledge of the errors introduced because of the simplification. The step represents the engineering or art in the process. Once the physical model is established, the rest of the process can be relatively routine.

As examples of physical modeling, consider wind-tunnel tests, ship-hull tests in towing tanks, structural modeling (used extensively in structural engineering), and a host of thermodynamic modeling tests. Each is characterized by one or more physical relations that are either held constant or controlled. The physical function of the *model* characterizes the physical function of the *prototype*. The model is small if the prototype is large; conversely, the model is large if the prototype is small. Usually there is a size difference, but there are situations where the model and the prototype are the same size. For example, a full-size wing may be structurally tested, with only the wing-to-fuselage attachment being modeled; it would be a full-size model. In any case, because it is easier to measure the performance of the model than it is to measure the performance of the prototype, we have and use physical models. The model and prototype are physically similar, so the physical performance of the prototype can be predicted by measuring the physical performance of the model.

8.7 SIMILITUDE

The word *similitude* refers to similarity, resemblance, or likeness. As a principle, similitude deals with observing the behavior of one system to predict the behavior of another, a necessary step in the design and use of models. Similitude is based on dimensional analysis and, in addition to providing principles of model design, helps reduce the number of variables and tests required in an experimental program. This process of simplification and generalization results in dimensionless terms, which have become commonplace in engineering analysis.

As a simple example of the use of similarity as a principle, let us consider the construction of an offshore drilling platform.

The platform will be subjected to tides and wave action, and we would like to construct it so that it can withstand the heaviest onslaught imaginable. Obviously, experimental tests are involved. Of necessity, we will have to prove that our design will withstand predicted wave action before it can be built. But how?

This problem involves only the free surface motion of a liquid, water, but on a very large scale. Wave velocity and wave height are the physical quantities we want to observe. Fluid density and gravity are the only physical properties influencing free surface flow and waves. We can duplicate this free surface flow in miniature by model tests in a tank of water, if we keep velocities and pressures similar. This similarity will be true similarity if

$$\frac{V^2{}_m}{L_m} = \frac{V^2{}_p}{L_p}$$

SHALLOW MOM → HIGH WAVES

V is wave velocity and L is any linear dimension, such as wave height and length. The subscript m represents the model, and the subscript p represents the prototype. This equation is really a form of the *Froude number* (named after the English engineer who first used a similarity principle in ship-model tests), but it is simplified because the gravitational constant is considered to be the same for both model and prototype. Proving the existence and deriving the Froude number are beyond this exercise, but in similitude, physical similarity requires some functional similarity such as the requirement that the Froude number of the model be equal to the Froude number of the prototype. This means that the flow conditions for the model will be kinematically and dynamically similar to the flow conditions of the prototype.

ALL DIMENSIONLESS
others e.g. mach #

More rigorously, the similarity expression should have been expressed

$$\frac{V_m}{\sqrt{g_m L_m}} = \frac{V_p}{\sqrt{g_p L_p}} = F$$

α in mod = α in prototype

which is the Froude number.

Algebraically, the previous equation will reduce to a single statement.

$$\frac{V_m}{V_p} = \sqrt{\frac{L_m}{L_p}}$$

If a model is constructed at $\frac{1}{16}$ scale, that is, $L_p = 16 L_m$, then $V_p = 4 V_m$. The model will have waves dynamically similar to those of the prototype at $\frac{1}{4}$ the wave velocity of the prototype and $\frac{1}{16}$ the height and length. This is a *physical model*; you would not expect this disproportion! What we have done is to establish the necessary physical similarity, or *similitude*, between model and prototype.

In aerodynamics, the *Mach number* is the ratio of velocity v to the velocity of sound c.

$$M = \frac{v}{c}$$

If you consider inertial and viscous forces, such as found in real fluids, and if you also require that flow conditions for the model are to be similar to flow conditions for the prototype, the *Reynolds numbers* will be required to be equal.

$$\text{Re} = \frac{\rho v d}{\mu}$$

where ρ = fluid density, v = velocity of the fluid, d = diameter or distance, μ = dynamic viscosity (resistance to relative fluid motion). If you include surface tension, equal *Weber numbers* will be required.

$$W = \frac{\rho L v^2}{\sigma}$$

where σ is the surface tension.

For dynamic similarity between a structural model and prototype, considering elastic and inertia forces, equal *Cauchy numbers* will be required. And so it goes.

If you include other parameters, such as heat transfer and temperature, you will have to include more than the three basic units of mass, length, and time. For this situation, a heat dimension and a temperature dimension would be required.

As you study the physical and engineering sciences, you will become familiar with numerous dimensionless terms, many of which were originally derived with the use of similitude. Their development is beyond the scope of this book, but the fact that these dimensionless parameters exist and are the bases on which physical modeling is done is not. It is central to our discussion of physical modeling. Fortunately, a dimensional approach to physical modeling exists, and it is a somewhat simple approach.

DIMENSIONLESS   ANAL.

8.8 DIMENSIONAL ANALYSIS

It is fundamental that any mathematical statement of a physical principle must be dimensionally correct. In other words, the dimensions on one side of an equation must be equivalent to the dimensions on the other. You cannot mix apples and cows, and you cannot have meters per second equal to kilograms per newton. (It would be more proper to say that this is axiomatic.)

It is possible to make all statements in mechanics using only four dimensions: force F, mass M, length L, and time T. Any three can be basic, and the fourth would be derived from the other three. You could either use force, length, and time as three basic dimensions and derive mass, or use mass, length, and time and derive force. By writing Newton's second law,

$$\mathbf{F} = m\mathbf{a}$$

and

$$F = \frac{ML}{T^2}$$

or

$$M = \frac{FT^2}{L}$$

If you consider something other than these three basic dimensions—for example, temperature—you would have to recognize temperature as a new basic dimension. We will consider only the four dimensions of force, mass, length, and time. Table 8.1 lists some common examples using these four dimensions.

These two thoughts are all that is needed to arrive at dimensionless physical parameters using dimensional analysis. The procedure was laid down by E. Buckingham in 1914 and is known as Buckingham's π theorem.

Let $P_1, P_2, P_3, P_4, \ldots, P_i$ be physical quantities involved in some physical statement. Buckingham states that there is some functional relationship for these physical quantities, such that

$$f(P_1, P_2, P_3, \ldots, P_i) = 0$$

Next, each of the physical quantities can be expressed in the *three* basic dimensions, mass M, length L, and time T. Buckingham's π theorem is that this physical statement can be restated in terms of $i - 3$ dimensionless numbers, called π, which is what Buckingham called them, presumably because they are, like π, a dimensionless ratio.

$$g(\pi_1, \pi_2, \pi_3, \ldots, \pi_{i-3}) = 0$$

This theorem can be proved, but the proof is quite complicated. Our task is simply to find any dimensionless numbers. The proof that they exist does not aid us in our search.

As an example of the use of the π theorem, consider the free surface motion of a liquid. The physical quantities involved, again, are surface wave velocity V, fluid density ρ, the acceleration of gravity g, and the wave height and length. For these four

[handwritten margin notes: "i phys. quant.", "π dimensions", "for # of Dimensionless terms needed"]

TABLE 8.1
Dimensions of Basic Engineering Quantities

	Basic L-T-M	Basic L-T-F
Dimensions		
Length, L	L	L
Mass, m	M	FT^2L^{-1}
Time, T	T	T
Force, F	MLT^{-2}	F
Dimensional quantities		
Area, A	L^2	
Volume, V	L^3	
Linear velocity, v	LT^{-1}	
Linear acceleration, a	LT^{-2}	
Angular velocity, ω	T^{-1}	
Angular acceleration, α	T^{-2}	
Kinematic viscosity, μ/ρ	L^2T^{-1}	
Torque or moment, FL	ML^2T^{-2}	FL
Density, ρ	ML^{-3}	FT^2L^{-4}
Specific weight, γ	$ML^{-2}T^{-2}$	FL^{-3}
Dynamic viscosity, μ	$ML^{-1}T^{-1}$	FTL^{-2}
Pressure or stress, p	$ML^{-1}T^{-2}$	FL^{-2}
Work, energy, or heat, LF	ML^2T^{-2}	LF
Power, P	ML^2T^{-3}	LFT^{-1}
Linear momentum, G	MLT^{-1}	FT
Angular momentum, H	ML^2T^{-1}	FLT
Dimensionless quantities		
Angular displacement (angles), θ		
Strain, ϵ		
Coefficient of friction, f		

$\overset{\circ}{Q}$ = HEAT FLOW

quantities, which all can be expressed in three dimensions, M, L, and T, there will be $4 - 3 = 1$, or one dimensionless π term. Buckingham's suggested procedure is to assume that the physical quantities are related by some power function, with unknown exponents for each physical quantity. Let a, b, c, and d be these exponents for the four physical quantities V, ρ, g, and L.

$$\pi = (V)^a(\rho)^b(g)^c(L)^d = M^0L^0T^0$$

Note that whatever the combination, because the π term is dimensionless, the combination of exponents must be such that the exponents of M, L, and T are all zero.

Introducing the three basic dimensions M, L, and T.

$$\pi = \left(\frac{L}{T}\right)^a\left(\frac{M}{L^3}\right)^b\left(\frac{L}{T^2}\right)^c(L)^d = M^0L^0T^0$$

dimensionless Quant.

WHAT IS IT FOR = WHY

There are three simultaneous equations that relate a, b, c, and d.

$$\text{for } M; \; b = 0$$
$$\text{for } L; \; a - 3b + c + d = 0$$
$$\text{for } T; \; -a - 2c = 0$$

There are four unknowns and three equations, which means that all of the exponents cannot be found explicitly; this will always be the case. Going on,

$$b = 0$$

and

$$c = -\frac{a}{2}\left(= d \;\right)$$

HOW? $\quad a(-3b) - \frac{a}{2} + d = 0$

$$d = -\frac{a}{2}$$

Recombining,

$$\pi = (V)^a (\rho)^0 (g)^{-a/2} (L)^{-a/2}$$

and setting $a = 1$, arbitrarily,

$$\pi = \frac{V}{\sqrt{gL}} \quad = \quad \frac{V^2}{gL}$$

for Graphs

This is a dimensionless parameter. You may recognize it as the Froude number. The zero exponent for density means that it does not affect the resulting dimensionless parameter, which can happen, particularly if you consider too many physical quantities. If the physical quantity should not be included, its exponent will develop to be zero.

After you obtain the dimensionless parameters π_1, π_2, π_3, . . . , and there may be more than one, physical modeling requires model and prototype to have identical dimensionless terms

$$(\pi_1)_m = (\pi_1)_p$$
$$(\pi_2)_m = (\pi_2)_p$$
$$(\pi_3)_m = (\pi_3)_p$$
$$\text{etc.}$$

Three final statements can be made about physical modeling. First, you may not be able to model all the π terms. This ought to be obvious, and when it happens compromise is the only solution, except, of course, the solution where you let $L_m = L_p$, which is full-size modeling.

Second, the π theorem cannot yield already nondimensional terms, which by themselves may need to be modeled. For example, coefficients of friction, Poisson's ratio, strain, angular displacement, and trigonometric and logarithmic functions are already dimensionless. The π theorem will not disclose that these influence physical functions.

If dimensional analysis is this limited, why do it? Why include it in a book on graphics and communication? The answer brings up the third statement on physical modeling *Dimensional analysis is fundamental to physical testing,* and the presentation of test information and data is almost always graphic. When you have developed two or more terms, the physical question is how do they relate to each other? The question can be answered graphically, and it usually is. In Figure 8.3*a*, $f(\pi_1, \pi_2) = 0$ is developed by showing π_1 as a function of π_2. In Figure 8.3b $f(\pi_1, \pi_2, \pi_3) = 0$ can be developed three-dimensionally. Unfortunately, we cannot graphically show more than three dimensions at any one time. This does not mean that we cannot use four-dimensional development; we cannot simply see it with three-dimensional geometry. The graphic presentation of relationships is the subject matter of Chapter 11.

8.9 ANALOG MODELS

behaves like real thing

An analog model is a system or device that is similar in behavior or function, but has no physical resemblance to the original. Current flowing in a wire is analogous to (or like) water flowing in a pipe. Voltage in an electrical system is analogous to pressure or force in a machine. The gear pair or a pulley system shown in Figure 8.4 can be used as a model of multiplication by a constant. Phrases such as *hard as a rock* and *light as a feather* are verbal analogies we use to communicate ideas and facts.

As you study the physical and engineering sciences, you will uncover many analogies between the various fields. These similarities are used in the teaching and understanding of concepts, as well as in the analysis and synthesis of systems and devices. When measurements or observations from one system can be used to predict the behavior of another system, the systems are analogous. In engineering, most analogs are identified from the fact that equations (mathematical models) that govern or model the two systems are of similar form. The simple equation

$$e = (P)c$$

where c = cause variable, (P) = system characteristics, and e = effect variable, is a common model for a number of simple systems where, given the nature of system (P) and the value of variable c, the resulting effect e can be obtained. Figure 8.4 presents several situations of the same mathematical form that you will encounter in your engineering studies.

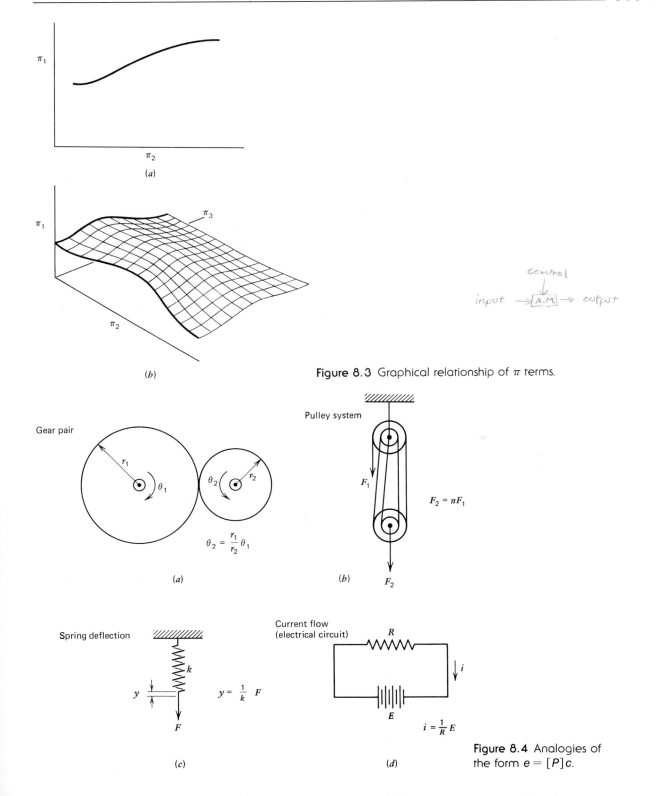

Figure 8.3 Graphical relationship of π terms.

Gear pair

Pulley system

$$\theta_2 = \frac{r_1}{r_2}\theta_1$$

$$F_2 = nF_1$$

(a)

(b)

Spring deflection

Current flow
(electrical circuit)

$$y = \frac{1}{k}F$$

$$i = \frac{1}{R}E$$

(c)

(d)

Figure 8.4 Analogies of
the form $e = [P]c$.

Pendulum

$$\frac{d^2\theta}{dt^2} + \frac{g}{\ell}\ \theta = 0$$

θ = angular displacement
g = gravitational constant
ℓ = pendulum length
t = time

Electrical

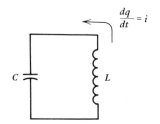

$$\frac{d^2q}{dt^2} + \frac{1}{LC}\ q = 0$$

q = electrical charge
C = capacitance
L = inductance
t = time

Column

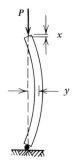

$$\frac{d^2y}{dx^2} + \frac{P}{EI}\ y = 0$$

y = deflection
x = deflection
P = force
E = elastic modulus
I = second moment of area

Elastic system

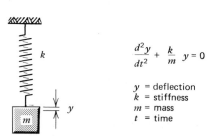

$$\frac{d^2y}{dt^2} + \frac{k}{m}\ y = 0$$

y = deflection
k = stiffness
m = mass
t = time

Figure 8.5 Analogies related by a differential equation.

As the complexity of the mathematical model of a system increases, the analogous system may be less obvious and therefore, when identified, can be a useful model. Four analogous systems linked with a second-order differential equation are represented in Figure 8.5. The details of these systems and their associated models will be treated in your engineering studies. Note how the pendulum, electrical circuit, column, and elastic system appear entirely different, but the mathematical behavior of the systems is similar.

Analog computers are a computational tool whose operation is based on analogies. Electrical circuits behave in the same way as many other systems. For example, you will discover that the general equation for the dynamic behavior of a mechanical system is the same, with the exception of symbols used, as the mathematical description of an electrical system. Therefore, an electrical system (the analog computer) can be used to model the behavior of another system having similar governing equations.

SAMPLE PROBLEM 8.15

Using dimensional analysis, determine the maximum stress s in the cross section of a rectangular beam of area A and height h acted on by a moment (bending) M.

Solution
Assume the relationship

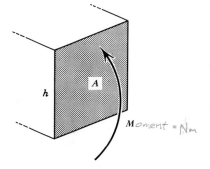

$$f(s, A, h, M) = 0$$

[4 phys. quant. (ie parameter)]

$\frac{F}{L^2}$ L^2 L $F \cdot L$

There are two basic dimensions, length L and force F, so there will be $4 - 2 = 2$, or two dimensionless π terms, and

$$g(\pi_1, \pi_2) = 0$$

Now, selecting any three quantities, for example,

$$\pi_1 = (s)^a (A)^b (h)^c = \left(\frac{F}{L^2}\right)^a (L^2)^b (L)^c = F^0 L^0$$

which gives the following relationships:

$$\text{for } F; \quad a = 0$$
$$\text{for } L; \quad -2a + 2b + c = 0$$

Setting $b = 1$, arbitrarily, $c = -2$ and

$$\pi_1 = \left(\frac{A}{h^2}\right)^{b=1}$$

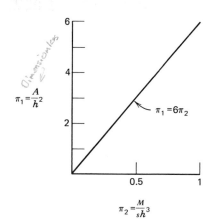

$$\pi_1 = \frac{A}{h^2}$$

$$\pi_2 = \frac{M}{sh^3}$$

$$\frac{A}{h^2} = 6 \cdot \frac{M}{sh^3}$$

For the second π term, taking another set of three quantities,

$$\pi_2 = (s)^a (h)^c (M)^d = \left(\frac{F}{L^2}\right)^a (L)^c (FL)^d = F^0 L^0$$

which gives additional relationships:

$$\text{for } F; \quad a + d = 0$$
$$\text{for } L; \quad -2a + c + d = 0$$

There are three unknowns and two equations which means, again, that we cannot find a, c, and d directly. Setting $d = 1$, arbitrarily, then $a = 1$ and $c = -3$. Therefore

$$\pi_2 = \frac{M}{sh^3}$$

If we were to use these dimensionless terms to find out experimentally how π_1 and π_2 relate to one another, we would find that $\pi_1 = 6\pi_2$ and

$$s = \frac{6M}{Ah} = \text{stress}$$

which is a relationship you will see in your study of strength of materials. $= s \alpha 6M \qquad s \alpha \frac{1}{h}$

$$s \alpha \frac{1}{A}$$

PROBLEMS

8.16 Show that the following numbers are dimensionless.

    a. Reynolds number, $\dfrac{\rho v L}{\mu}$

    b. Froude number, $\dfrac{v^2}{gL}$

    c. Euler number, $\dfrac{\rho v^2}{p}$

    d. Weber number, $\dfrac{\rho v^2 L}{\sigma}$

Using dimensional analysis, develop the form of the following relationships.

8.17 $s = f(g, t)$, distance traveled by free-falling body.

Answer: $s = \frac{1}{2}gt^2$

8.18 $v = f(g, h)$, velocity of free-falling body.

Answer: $v = \sqrt{2gh}$

8.19 $KE = f(m, v)$, kinetic energy of a particle.

$$Answer: KE = \tfrac{1}{2}mv^2$$

8.20 $v = f(E, \rho)$, velocity of sound in a material.

$$Answer: v = \sqrt{\frac{E}{\rho}}$$

8.21 Experiments are to be conducted on gas-filled balloons rising in still air. Consider a balloon of diameter D, rising in air of density ρ, rising with a velocity v, in a gravity field g. Find a set of parameters for organizing your experimental results for balloons of differing diameters. Ignore the weight of the balloon.

$$Answer: \frac{Dg}{v^2}$$

8.22 An ocean vessel 500 ft long is to travel at a speed of 15 knots (nautical miles per hour). At what speed should a 5 ft long model be towed, in seawater, for surface-wave resistance similarity?

$$Answer: v_m = 1.5 \text{ knots}$$

8.23 The theory of the wind-induced vibration of electric transmission lines states that the frequency of vibration is a function of wind velocity and the diameter of the cable, but not the mass. Show that this is true. What dimensionless combination must be held constant?

$$Answer: \frac{fD}{v}$$

8.24 Testing a prototype to failure can be very, very expensive, so one use of a structural model is to test the model physically for buckling failure. Considering force P, length, and the elastic modulus E, determine what dimensionless parameter or parameters must be held between model and prototype.

$$Answer: \frac{P}{EL^2}$$

8.25 Determine the possible combination of parameters affecting friction on a lubricated surface. Include the tangential friction force R, the normal force P, the velocity v, the surface area A, and the dynamic viscosity of the lubricant μ.

$$Answer: \frac{R}{P}, \frac{v\mu\sqrt{A}}{P}$$

8.26 An open cylindrical tank of 2 m diameter contains water to a depth of 1.5 m. The tank is to be rotated at 100 rev/min about its axis (which is vertical). A quarter-size model of this tank is to be made with water as the fluid. At what speed must the model be rotated for similarity? What is the ratio of the pressure at corresponding points in the liquids?

Answer: 200 rev/min, $\frac{p_m}{p_p} = \frac{1}{4}$

8.27 Airflow through a very large pipe can be modeled by water flow through a much smaller pipe. The velocity of water that will model the air velocity depends on the ratio of pipe diameters and the ratio of the kinematic viscosity of air to water, which at normal temperatures and pressures is 15:1, water being the smaller. If the prototype pipe diameter is 10 times larger than the model pipe diameter, what water velocity will model an air velocity of 3 m/sec?

Answer: 2 m/sec

8.10 ICONIC MODELS

Iconic models look like the original system or component. They are a visual representation. The word *iconic* comes from the Greek word *eikon*, which means an image or likeness. Iconic models can be in two or three dimensions and can be a smaller version of the original, an enlargement of the original, or the same size as the original. Iconic models, just like all types of models, leave out many details and only incorporate the details of interest or significance. Look at Figure 8.6. Is there enough detail for you to see immediately what is being represented? If you had trouble seeing what is in the picture, is the problem too much or not enough detail? Are you considering the correct subject matter? The point we are making is that what is clear to one person may not be clear to others. This is true not only for iconic modeling but also for many other types of modeling and communication.

Geometrically similar three-dimensional models are used to communicate their real counterparts. Figures 8.7 and 8.8 are examples of scaled-down and scaled-up three-dimensional models, respectively. Both cases are representations of how the originals appear but do not include much of the actual detail. The molecular model is a perceived geometric configuration which people have agreed to use for investigative purposes.

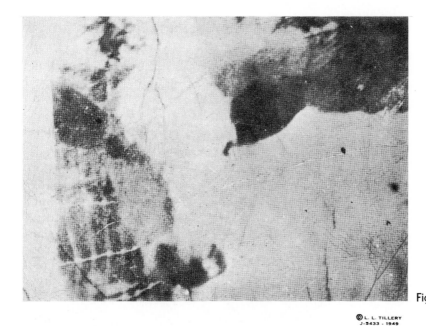

COW

Figure 8.6 Can you see what it is?

© L. L. TILLERY
J-5433 - 1949

A mock-up as shown in Figure 8.7 gives you a mental image of the object by just representing those parts you see. The next time you attend a movie, ask yourself about the authenticity of the buildings you are seeing.

The well-known and often used saying "a picture is worth a thousand words" does define the power of two-dimensional

Figure 8.7 Scaled-down three-dimensional model. (Courtesy JPL and NASA)

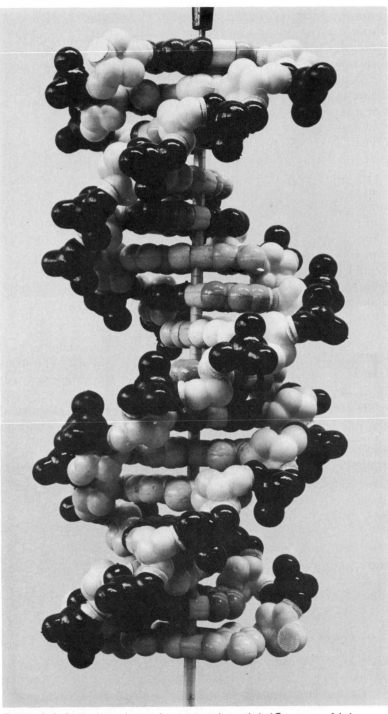

Figure 8.8 Scaled-up three dimensional model. (Courtesy of John Oldenkamp/Psychology Today Magazine.)

iconic models. The primary problem we face in graphic communication is our desire to communicate and record three-dimensional objects using a two-dimensional medium. Until someone invents three-dimensional paper, there will be a need for books such as this and courses in graphic communication. Photographs are detailed two-dimensional iconic models. Maps are a common form of two-dimensional iconic models. Recall that models exclude unimportant details and include or emphasize important details. A typical map emphasizes the details that will help guide your travels.

Graphite, lines, and paper are other important media for the preparation of two-dimensional models. An engineering drawing is a two-dimensional iconic model where geometric and material details are emphasized. Informal sketches are also a useful two-dimensional iconic model. The study and application of graphics, in large part, is the study of two-dimensional iconic modeling. Therefore, many of the principles and practices covered in the previous chapters are meant to help you make better iconic models.

8.11 THE DESIGN MODEL

A generation ago, a scale model was a replica of a plant or system that had already been constructed. It was an "after-the-fact" reproduction, most likely used for public relations or training purposes. Scale models resided in the foyers of office buildings, under glass. They could be viewed or admired, whichever was more appropriate, but they were not to be touched. Certainly, these scale models had no purpose in engineering design.

The engineering use of scale models began as a check on design, particularly in those instances where the constructor and the owner were not the same. The owner wanted to "see" what was being constructed, and so a model was built. Comparisons could be made. As plans for future changes arose, conflicts with the existing design could be discovered and averted. It became apparent that scale models could serve a useful purpose in reducing interferences and conflicts for space. As the uses and advantages of scale models became more obvious, models became bigger and were constructed with more precision.

Today, scale models are constructed with great precision before final working drawings are released or even started. In some cases, assembly drawings are eliminated altogether. A complete three-dimensional scale model, accurate in most details, is constructed, and component dimensions are scaled up from the model. This is a *design model*.

It is felt that using design models is a much better method for some projects than using assembly drawings. Many opinions and much expertise can be registered if a design model is available—a cumbersome process with drawings. How many people can view a drawing at one time? How many can view a model? Remember, a drawing has a preferential direction for viewing. A design model does not. With a design model, costly interferences are virtually eliminated. Interface errors are avoided. In complicated designs, there may be four or five candidates for the same space, such as air ducting, electrical circuits, water, and other fluid piping and structural members. All these candidates are complete systems. Calendar time required to design plants is reduced as a direct result of the reduced coordinating time required. Decisions between engineering and other disciplines can be reached in a fraction of the time when viewing the model as opposed to poring over stacks of drawings.

Early design models were mostly wood with pipes and piping components shown in brass wire, brass discs to check clearance, and copper nails. Each pipe was then hand-painted in a particular color to depict a certain system.

Today's design models are 99 percent plastic, with components and fittings either extruded or injection molded in many different colors. Structural shapes are available in a complete range of types, sizes, and colors. Components for fabricating any piece of mechanical equipment imaginable are now commercially available.

The design model begins early in the life of the project. Model design involves two stages: (1) preliminary design, and (2) final design. Both stages are embodied in the principle of model design, and both are of major importance in job cost and planning.

The preliminary design model is a block-type model, built to allow quick changes and to try things—to establish the first estimates of sizes, plot plans, and capital expenditures. The first model arrangement is assembled from rough freehand sketches. The equipment pieces are usually made from styrofoam blocks, cardboard, and tape. Components are simple, and it is important that they be kept simple. At the beginning of a design, it must remain flexible. It is also important that the parts be movable; you do not want to fix the location of components in a preliminary model. The preliminary model scale should be small. A later model can and will be larger and more accurate. The preliminary model should only be large enough to permit internal rearrangement of components and viewing. The re-

Figure 8.9 Preliminary design model
(Courtesy The Ralph M. Parsons Co.).

view of the preliminary model is most important. Figure 8.9 is a preliminary design model. The engineering design model or design model to follow will be much larger and more accurately constructed.

Design models are constructed in modules. This means that sections can be separated. Engineers and model designers can get right inside the model to study particular problems or make changes.

The size of the design model is determined by workability and portability. The smaller models are more portable. Scales range from $\frac{1}{4}$ in. to 1 ft (1:48) to $\frac{3}{4}$ in. to 1 ft (1:16), with probably the most popular scales being $\frac{3}{8}$ in. to 1 ft (1:32) or $\frac{1}{2}$ in. to 1 ft (1:24). In the 1:32 scale, a 6-ft man is modeled at $2\frac{1}{4}$ in. which is toy-size, and many toy trucks, cranes, etc., can be used to check clearances and headroom.

Modules seldom exceed 30 in. wide, 60 in. long, and 80 in. high. These are built on desk-height tables, and a complete model can require as many as 75 tables and 1000 ft². The design model is built in layers or trays cut horizontally at each major elevation. These trays are taken out of the model and photographed in the plan and elevation views. Photodrawings are intended to replace handmade drawings. A design model is a serious effort. Figure 8.10 is a design model of a nuclear reactor. The total height of this model is about 1 m. At some point in the model design, progress photodrawings are made of the model.

Figure 8.10 Design model (Courtesy Engineering Model Associates, Inc.).

Next to using the scale model as a primary engineering tool that eliminates conventional paper design methods, the most significant benefit is the production of isometric pipe-fabrication drawings. Information from the model is verbally transferred to a computer memory. The model is dimensioned by model technicians and designers to show dimensions of pipe and other components from the nearest column line. The computer converts these dimensions to plant-origin coordinates.

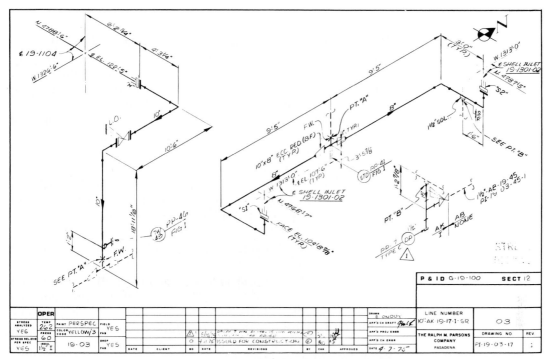

(a)

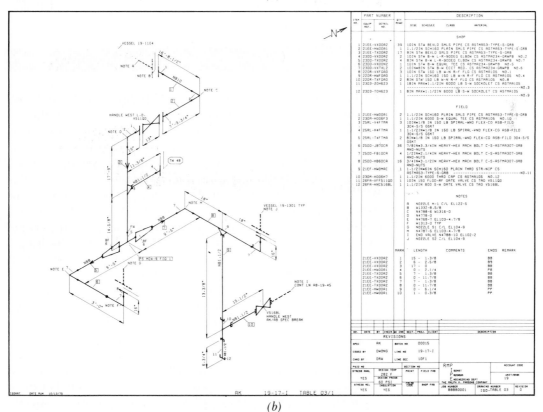

(b)

Figure 8.11 Piping drawing. (a) Hand drawn isometric drawing. (b) Computer drawn isometric drawing. (Courtesy the Ralph M. Parsons Co.)

Mass points and working points are also identified for stress-analysis purposes. An isometric drawing with unique identifying dimensions for each individual component is produced from this information. Figure 8.11 is such a drawing for a piping system; a bill of material for fittings and pipe sizes and lengths is also produced. This process alone creates huge savings of time and money, and it is relatively simple to do.

The unique aspect of the engineering design model is that the plant layout is done on the model, not on a series of composite layout drawings. The designers continue to design just as they would using the conventional drawing approach. Solutions to layout problems are resolved on a three-dimensional framework with interaction among all the design disciplines. A set of layout drawings showing individual commodities is developed from the model for record purposes and for use by construction. The design model is an extremely valuable tool, and its value is growing as more people put their imaginations to work. The improved communications avoid costly problems of the past, optimum design is achieved, team spirit between design and construction is greatly multiplied, and ultimately a better design is achieved at lower cost.

8.12 CHOICE OF MODELS

You have now been introduced to a wide variety of models. We use these models to communicate, analyze, and synthesize. Verbal models utilize language with all its imperfections. Iconic models are look-alike representations. Diagrams represent structure or relationships of the elements in a system. Symbols are shorthand. Mathematics is a game with precise rules that can be used to represent and manipulate physical systems, relationships, and concepts. Analog models behave like the original. And so on. Therefore, which model should be used, and when?

Choosing a model is just like giving a speech or writing a paper or report. It is extremely important to know the members of your audience. Do they understand the language and terms you are using? Can they read the drawings or sketches you are preparing? Are the symbols and nomenclature familiar? Do they have the mathematical background to follow what you are doing? How interested are they in receiving what you are giving them?

In engineering, we are continually concerned, or should be, about accuracy. Some iconic and mathematical models can be extremely accurate. That is, they communicate many specific facts about the device or system being modeled. On the other hand, in this age of electronic calculation, it is easy to obtain higher mathematical accuracy than necessary. A good engineer works to the accuracy appropriate for the job. Don't measure it with a micrometer, mark it with a soft lead pencil, and then cut it off with an ax!

The choice of model type takes experience, but the first step is to try to consider all the alternatives available to you. When you discuss the making of a part with a machinist, you could verbally describe the part, prepare mechanical drawings, or maybe make a model from a material you can form and work. The choice here is obvious, since the mechanical drawing will give the machinist all the necessary details needed to manufacture the component. If you are discussing the results of an experiment with a co-worker, you could verbally describe what you observed, present the data in tabular form, or reduce the data to charts and graphs or even to mathematical models. The choice here would be influenced by whether you wanted to communicate the general trends or the accuracy and spread of the data. The choice of the model is influenced by the degree of abstraction, the audience, and the accuracy.

PROBLEMS

8.28 Make a map that shows how to walk from your Registrar's Office to the classroom in which this course meets.

8.29 Using a scaled model, determine the maximum length L of a 4-in.-wide beam that can be carried through the hallway and right-angle corner.

Answer: 115 in.

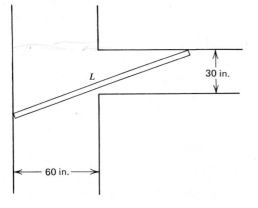

8.30 Using the given portion of the road map, prepare the following.

    a. A verbal description of how to go from West Sacramento (lower left) to Pilot Hill (upper right).

    b. A map (iconic model) of how to go from Sacramento Metro Airport (left center) to El Dorado Hills (right center).

    c. A verbal description of how to go from Rocklin (upper center) to Mather Air Force Base (lower center).

    d. A map that includes primary landmarks for a canoe trip from Folsom Lake to the Sacramento River.

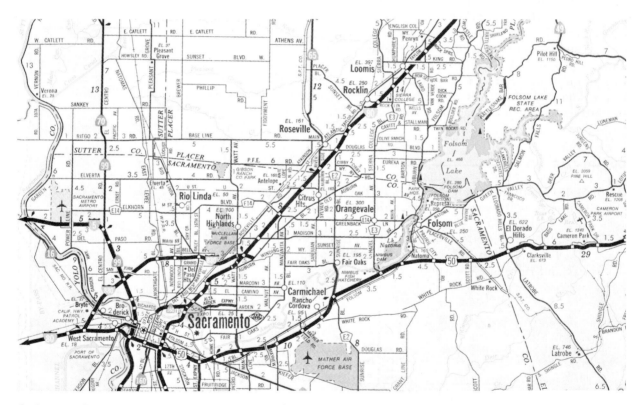

Prob. 8.30 (Map copyrighted 1979 by the California State Automobile Association. Reproduced by permission).

8.31 Using a scaled model, determine the maximum length and width combinations of a rectangular object that can be carried, pushed, or shoved through the corner described in Problem 8.29.

8.32 Using a scaled model, determine the largest pipe assembly that can successfully pass through the doorway. The pipe is 40 mm in diameter, and all 10 straight sections are of equal length.

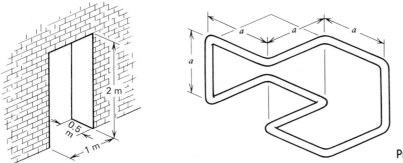

Prob. 8.32

8.33 Using a scaled model, determine the largest assembly that can pass through the doorway described in Problem 8.32. The three orthogonal square planes are relatively thin.

Answer: $b = 1.36$ m

Prob. 8.33

8.34 Using commercially available modeling materials, make an iconic model of a solar hot-water heater and storage system to be installed on the roof of a one-story, one-family home.

8.35 Using commercially available modeling materials, make an iconic model of a hot tub and its equipment to be used in a one-family home.

8.36 Build a design model of a hot tub and its supporting systems to be constructed in the back of a pickup truck.

8.37 Build a design model for a self-contained refrigerator for camping. Provide for 2 ft³ of cooled volume.

8.38 Build a design model for a self-contained, solar-heated, portable shower.

8.39 Build a design model of a device that elevates an automobile 3 ft for the purpose of working on the engine or the chassis from underneath. Assume that the device is to be obtained from a rental agency; therefore, it must be portable.

8.40 Using two toy cars, with a scale of $\frac{1}{4}$ in. to 1 ft, make a design model of a one-car garage for two cars. There is no limit on garage height, but the floor must remain unchanged.

8.41 Build a design model of a home greenhouse. Include shelf space for plants, a work area, a temperature-control system, and a watering system.

8.42 Build a design model of a low-volume still to make industrial alcohol to combine with gasoline to make your own gasohol using farm refuse as your raw material.

9
GRAPHIC MATHEMATICS

9.1 FUNCTIONS

Problems in engineering, as well as problems in mathematics, physics, and chemistry, can often be expressed in very simple mathematical terms. If one variable is known, its value will determine others. If two variables are so related that when one is known the other can be determined, the second variable is called a *function* of the first. The first is called the *independent variable* and the second is called the *dependent variable*. The statement

$$y = f(x) \tag{9.1}$$

is the mathematical expression that the dependent variable y is a function of the independent variable x. Algebraic, trigonometric, inverse trigonometric, logarithmic, exponential, hyperbolic, and inverse hyperbolic functions all use this definition, and they are usually stated in terms of one variable.

The idea that one variable can be a function of another is not limited. There may be three, four, or more variables, and one or more of these may be independent.

$$z = f(x, y) \tag{9.2a}$$

$$w = f(x, y, z) \tag{9.2b}$$

are the mathematical expressions that the dependent variable

H

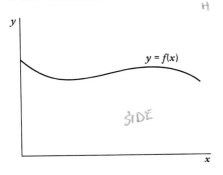

$y = f(x)$

SIDE

(a)

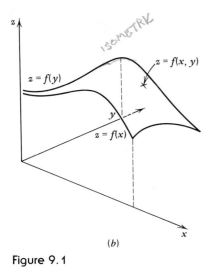

ISOMETRY

$z = f(x, y)$

$z = f(y)$

$z = f(x)$

(b)

Figure 9.1

z is a function of independent variables x and y, and the dependent variable w is a function of independent variables x, y, and z.

A graph is another way to express one variable as a function of another. In two dimensions, the dependent variable is plotted as the ordinate y, and the independent variable is plotted as the abscissa x. This is conventional, although there is no reason why they could not be reversed. In three dimensions, one dependent variable can be plotted as a function of two independent variables. Three-dimensional space is the limit of our graphic expression, but it is not the limit of our mathematics. Mathematically, n-dimensional space exists, but graphics cannot show it.

In Figure 9.1a the variable $y = f(x)$. In two-dimensional space, this function is described by a line. In Figure 9.1b $z = f(x, y)$. In this case the function is described as a surface.

9.2 ALGEBRAIC EQUATIONS

One of the simplest algebraic equations expresses the linear relation between two variables. Let $y = f(x)$ such that

$$y = mx + b \tag{9.3}$$

This is a linear first-degree equation, where the dependent variable y is related to the independent variable by a constant b and a constant coefficient m. The value of x for which $y = 0$ is

$$x = -\frac{b}{m} \tag{9.4}$$

and is known as the *root* of the equation. Graphically, the root is the location of the x-intercept of the graph. The location of the y-intercept is the constant b, and the coefficient m is the slope. All this is shown in Figure 9.2.

When two linear equations are expressed simultaneously, a specific value of x and a specific value of y satisfy both equations. Graphically, the solution is the intersection of two lines.

Let $y = m_1x + b_1$ be one linear relation between x and y, and let $y = m_2x + b_2$ be the other. These two equations would have an intersection unless they were parallel or coincident. The value of x and y at the point of intersection would simultaneously satisfy both equations, as shown in Figure 9.3, where the simultaneous solution is at $y = r$ and $x = s$. The values of r and s satisfy both equations. For these two lines to be parallel, $m_1 = m_2$. To be coincident, the equations would have to be identical.

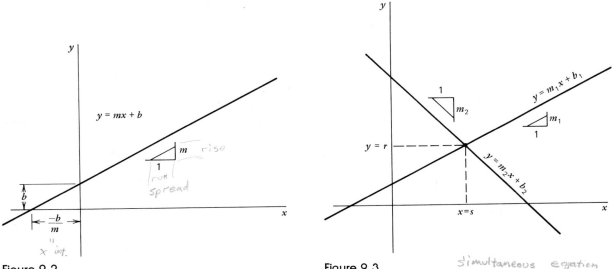

Figure 9.2 Figure 9.3

Polynomials are algebraic functions of higher degree. They are single-valued, continuous functions with any number of continuous derivatives. Both the function and its derivatives have physical significance in engineering. The proof of continuity is beyond the scope of this work, but the important fact is that we may assume that there is a *continuous* curve for all functions that are polynomials in x. The degree of the equation is the highest exponent of the variable x. Equations of the first, second, third, and fourth degrees are called linear, quadratic, cubic, and quartic equations, respectively. For example,

$$y = a_0 + a_1x + a_2x^2 \ldots a_nx^n \qquad (9.5)$$

This equation is a polynomial in x, of degree n, and it will have n roots. These roots may be real or imaginary.

The polynomial may also be expressed in terms of its factors.

$$y = (x - x_1)(x - x_2)(x - x_3) \ldots (x - x_n) \qquad (9.6)$$

Since $y = 0$, whenever $x = x_1$, $x = x_2$, or $x = x_n$, $x_1, x_2, x_3, \ldots, x_n$ are the roots of the polynomial.

In drawing a graph of a polynomial, a given number of values of y are plotted as a function of x, and a smooth curve is passed through these points. How well the polynomial is defined is a matter of how many points are selected. Figure 9.4 is a graph of the polynomial $y = x^3 - 3x + 2$. The roots are $x_1 = -2$ and $x_2 = x_3 = 1$. x_2 and x_3 comprise a *double root*. In order to define any curve there must be at least one more data point than the degree of the polynomial. This particular curve was defined by 12 points.

$y = f_1 - f_2$

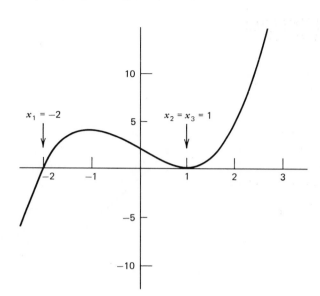

Figure 9.4

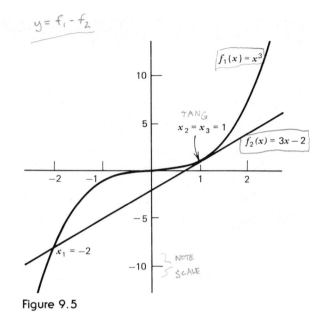

$f_1(x) = x^3$

TANG
$x_2 = x_3 = 1$

$f_2(x) = 3x - 2$

$x_1 = -2$

NOTE
SCALE

Figure 9.5

SET $y(x) = \phi$

$x^3 - 3x + 2 = \phi$

plot $y = f(x)$

find x intercepts

blow up intersection (w. straight
line)

enlarge again

An alternate way to solve for the roots of a polynomial would be to divide the function into two functions, $f_1(x)$ and $f_2(x)$. In the case of the polynomial $y = x^3 - 3x + 2$,

$$f_1(x) = x^3$$
$$f_2(x) = 3x - 2$$

and

$$f_1 - f_2 = \phi$$

$$y = f_1(x) - f_2(x) = 0 \tag{9.7}$$

or

$$y = x^3 - 3x + 2 = 0$$

This solution has one advantage in that $f_1(x)$ is very easy to graph. Plotting these as separate graphs on the same set of co-ordinates, $f_1(x)$ and $f_2(x)$ will intersect at $(-2, -8)$ and be tangent at $(1, 1)$ (see Figure 9.5). The x-coordinates of these points are the roots of the original quadratic equation since $y = 0$ when $f_1(x) - f_2(x) = 0$. If the two functions do not intersect, there are no real roots to the equation. This concept, splitting the function $y(x)$ into two functions and solving for the intersection of the two functions, is very powerful. Later in the chapter it will be used to solve nonalgebraic equations.

9.3 RATE PROBLEMS

Algebraic

As we have seen, some concepts in algebra are more easily understood when presented graphically rather than analytically. This is especially true for problems involving rates.

As an example, let us consider a familiar problem. If A can dig a hole in 6 h, and B can dig the same hole in 4 h, how long will it take if A and B dig together? The answer is 2.4 h.

If x is the time that it takes A to dig the hole, then $1/x$ is the rate at which A digs the hole. If y is the time that B takes, then $1/y$ is the rate at which B digs the hole. If they dig together, their rates are added, and the problem is expressed analytically,

$$\frac{1}{x} + \frac{1}{y} = \frac{1}{z} \tag{9.8}$$

where z is the time they take to dig the hole together, and $1/z$ is their combined rate.

Graphically, this problem can be expressed quite simply. It is described in Figure 9.6. Note that the coordinates for the graph are holes and time. Line A represents A's digging progress, one hole in 6 h, two holes in 12 h, etc. Line B is B's digging progress, one hole in 4 h, two holes in 8 h, three in 12 h, etc. This is all presuming that A and B can dig at a constant rate and that they will not interfere with each other. Line $A + B$ is their combined rate, $1\frac{2}{3}$ holes in 4 h, 5 holes in 12 h, etc. Line $A + B$ crosses the one-hole line at 2.4 h, which means that together they would dig one hole in 2 h and 24 min.

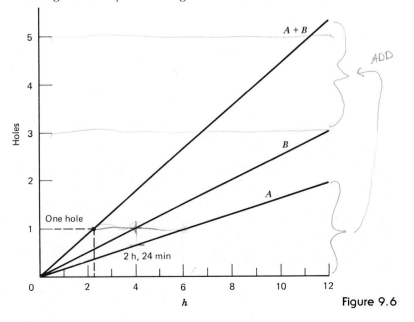

Figure 9.6

These problems have a variety of form. If a tank can be filled in 2 h and drained in 6 h how long would it take to fill the tank with the drain open? The problem involves the difference between two rates. If car A leaves one city and moves along a road toward a second city with a constant velocity of 60 km/h, and B leaves $\frac{1}{2}$ h later from the second city and travels toward the first, traveling at 80 km/h, how long will it be before B passes A? This could be rephrased to ask how far would A and B have traveled? Also, A and B could leave from the same city. You have seen algebraic rate problems many times. The advantage of a graphic solution is best seen by considering examples.

SAMPLE PROBLEM 9.1

Determine the value of y and x that will satisfy the two simultaneous equations

$$2y = 11 - x$$
$$y = 2x - 2$$

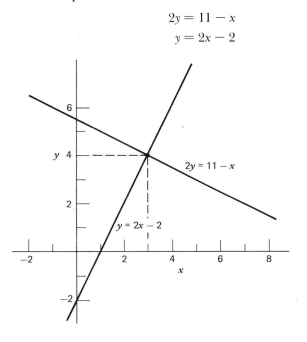

Solution
When the two curves are plotted on the same set of coordinates, they have one intersection at $y = 4$, $x = 3$. A check calculation will show that these values do satisfy the equations.

SAMPLE PROBLEM 9.2

The time of flight for a projectile in free-fall is defined by the equation

$$h = -\frac{gt^2}{2} + v_0 t$$

where h = vertical displacement at time t
v_0 = initial vertical velocity
g = acceleration of gravity, 32.16 ft/sec²

If a stone is thrown at a velocity of 60 ft/sec, upward at an angle of 30 deg to the horizontal from a 200-ft cliff (a), how long will it be before it strikes the ground?

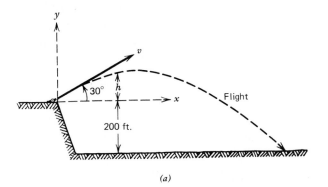

(a)

Solution

The horizontal and vertical coordinates are separable. The time of flight depends only on the initial vertical velocity, the acceleration of gravity, and the vertical height of the cliff. The x–y coordinates for this problem would be as shown (a). The vertical displacement at time t is -200 ft. Note that the negative sign means that the stone will come to rest 200 ft below its starting position.

$$v_0 = v \sin \theta$$
$$= 60 \sin 30° = 30 \text{ ft/sec}$$

and

$$-200 = \frac{-(32.16)t^2}{2} + 30t$$

This equation is a quadratic in t.

$$-12.438 = -t^2 + 1.866t$$

Plotting $y(t) = t^2 - 1.866t - 12.438$, the function $y(t)$ has two roots (b). One lies between $t = 4$ and $t = 5$ sec. There is a second root between $t = -2$ and $t = -3$ sec, but a negative root is mean-

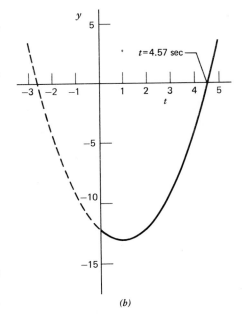

(b)

ingless. Interpolation of the curve between $t = 4$ and $t = 5$ sec yields the value of

$$t = 4 + \frac{3.952}{(3.182 + 3.952)} = 4.57 \text{ sec}$$

This value can be checked by the quadratic formula. The exact value is $t = 4.576$ sec. Note that $h(x)$ and $y(t)$ are entirely different quantities. $h(x)$ is vertical displacement in feet. The units for $y(t)$ are seconds squared.

SAMPLE PROBLEM 9.3

In water, a solid steel hemisphere will sink, but a hemispheric steel shell will float. For a hemisphere with an outside diameter of 20 cm, determine the maximum thickness of the shell that will allow the hemisphere to float (a). The density of water is 1 g/cm³ and the density of steel is 7.85 g/cm³.

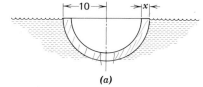

(a)

Solution

The volume of a hemisphere is $(\frac{2}{3}) \pi r^3$. Calling x the thickness of the steel hemispherical shell, the shell would have a mass of

$$m_s = (7.85) \left[\frac{2}{3} \pi 10^3 - \frac{2}{3} \pi (10 - x)^3 \right]$$

The mass of the displaced water is

$$m_w = (1) (\frac{2}{3} \pi 10^3)$$

Equating the mass of the steel hemisphere to the mass of the displaced water,

$$(7.85) \left[\frac{2}{3} \pi 10^3 - \frac{2}{3} \pi (10 - x)^3 \right] = (1) \frac{2}{3} \pi 10^3$$

This equation is a cubic equation in x and reduces to

$$x^3 - 30x^2 + 300x = 127.4$$

Plotting $y(x) = x^3 - 30x^2 + 300x - 127.4$, the function $y(x)$ is observed to have a root between $x = 0$ and $x = 1$ since the function $y(x)$ changes sign between these two values of $x(b)$. (Note that the units for $y(x)$ are cubic centimeters.)

$$y(0) = -127.4$$
$$y(1) = 143.6$$

There are no other real roots between $x = 0$ and $x = 10$.

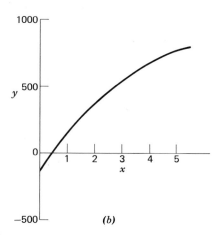

(b)

Expanding the x-coordinate of the graph (c), the root of the function can be found between $x = 0.4$ and $x = 0.5$

$$y(0.4) = (0.4)^3 - 30(0.4)^2 + 300(0.4) - 127.4 = -12.14$$
$$y(0.5) = (0.5)^3 - 30(0.5)^2 + 300(0.5) - 127.4 = +15.23$$

Interpolating between these two values

$$x = 0.4 + (0.5 - 0.4)\frac{(12.14)}{(12.14 + 15.23)} = 0.444$$

A more exact value would be $x = 0.44407$, correct to five significant figures. The other roots of this polynomial are a pair of conjugate complex numbers $29.56 \pm 16.55i$.

There are a number of better methods for determining the roots of a polynomial by approximation. Newton's method and the interpolation of roots by using chords instead of straight lines are two. In most cases, the need for a numerical answer can be fully satisfied by using graphic techniques and straight-line interpolation.

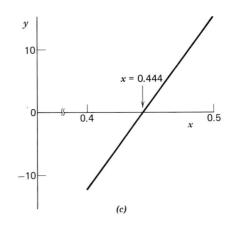

(c)

USUALLY
DISTANCE
TIME

SAMPLE PROBLEM 9.4

A car leaves city A and travels at a constant speed of 60 km/h toward city B 300 km away. One hour later, a second car leaves city B and travels at a speed of 80 km/h toward city A along the same road. When and where will the cars pass?

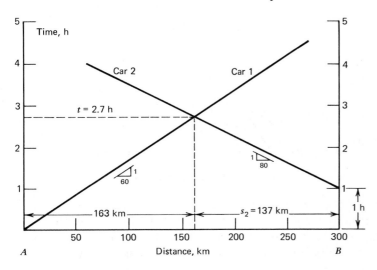

Solution

The coordinates for a graphic solution will be time and distance traveled. The line designated as car 1 is the graph of the distance traveled by car 1 along the road from A to B. The line designated as car 2 is the graph of the distance traveled by car 2. The slopes of these two lines are established by the respective rates of travel, 60 km/h for car 1 and 80 km/h for car 2. Note that the displacement-time curve for car 2 has a negative slope, which simply means that if the direction of line AB is positive, car 2 is traveling in a negative direction. The line representing its travel also passes through the point $s = 300$ km, $t = 1$ h.

These two straight lines intersect at $t = 2.7$ h at a point 163 km from A and 137 km from B.

SAMPLE PROBLEM 9.5

A topsoil batching plant sells two grades of topsoil mix. One contains 60 percent of fir bark and another contains 30 percent fir bark, by mass. The specifications for a particular landscaping project require a mix containing 50 percent fir bark. How many kilograms of each mix are required for each 1000 kg of the total mix?

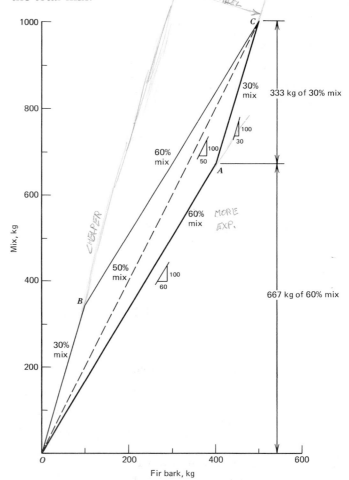

Solution

Each batch of topsoil mix contains a certain mass of fir bark. In the pile of 60-percent mixture there are 600 kg of fir bark in every 1000 kg of mixture. Likewise, there should be 60 kg of fir bark in 100 kg of mixture, and 6 kg in 10 kg of mixture. All this can be expressed graphically by the line labeled 60 percent mixture on the accompanying graph. The coordinates are kilograms of fir bark and kilograms of mixture. In the 30-percent mix, 300 kg of each 1000 kg is fir bark. The line marked 30 percent represents the quantity of fir bark in each quantity of that mix, 30 kg of fir bark in 100 kg of mixture, 3 kg in 10 kg of mixture, etc.

If a 50-percent mix were available, the dashed line marked as a 50-percent mixture would represent the proportion of fir bark by mass in that mix. Obviously, the 60-percent mix has too much fir bark, and the 30-percent mix has too little. In order to make the correct mixture, suppose we start by taking from the pile containing the 60-percent mixture and stop when we have enough of that mix. At that time, we start taking from the pile of the 30-percent mix. To do this, we would follow line OA to point A and then AC to point C. We could also reverse the procedure, taking first from the 30-percent mix. This would follow line OB to point B and then BC to point C. In either case, we are mixing 667 kg of the 60-percent mixture with 333 kg of the 30-percent mixture. $OACB$ is a parallelogram. Why?

PROBLEMS

Solve the following systems of equations for x and y.

9.6 a. $y = 5 + x$; $2y = 4 - x$.

  b. $2y = x + 3$; $x^2 - 2x = y$.   *Answer: $x = +3$, $y = +3$*

    $x = -\frac{1}{2}$, $y = +\frac{5}{4}$

9.7 a. $4y = 3x + 8$; $y = 2x - 3$.

  b. $3x - y = 8$; $y^2 = 4x$.   *Answer: $x = +4$, $y = +4$*

    $x = +\frac{16}{9}$, $y = -\frac{8}{3}$

9.8 a. $4y = x + 2$; $x + y = 3$.

  b. $xy = 5$; $x^2 + y^2 = 25$.   *Answer: $x = +1.021$, $y = +4.895$*

    $x = +4.895$, $y = +1.021$

    $x = -1.021$, $y = -4.895$

    $x = -4.895$, $y = -1.021$

9.9   a. $6y + 6 = x; \quad 2y + x = 2.$

      b. $\dfrac{4}{x} - \dfrac{3}{y} = 1; \quad 4 - 2x - x^2 = y.$

> *Answer:* $x = +1, y = +1$
>
> $x = +4.531, y = -25.59$
>
> $x = -3.531, y = -1.407$

9.10 It is estimated that the product-development cost for a new bathroom mixing valve will be $9500. This is a fixed cost that cannot be lowered. The best estimate for the manufacturing cost is $16.55 per unit. A market price of $29.95 per unit has been set by marketing surveys. How many units must be sold before a profit is realized? This is known as the break-even point.

> *Answer:* 710 units

9.11 A salesman uses his private car for business and is reimbursed for his expenses. Company policy allows him to claim either his actual expenses or a straight 30¢ per mile, up to 9000 miles, and 20¢ per mile thereafter. He calculates his fixed expenses as $2500 per year for depreciation, license, and insurance. Service and fuel costs combined are estimated to be 9¢ per mile. At what point is it to his advantage to claim the straight rate rather than actual expenses? Base your answer on miles traveled per year.

> *Answer:* 14,545 mi/year

Graphically, find any real roots of the following equations.

9.12 $y = x^3 + 3x - 20$

9.13 $y = x^3 + 4x - 7$

9.14 $y = 2x^3 + 3x^2 - 4x - 10$ *Answer:* $x = +1.625$

9.15 $y = x^4 - 3x - 1$ *Answer:* $x = +1.540$

> $x = -0.329$

9.16 By increasing the radius of a sphere by 1 in., the volume of the sphere is increased by 20 in.^{3}. What was the diameter of the original sphere? Solve the problem graphically.

> *Answer:* 1.456 in.

9.17 In riveted lap joints, pitch p is defined as the distance between rivets. For a quadruple-riveted double-strap butt joint,

$$p = 10\frac{d^2}{h} + d$$

h is the thickness of the shell and d is the diameter of the rivet. For $h = 1$ cm and $p = 5$ cm, what is the diameter of the rivet?

Answer: 6.59 mm

9.18 A sphere of ice 1 m in diameter is floating in water. Determine the height of the sphere of ice exposed above the water surface if the density of water is 1.00 and the density of ice is 0.90.

9.19 An automobile starts at 1 P.M. on a 120-km trip, at a speed of 80 km/h. At 2 P.M. the automobile reaches the outskirts of a city, stops for service for 15 min, and resumes the trip at a speed of 40 km/h. A second car starts on the same trip at the same time, begins with a velocity of 70 km/h, but the driver has to reduce the velocity by 10 km/h every 30 km. Which car reaches the destination first, and how long afterward does the other car arrive?

Answer: Car 1; car 2 arrives 1.7 min later.

9.20 A tank can be filled by pipes A and B in 2 h. If a discharge valve C is open, A alone can fill the tank in 4 h. If discharge valve C is open, B alone can fill the tank in 8 h. How long would it take for discharge valve C to empty the tank with both A and B closed?

Answer: 16 h

9.21 A coiled spring is described by its modulus k, which is the force required to displace the spring one unit. When two springs are placed in series, their equivalent modulus is

$$\frac{1}{k_e} = \frac{1}{k_1} + \frac{1}{k_2}$$

If $k_1 = 97$ lb/in. and $k_2 = 142$ lb/in., what is k_e? Determine graphically.

9.22 A stone is dropped into a deep well. Four seconds later the sound of a splash is heard. How far below the ground surface is the water surface in the well? Assume that the velocity of sound is 1100 ft/sec. (Hint. Refer to the equation for free-fall in Sample Problem 9.2.)

Answer: 231 ft

9.23 One bar of an alloy of silver is 40 percent pure silver. A second bar is 12 percent silver. How much of the first alloy and how much of the second are needed to produce 50 oz of alloy that is 20 percent silver.

Answer: 14.3 oz of 40-percent alloy
35.7 oz of 12-percent alloy

9.4 TRANSCENDENTAL EQUATIONS

A function that is not algebraic is called *transcendental*. Transcendental functions include trigonometric, logarithmic, exponential, hyperbolic, Bessel, inverse trigonometric, and other functions. Equations that contain these functions are called *transcendental equations*, and they abound in engineering.

Formulas for the solution of transcendental equations are not available. Some transcendental equations are easy to solve because the numerical value of the transcendental functions in the equations is known. For example, consider

$$\sin x = 1$$

which is a transcendental equation. It is trigonometric but it is *not* algebraic. The solutions are $x = \pi/2,\ 5\pi/2,\ 9\pi/2,\ \ldots,\ (4n - 3)\pi/2$. The value of $\sin x$ can be found for any x from trigonometric tables. $\sin x$ is known because of our knowledge of trigonometry.

Now consider the equation

$$x \sin x = 1$$

This is also a transcendental equation, but our knowledge of trigonometry does not extend to its solution. The equation has solutions, just as an algebraic equation has solutions, but the solutions are not found easily. We do not have a table for the function $x \sin x$, but we could make one if we needed it. The best methods to determine the solutions of transcendental equations are by successive approximations, which is trial and error, or by using a graphic solution.

To solve these equations graphically, the equation must first be expressed in another form. Rearranging terms and expressing it as $y(x)$,

$$y = \sin x - \frac{1}{x}$$

Two separate functions are now created as in Section 9.2.

$$f_1(x) = \sin x$$

and

$$f_2(x) = \frac{1}{x}$$

These functions are plotted. The intersection of the graphs of the two functions is the solution to the original equation. The graphs also show that there could be more than one intersection.

Intersections would occur when $f_1(x) = f_2(x)$ or

$$y = f_1(x) - f_2(x) = 0 \qquad \text{(9.7 repeated)}$$

Note that the independent variable x must be a pure number. It cannot have dimensions.

SAMPLE PROBLEM 9.24

What are the roots of the transcendental equation $x \sin x = 1$?

$$f = f_1 - f_2 = \emptyset$$

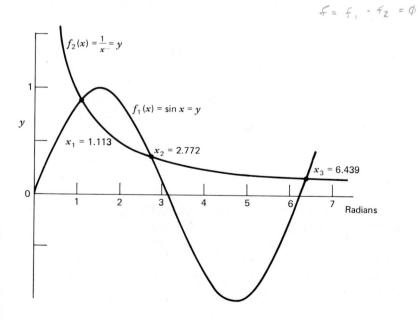

$f_2(x) = \frac{1}{x} = y$

$f_1(x) = \sin x = y$

$x_1 = 1.113$

$x_2 = 2.772$

$x_3 = 6.439$

Radians

1.000326
$.999774$

Solution

Rearrange the transcendental equation as two equations, one algebraic and one trigonometric, both of which are known.

$$f_1(x) = \sin x$$

$$f_2(x) = \frac{1}{x}$$

Now, plot as two separate graphs on the same set of coordinates.

The first three roots of these equations are

$$x_1 = 1.113$$

$$x_2 = 2.772$$

$$x_3 = 6.439$$

Beyond the first two, succeeding roots are approximately $x_n = (n-1)\pi$, and the third is in error by only 2.5 percent if it is taken as $x_3 = 2\pi$ instead of $x_3 = 6.439$.

SAMPLE PROBLEM 9.25

What is the solution to the equation $y = e^x + x - 6$?

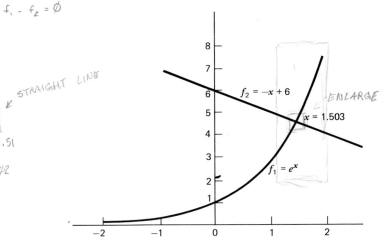

Solution

This is a transcendental equation because it contains the exponential e^x. We cannot solve for x without knowing e^x, and we cannot solve for e^x without x.

Rearrange these three terms into two equations,

$$f_1(x) = e^x$$

and

$$f_2(x) = -x + 6$$

The graphic solution of these two equations yields the single root $x = 1.503$, which satisfies the equation

$$e^{1.503} + 1.503 - 6 = 0$$

PROBLEMS

9.26 Determine the solution of the equation $y = x(\ln x) - 1$.

9.27 What is the solution (or solutions) of the equation $y = e^x + x - 2$?

Answer: 0.443

9.28 Determine the first two solutions to the equation $y = x \tan x - 1$.

Answer: 0.860
3.437

9.29 Determine the first two solutions of the equation

$$\cosh \beta l \cos \beta l = 1$$

This equation is used to solve for the natural frequencies of a uniform beam built in at both ends. β is a function of frequency and the flexural constants of the beam, l is the length of the beam, and βl is dimensionless.

Answer: $\beta_1 l = 4.730$, $\beta_2 l = 7.853$

9.30 Determine the roots of the equation $\tan \beta l = \tanh \beta l$. This equation is used to solve for the several frequencies of a long rod suspended as a pendulum. βl is dimensionless.

9.31 A transmission line with a span of 300 m between two towers of the same height has a mass of 3 kg/m and a sag of 10 m. The tension in the cable is determined by the transcendental equation

$$\frac{0.294}{T_0} = \cosh \frac{4.413}{T_0} - 1$$

T_0 is measured in kilonewtons. What is T_0?

Answer: 33.06 kN

9.32 The logarithmic mean temperature difference ΔT_m in a parallel-flow heat exchanger is defined as

$$\Delta T_m = \frac{\Delta T_2 - \Delta T_1}{\ln\left(\dfrac{\Delta T_2}{\Delta T_1}\right)}$$

for a final temperature difference $\Delta T_2 = 20$ K, determine ΔT_1 if it is required that $\Delta T_m = 15$ K.

Answer: $\Delta T_1 = 10.912$ K

9.33 The linear acceleration of a piston in an internal combustion engine is given as

$$a = r\omega^2\left(\cos\theta + \frac{r}{l}\cos 2\theta\right)$$

where θ = angular displacement of crankshaft
   r = radius of crank arm
   l = length of connecting rod
   ω = crank speed, radians/sec

Determine the angular displacement of the crankshaft at which the piston acceleration is zero, if $\dfrac{r}{l} = \dfrac{1}{4}$.

Answer: $\theta = 77{,}283$ deg

9.34 In the flow of fluid through a smooth pipe, the friction factor f has been related to the Reynolds number (Re) by Nikuradse.

$$\frac{1}{\sqrt{f}} = 2\log_{10}\mathrm{Re}\sqrt{f} - 0.8$$

This can be used to determine friction factors for Re > 5000. For $R = 10^5$, determine the friction factor f.

Answer: $f = 0.0181$

9.35 For an airplane, the difference between the power available and the power required determines its rate of climb. When the curves of power required and power available become tangent to each other, there is only one speed at which the airplane can fly level and the rate of climb is zero. The corresponding altitude is the absolute ceiling H; it can be determined from the curve of maximum rate of climb as a function of altitude. The time required to climb to any altitude h is determined as

$$t = 2.303\frac{H}{r_0}\log_{10}\left[\frac{H}{H-h}\right]$$

where r_0 is the rate of climb at sea level. With an initial rate of climb of $r_0 = 1500$ ft/min, an airplane can climb to an altitude of $h = 10,000$ ft in 13 min. What is its absolute ceiling H?

Answer: $H = 12,780$ ft

9.36 For an electrical cable to carry the highest voltage with minimum insulation safely, the ratio $x = \dfrac{r_0}{r_i}$ of the outside to inside radius of the insulation must satisfy the following transcendental equation.

$$\ln x = 1 - \frac{1}{x^2}$$

Other than $x = 1$, determine the ratio $x = \dfrac{r_0}{r_i}$ that satisfies this equation.

Answer: $x = 2.218$

9.5 THE DERIVATIVE

The derivative is one of two basic concepts of the calculus that allow us to study how functions behave. The second basic concept is the integral. Both cases have geometric meaning and both can be determined graphically.

In Figure 9.7 A is any point (x, y) on the curve described by the function $y = f(x)$. B is another point (x_1, y_1) on the same curve. The slope of line AB is

$$m = \frac{y_1 - y}{x_1 - x} \tag{9.9}$$

Using the symbol Δ for the difference or incremental change of the variables x and y

$$\Delta y = y_1 - y$$
$$\Delta x = x_1 - x \tag{9.10}$$

Slope m, in terms of the incremental changes, is expressed more simply as

$$m = \frac{\Delta y}{\Delta x} \tag{9.11}$$

Line AB may either be a chord or the tangent of the curve depending on the curvature of the graph. However, if A remains fixed, and the second point B is considered to move

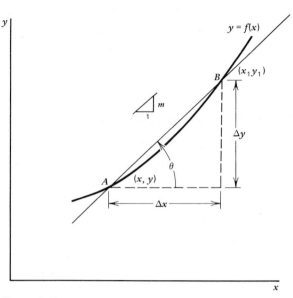

Figure 9.7

closer to point A, the slope of line AB will approach a constant limiting value. B cannot move closer than being coincident with A. When this happens, line AB will be the tangent to curve $y = f(x)$ at A, and

$$m = \tan \theta \tag{9.12}$$

Using increment Δx, slope m may also be written as

$$m = \frac{\Delta y}{\Delta x} = \frac{y(x + \Delta x) - y(x)}{\Delta x} \tag{9.13}$$

Now, holding x fixed, as Δx becomes smaller and smaller, the limiting value of slope m is defined as the derivative of y with respect to x. Mathematically, this is stated in the following way.

$$m = \lim_{\Delta x \to 0} \frac{\Delta y}{\Delta x} = \lim_{\Delta x \to 0} \frac{y(x + \Delta x) - y(x)}{\Delta x} = \frac{dy}{dx} \tag{9.14}$$

which means also that $\tan \theta = dy/dx$.

The derivative exists and can be determined, but the concept of the limit is needed in the definition of the derivative. If we make Δx equal to zero, the ratio $\Delta y/\Delta x$ is indeterminate and meaningless. We define the derivative as the value of $\Delta y/\Delta x$ as Δx approaches but does not equal zero. We avoid the problem of indeterminacy by not letting it happen. Since Δy and Δx are always finite and definite, the expression dy/dx is the limiting value of a fraction but not the fraction.

Both Leibniz and Newton understood this concept and systematically used limits to determine derivatives. One further piece of nomenclature is that dy/dx is often written as $y'(x)$ or simply y'. This is just a shorter way to indicate the derivative. All are equally acceptable.

9.6 GRAPHIC DIFFERENTIATION

The derivative dy/dx has been shown to be the slope of the tangent to the curve representing the function $y = f(x)$ at a specific point. That is, $y'(x_1) = dy/dx]_{x=x_1}$, $y'(x_2) = dy/dx]_{x=x_2}$, etc. If we determine the derivative of $y = f(x)$ at enough points, we can plot derivative $y'(x) = dy/dx$ as a continuous curve. A very simple graphic method exists to do this.

The first step is to plot $y(x)$ carefully using convenient scales for both the x- and y-coordinates. It is not important for these scales to be equal, but they will influence the determination of y and x, so they must be convenient. Using a convenient scale means that the smallest division should be 1, 2, 5, 10, 20, 100, etc.—a number that is easy to plot and convenient for calculation. At the same time, a second set of axes is plotted to be used for the curve of the derivative. In Figure 9.8a the known curve of function $y(x)$ is plotted. The second set of axes is placed directly below. Note that the abscissas are identical.

The second step is to locate an arbitrary pole P, placing it at a pole distance p, to the left of the origin of the new set of axes to be used for the curve of the derivative. It should be on the same line as the x-axis, but the distance p is not negative. It is simply a geometric construction placed at the left of the origin. The units for measurement of the pole distance will be the same as the units for the x scale. The pole distance itself is arbitrary. In Figure 9.8b pole P is placed two units to the left of the origin.

After the pole is located, the scale for the ordinate of the derivative can be set. On the curve of $y(x)$, draw a line from the origin with a selected slope such as 1:1, 5:1, 10:1, 1:10, etc. The slope to be selected will depend on the scale of the ordinate for $y(x)$. If the numbers for $y(x)$ are large, it may be necessary to condense the scale for $y(x)$ and select a slope such as 5:1, 10:1, or higher. The reverse is true if the numbers for $y(x)$ are small numbers (i.e., decimals). In Figure 9.8b a slope 1:1 has been selected. Now draw a second line from pole P that intersects the ordinate of the derivative and is parallel to the first line. The intersection of this second line with the ordinate of the derivative locates the unit scale for the derivative axis. In Figure 9.8b we have a 1:1 slope, so the intersection locates one unit on the

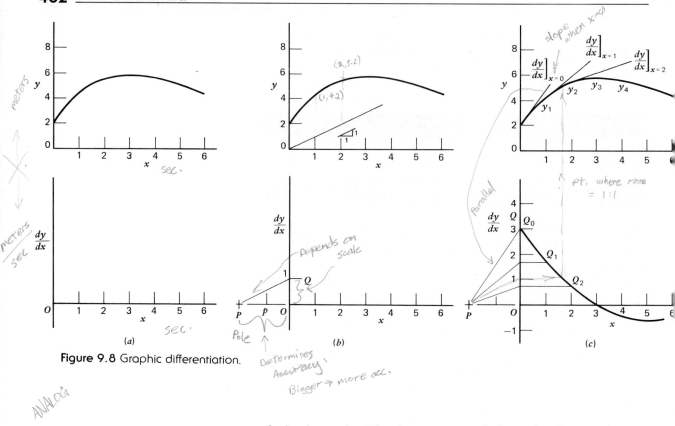

Figure 9.8 Graphic differentiation.

derivative axis. The importance of the pole distance is now apparent. A larger pole distance will expand the ordinate of the derivative. A shorter pole distance will contract it.

The third and last step (Figure 9.8c) is to draw the tangent to curve $y(x)$ at selected points. Lines parallel to each of these tangents are drawn through pole P to the ordinate of the derivative curve. The intercept of these lines is the direct measure of derivative dy/dx.

Consider the tangent to the curve at $x = 0$. Line PQ is parallel to that tangent. From geometric similarity,

$$\frac{\Delta y}{\Delta x}\bigg]_{x=0} = \frac{\overline{OQ}}{\overline{OP}}$$

and

$$\overline{OQ} = \frac{\Delta y}{\Delta x}\bigg]_{x=0} \cdot \overline{OP} = \frac{\Delta y}{\Delta x}\bigg]_{x=0} \cdot p \tag{9.15}$$

Thus, linear distance OQ is the value of the fraction $\Delta y/\Delta x$ multiplied by pole distance p.

Point Q has been located on the y'-axis. Call it Q_0, meaning slope y' at $x = 0$. For tangents to the curve at other points such as $x = 1$, $x = 2$, etc., other lines are drawn through P, each parallel to a respective tangent. For these points, the new Q is carried horizontally to Q_1, Q_2, Q_3, etc., and plotted. Thus point Q_1 is the intersection of the horizontal line $y' = Q_1$ and the vertical line $x = 1$. Point Q_2 is the intersection of the horizontal line $y' = Q_2$ and the vertical line $x = 2$. A smooth, continuous curve is now drawn through these points. This is the curve of the derivative $y' = dy/dx$.

9.7 NUMERICAL DIFFERENTIATION

Many times, instead of finding the best curve that would fit given data, and then differentiating graphically, it is possible to use the data numerically and obtain the derivative directly. This is particularly true if the data are calculated and fall on a regular curve but the curve itself is not known analytically.

Figure 9.9 shows a regular curve for which points y_0, y_1, y_2, y_3, . . . , are known. The incremental change in y at point y_1 would be

$$y_2 - y_0 = \Delta y \qquad {\scriptstyle x_2 - x_0 = 2h}$$

and for an incremental change $\Delta x = h$

$$\frac{\Delta y}{\Delta x} = \frac{y_2 - y_0}{2h} \quad \longleftarrow \quad {\scriptstyle ;\ y_2 - y_0}$$

If the interval $\Delta x = h$ is small,

$$\left.\frac{dy}{dx}\right]_{x=n} \approx \left.\frac{\Delta y}{\Delta x}\right]_{x=n} = \frac{y_{n+1} - y_{n-1}}{2h} \qquad (9.16)$$

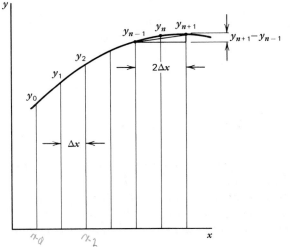

Figure 9.9

Note that we have had to take the value of the ordinate y_{n+1} immediately ahead of y_n, and y_{n-1} immediately behind y_n, in order to find the average slope at the point we seek. For this reason, we had to use $2\Delta x = 2h$. We could have used

$$\left.\frac{\Delta y}{\Delta x}\right]_{x=n+1/2} = \frac{y_{n+1} - y_n}{h}$$

or

$$\left.\frac{\Delta y}{\Delta x}\right]_{x=n-1/2} = \frac{y_n - y_{n-1}}{h}$$

but these find the value of the slope halfway in between two points and not any specific point. They are not convenient.

These difference calculations yield $\Delta y/\Delta x$ for all points except the first and last, y_0 and y_n, which can be found by extending a parabola through three given points. Extending y_2 and y_1 to y_0 backward,

$$\left.\frac{\Delta y}{\Delta x}\right]_0 = \frac{-3y_0 + 4y_1 - y_2}{2h} \tag{9.17}$$

Extending y_{n-2} and y_{n-1} to y_n forward,

$$\left.\frac{\Delta y}{\Delta x}\right]_n = \frac{y_{n-2} - 4y_{n-1} + 3y_n}{2h} \tag{9.18}$$

At times, these can be very useful.

SAMPLE PROBLEM 9.37

The following data give the value of y as a function of x. Plot $y(x)$ carefully and draw a smooth curve through these points. Graphically differentiate this curve and plot dy/dx. What is dy/dx at $x = 1$, $x = 2$, and $x = 5$?

x	y
0	5
1	14
2	22
3	27
4	29
5	28
6	25
7	22
8	19
9	17
10	16

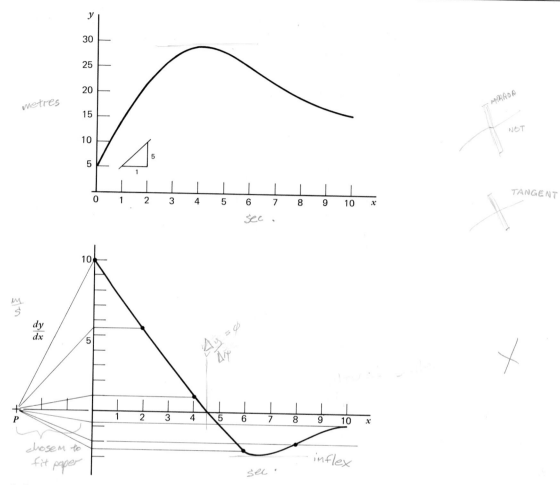

Solution

As in Figure 9.8, the graph of $y(x)$ is placed directly above the axes for the proposed graph of the derivative. Both sets of axes have the same abscissa.

A pole P is set three units to the left of the origin of the derivative curve. This is an arbitrary distance, but $p = 3$ produces a full scale on the derivative curve. A slope of 5:1 is now drawn on the $y(x)$ curve, and a line parallel to it is drawn on the dy/dx curve to establish the scale of the dy/dx curve. Where the line drawn from pole P parallel to the 5:1 slope intersects the dy/dx ordinate, mark $dy/dx = 5$. Now scale the balance of the dy/dx ordinate.

Tangents to the $y(x)$ curve are drawn at $x = 0, 2, 4, 6, 8, 10$. Parallel lines drawn from P intersect the dy/dx ordinate. As examples, at $x = 0$, $dy/dx = 10$; at $x = 2$, $dy/dx = 5.5$; at $y = 4$, $dy/dx = 1$; at $t = 6$, $dy/dx = -2.5$. Note that this is negative and is also near a point of inflection.

Fairing the curve, we can read that $dy/dx = 7.8$ at $x = 1$, and $dy/dx = -0.8$ at $x = 5$.

SAMPLE PROBLEM 9.38

In the position shown (a) the spring is unextended. It is known to have a modulus of 1000 N/m, which means that for every millimeter it stretches, the spring force increases by 1 N. A mass of 100 kg is now hung by a small hook in the middle of the string that leads from the spring around pulley B to A (b). The distance AB is 3 m. The string now falls under the weight of the mass stretching the spring (c). Let x be the stretch in the spring and y be the vertical displacement of the 100-kg mass. The total energy V in the system is

$$V = \frac{1}{2}(1000)x^2 - 980.5y$$

Note that the 100-kg mass would weigh 980.5 N. At static equilibrium, $dV/dx = dV/dy = 0$. Determine what x and y would be at static equilibrium.

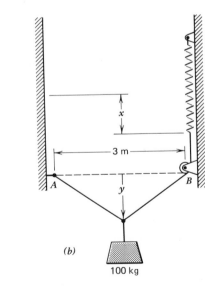

(a)

(b)

100 kg

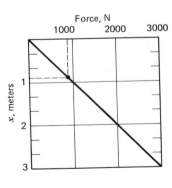

(c)

Solution

This problem can be solved either analytically or graphically. x and y are related to each other through the geometric constraint

$$x = 2\sqrt{(1.5)^2 + y^2} - 3$$

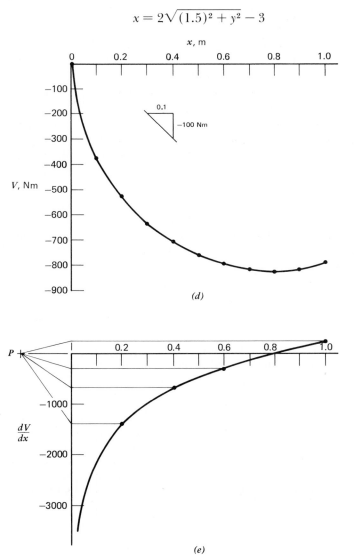

(d)

(e)

Analytically, this would mean that

$$V = \frac{1}{2} 1000[2\sqrt{(1.5)^2 + y^2} - 3]^2 - (980.5)y$$

Differentiating this expression would lead to the minimum value for V. The problem is that it will lead through some cumbersome calculus. Why not try a graphic solution?

The following table shows x, y, and V as x varies from 0 to 1 m.

x	y	V
0	0	0
0.1	0.391	−377.9
0.2	0.556	−525.9
0.3	0.687	−629.0
0.4	0.800	−704.4
0.5	0.901	−758.8
0.6	0.995	−795.6
0.7	1.083	−816.7
0.8	1.166	−823.4
0.9	1.246	−816.7
1.0	1.323	−797.1

Obviously, there is a minimum somewhere around $x = 0.8$ m (d).

A pole P is set two units to the left of the origin of the derivative curve (e). A slope of -1000 N·m/m is a 45-deg line down and to the right on the V–x curve. This means that if a 45-deg line is drawn downward from pole P, that line will intercept the ordinate of the dV/dx scale at -1000 N·m/m. Each unit on the dV/dx scale will be -500 N·m/m. This could have also been found by dividing -1000 N·m/m by the pole distance $p = 2$.

Tangents to the $V(x)$ curve are drawn at $x = 0.1, 0.2, \ldots$, through $x = 1$ m. For each tangent, parallel lines are drawn from P intersecting the dV/dx ordinate. dV/dx is then recorded for each x. Fairing a curve through these points shows that dV/dx is an increasing function and crosses the x-axis a little before $x = 0.8$. Actually, taking a smaller interval would show the minimum to be at $x = 0.799$ m, and $y = 1.165$ m.

SAMPLE PROBLEM 9.39

Using the data given in Sample Problem 9.38, determine dV/dx numerically. Check these results with the results obtained graphically.

Solution

In preference to obtaining the whole curve with a wide spacing, numerically let us examine the area from $x = 0.6$ m to $x = 1.0$ m, at intervals of $\Delta x = h = 0.01$ m. The following table presents the calculations of dV/dx, where

$$\frac{\Delta V}{\Delta x} = \frac{V_{n+1} - V_{n-1}}{2h}$$

x, m	y, m	V, N·m	$\dfrac{V_{n+1} - V_{n-1}}{2(0.01)}$, N·m/m
0.60	0.995	−795.6	—
0.61	1.004	−798.4	−271.4
0.62	1.013	−801.0	−256.0
0.63	1.022	−803.5	−240.8
0.64	1.031	−805.8	−225.7
0.65	1.040	−808.0	−210.7
0.66	1.048	−810.0	−195.9
0.67	1.057	−811.9	−181.1
0.68	1.066	−813.7	−166.5
0.69	1.074	−815.3	−152.0
0.70	1.083	−816.7	−137.6
0.71	1.091	−818.0	−123.3
0.72	1.100	−819.2	−109.1
0.73	1.108	−820.2	−95.0
0.74	1.117	−821.1	−81.0
0.75	1.125	−821.8	−67.1
0.76	1.133	−822.4	−53.3
0.77	1.142	−822.9	−39.5
0.78	1.150	−823.2	−25.9
0.79	1.158	−823.4	−12.3
0.80	1.166	−823.4	1.3
0.81	1.174	−823.4	14.7
0.82	1.182	−823.2	28.1
0.83	1.190	−822.8	41.4
0.84	1.198	−822.3	54.6
0.85	1.206	−821.7	67.8
0.86	1.214	−821.0	80.9
0.87	1.222	−820.1	93.9
0.88	1.230	−819.1	106.9
0.89	1.238	−818.0	119.9
0.90	1.246	−816.7	132.7
0.91	1.254	−815.3	145.6
0.92	1.262	−813.8	158.3
0.93	1.269	−812.1	171.1
0.94	1.277	−810.4	183.7
0.95	1.285	−808.5	196.4
0.96	1.292	−806.4	208.9
0.97	1.300	−804.3	221.5
0.98	1.308	−802.0	234.0
0.99	1.315	−799.6	246.4
1.00	1.323	−797.1	—

These calculations were done on a hand calculator. Four significant figures for V are needed to determine the slope $\Delta V/\Delta x$ accurately. From these numerical data, $dV/dx = 0$ at $x = 0.799$ m.

PROBLEMS

9.40 Show by graphic differentiation that

$$\frac{d}{dx}(e^x) = e^x$$

9.41 Show by graphic differentiation that

$$\frac{d}{dx}(\ln x) = \frac{1}{x}$$

9.42 Show by graphically differentiating $y = \arctan x$ that the derivative dy/dx is identical to $f(x) = 1/1 + x^2$).

9.43 Plot the following data with y as the ordinate and x as the abscissa. Draw a fair curve through the given points and differentiate this curve.

x	y
−2	−2.0
−1	0
0	−1.0
1	−1.5
2	1.0
3	0
4	1.6
5	2.8
6	4.0
7	4.8
8	5.5

9.44 Physically, the incremental rate for a power plant is the amount of additional input energy required to produce an added unit of output. Mathematically, this is the slope of the input-output curve $\frac{dI}{dL}$. The data in the following table are for a 20 MW station. I is in millions of Btu/h and L is in megawatts. Plot I as a function of L. Graphically differentiate this curve to determine dI/dL. What is the incremental energy rate $\frac{dI}{dL}$ at 8 MW of output?

Answer: 12.82×10^6 Btu/MWh

Output L, MW	Input I, 10⁶ Btu/h
0	20.0
1	21.2
2	23.7
3	27.6
4	33.0
5	40.0
6	48.6
7	58.8
8	70.7
9	85.4
10	100.0
11	117.5
12	136.7
13	158.3
14	181.8
15	207.5
16	235.4
17	265.5
18	297.9
19	332.7
20	370.0

9.45 In a reciprocating engine, a piston moves vertically in a cylinder, driving its crankshaft by means of a connecting rod. Taking x as the displacement from top dead center (the highest position of piston), plot x as a function of the angular position θ of the crank. Do this graphically by laying out the crank and slider mechanism to scale. Now, graphically differentiate x as a function of θ. At what position of θ is $dx/d\theta$ a maximum?

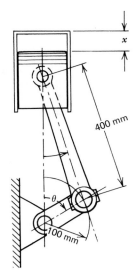

IMAGRUNT

9.8 THE INTEGRAL

In mathematics, there are many examples of mutually reversible operations such as addition and subtraction, multiplication and division, powers and roots, and logarithms and antilogarithms. For each example, one is defined as being the inverse of the other. Consider the following examples,

$$y = x + a \qquad x = y - a$$

$$y = \frac{x}{a} \qquad x = ya$$

$$y = e^x \qquad x = \ln y$$

$$y = \sin x \qquad x = \sin^{-1} y$$

Using this description, integration is the inverse of differentiation and is sometimes referred to as antidifferentiation.

From differential calculus, we have the derivative $dy/dx = y'$, or in differential form $dy = y' \, dx$. The problem in integral calculus is to invert this operation. Given the differential of a function, find the function itself. The function y is called the integral of the differential expression $y' \, dx$, and the operation is indicated by the integral sign $\int$, a long distorted script s.

$$y = \int y' \, dx \qquad (9.19)$$

In general, integration is a more difficult operation than differentiation. As a matter of fact, the integral of so simple a function as $y = \sqrt{x} \sin x$ cannot be found.

9.9 GRAPHIC INTEGRATION

Using the concept of integration as antidifferentiation, let us start with a given derivative $y'(x)$, from which the function $y(x)$ is required.

In Figure 9.10a the derivative y' is known. The ordinate of each x-coordinate on the given curve is the slope y' of a function y at that coordinate. A large value of y' means that $y(x)$ would have a steep slope, and a small value means a shallow slope. Negative values of y' indicate a negative slope. Again, convenient scales should be used for both the x- and y'-coordinates.

The integral curve is to be the graph of y expressed as a function of x. It is placed directly above the known curve of the derivative y', which is the inverse of the operation used to determine the derivative graphically. The abscissas are again identical. The first step in determining the function $y = \int y' \, dx$ is to draw the coordinate axes for each and to draw the curve of the derivative accurately.

The second step is to locate an arbitrary pole P, placing it at a pole distance p to the left of the origin of the known curve. Now locate point Q on the y'-axis. Distance $\overline{OQ}$ is y', the slope of the function y at $x = 0$

$$\overline{OQ} = y']_{x=0}$$

Our objective here is to reconstruct the integral curve for which the known curve is the derivative. Connecting point P to point Q constructs the initial slope. Using triangle OPQ.

$$\frac{\overline{OQ}}{\overline{OP}} = \frac{1}{p} \cdot y' \Big]_{x=0} = \frac{1}{p} \cdot \frac{\Delta y}{\Delta x} \Big]_{x=0} \qquad (9.20)$$

The difficulty is that a number of lines in Figure 9.10b would be equally valid as the slope of the function y, for an infinite number of curves could have the slope $y' = \Delta y/\Delta x]_{x=0}$.

find: units, starting pt.

abscissa the same

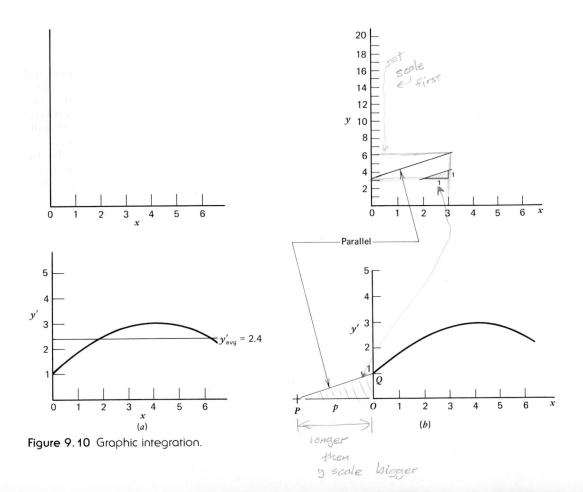

Figure 9.10 Graphic integration.

set scale first

Parallel

longer then y scale bigger

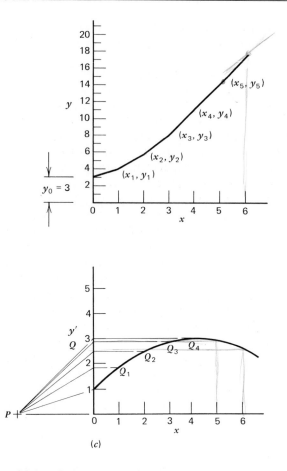

Figure 9.10 (cont.)

(c)

Which of these functions would be the required function still must be determined. To determine which is the correct curve, we must first have a starting condition. The proper curve can be determined quite easily, if only one point on the integral curve is known, either graphically or mathematically. Usually the correct curve is determined by a boundary condition.

The location of pole P also establishes the scale of the integral curve. The units for measurement of pole distance p will be the same as the units for the x scale. For $y' = 1$, line PQ is parallel to a 1:1 slope on the integral curve. This is the change in the variable y for a corresponding change in the variable x. For $y' = 2$, line PQ is parallel to the 2:1 slope on the integral curve; for $y' = 5$, line PQ is parallel to the 5:1 slope; etc. For a pole distance $p = 1$ unit, the x and y scales would be the same. In Figure 9.10b pole P is placed 3 units to the left of the origin. This means that the unit distance on the y scale will be reduced to one-third the unit distance of x scale.

Starting with the point x_0, y_0, the approximate point x_1, y_1 can then be found easily.

$$\frac{y'}{p} = \frac{\Delta y}{\Delta x} = \frac{y_1 - y_0}{x_1 - x_0}$$

(9.21)

$$y_1 = \frac{y'}{p} \Delta x + y_0$$

This is a straight line with an intercept at $x = x_0$, $y = y_0$ and a slope of y'/p, or dy/dx divided by pole distance p. Note that for correct units y' would have to be divided by p. Of course, in order to find y_1, the incremental change $\Delta x = x_1 - x_0$ must be known or arbitrarily taken. We will go into this in more detail.

Assume that $\Delta x = 1$ unit. In Figure 9.10c the initial value of y is $y = 3$ at $x = 0$. On the y scale this is 1 graphic unit. Remember, the unit distance has been reduced by a factor of 3. Graphically, with $\Delta x = 1$ unit, y_1 would be $1\frac{1}{3}$ graphic units or $y_1 = 4$ for its true value.

With y_1 at x_1 determined, the slope of the integral curve at x_1, y_1 can now be found. Drawing a horizontal line QQ_1 to the vertical axis from point $y'(x_1)$, the new slope at x_1 can be found. It will be the new line PQ. Drawing a line parallel to this new slope from point x_1, point x_2 and y_2 can be established, as in Figure 9.10c. In turn, other points on the integral curve can be established. For each point y_n, the slope of the integral curve y'_n is used to determine the next point y_{n+1} on the integral curve. This is done in succession until the integral curve $y = \int y' \, dx$ is completed. The method is known as chordal approximation. The integral curve is approximated by a series of chords. Obviously, the accuracy of this method depends on the number of points used and how close they are selected to one another.

There are two important observations. First, pole distance p determines the graphic scale of the integral curve. Increasing the pole distance will make the slope of the integral curve shallow. Decreasing the pole distance will make the slope steep. Second, the initial value of y_0 is arbitrary and must be established from external conditions. The value of y at $x = 0$ is in fact the constant of integration.

9.10 GEOMETRIC INTERPRETATION OF THE INTEGRAL

Historically, integration has been viewed as a summation process that is the original reason for the script s indicating a sum. In fact, the integral was invented as a means of determining the area bounded by curves, and this still remains an important use for the integral.

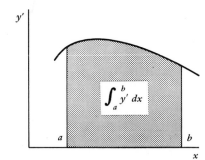

Figure 9.11

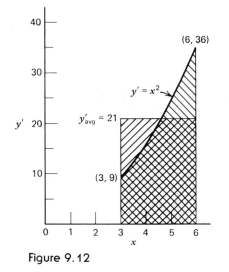

Figure 9.12

CAN'T WIN 'EM ALL;

SOMETIMES CAN'T WIN ANY

The total change in the variable $y(x)$ from $x = a$ to $x = b$ is the area under the curve $y'(x)$ from a to b in Figure 9.11.

$$y_b - y_a = \int_a^b y' \, dx = \text{area under } y'(x) \text{ curve} \qquad (9.22)$$

This form of the integral is known as a *definite integral*. It is a number, and it is more proper to say that the number represents the area under the $y'(x)$ curve. Definite integrals can be determined graphically if the graph of the function $y'(x)$ is known. Definite integrals can represent many things—for example, volume, surface area, and mass.

In Figure 9.12, the function $y' = x^2$ is known from $x = 3$ to $x = 6$. At $x = 3$ $y' = 9$, and at $x = 6$ $y' = 36$. The area under this curve is the value of definite integral.

$$y_6 - y_3 = \int_3^6 y' \, dx = \int_3^6 x^2 \, dx = \left(\frac{x^3}{3}\right)_3^6 = 72 - 9 = 63$$

This simple integration is easily verified.

The approximate value of the definite integral can be put to good use to estimate the limit of the integral curve. In Figure 9.10a the average value of the ordinate y' from $x = 0$ to $x = 6$ is $y' = 2.4$. This is not an exact value, but any approximate value also would be useful to establish the scale for the ordinate y of the integral curve. For example, what should be the limit of the ordinate? If the average were $y' = 2.4$, the change in y from $x = 0$ to $x = 6$ would be $(6)(2.4) = 14.4$. The maximum value for the ordinate y would then be $y_0 + 14.4 = 17.4$. For good measure, you could pick a slightly larger value such as 20.

9.11 NUMERICAL INTEGRATION

When the function $y'(x)$ is not known analytically and therefore cannot be integrated analytically, or when graphic integration would be tedious, the approximate definite integral can always be obtained numerically if the function $y'(x)$ can be expressed as tabular data or is known in graphic form. Three numerical methods are widely used.

The first is the rectangular rule in which each successive element is a rectangular area. In Figure 9.13a the graph of the function $y'(x)$ is known from $x = a$ to $x = b$. The segment from a to b is divided into a finite number of elements, each with a width Δx.

The successive ordinates for $y'(x)$ are

$$y'_0 = y'(a)$$
$$y'_1 = y'(a + \Delta x)$$
$$y'_2 = y'(a + 2\Delta x)$$
$$\vdots$$
$$y'_n = y'(a + n\Delta x)$$

The area of each element is simply the product of the width and height.

$$A_1 = y'_0 \cdot \Delta x$$
$$A_2 = y'_1 \cdot \Delta x$$
$$A_3 = y'_2 \cdot \Delta x$$
$$\vdots$$
$$A_n = y'_{n-1} \cdot \Delta x$$

The total area is the definite integral

$$\int_a^b y' \, dx = (y'_0 + y'_1 + y'_2 + \ldots + y'_{n-1}) \Delta x \qquad (9.23)$$

It is obvious that unless Δx is very small this method can be inaccurate. It is the simplest to use, however, and in the age of computers it is much easier to take a greater number of points, let a computer do the computational work, and regain any accuracy that the method would lose.

A second and more accurate scheme uses trapezoidal areas instead of rectangular areas (see Figure 9.13b). In this method, the successive ordinates are connected by straight lines or chords making each elemental area a trapezoid. Since the area of each elemental area is the product of the average ordinate and the width Δx, and assuming equal increments Δx,

$$A_1 = \frac{1}{2}(y'_0 + y'_1)\Delta x$$

$$A_2 = \frac{1}{2}(y'_1 + y'_2)\Delta x$$

$$A_3 = \frac{1}{2}(y'_2 + y'_3)\Delta x$$

$$\vdots$$

$$A_n = \frac{1}{2}(y'_{n-1} + y'_n)\Delta x$$

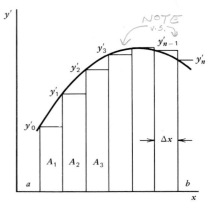

(a)

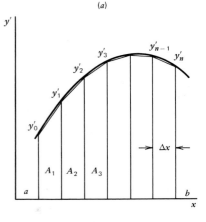

(b)

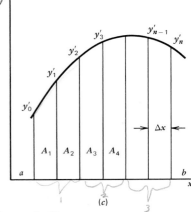

(c)

Figure 9.13

and the total area, and thus the definite integral

$$\int_b^a y' \, dx = \left(\frac{1}{2} y_0 + y_1 + y_2 + \ldots y_{n-1} + \frac{1}{2} y_n \right) \Delta x \qquad (9.24)$$

NOTE

A third scheme is to connect the ordinates of the curve by parabolic arcs instead of straight lines, summing the areas under these arcs. A parabola may be passed through any three points on a curve, and the series of parabolic arcs will fit the given curve more closely than a series of chords, as shown in Figure 9.13c.

Since three points determine a parabola, the function $y'(x)$ is divided into pairs of parabolic strips. The center ordinate is considered the vertical axis of the parabola. Thus

$$A_1 + A_2 = \frac{1}{3}(y_0 + 4y_1 + y_2)$$

$$A_3 + A_4 = \frac{1}{3}(y_2 + 4y_3 + y_4)$$

$$A_5 + A_6 = \frac{1}{3}(y_4 + 4y_5 + y_6)$$

$$\vdots$$

$$A_{n-1} + A_n = \frac{1}{3}(y_{n-2} + 4y_{n-1} + y_n)$$

A check of a text on analytic geometry will verify the formula for the area under a parabolic curve.

The definite integral is

$$\int_a^b y' \, dx = \frac{1}{3}(y_0 + 4y_1 + 2y_2 + 4y_3 + 2y_4 \ldots$$
$$2y_{n-2} + 4y_{n-1} + y_n)\Delta x \qquad (9.25)$$

This is classically known as Simpson's rule. The only qualification for its application is that there must be an even number of strips.

SAMPLE PROBLEM 9.46

Graphically determine the half cross-sectional area of the irregular tank. Draw calibration lines to indicate $\frac{1}{8}$, $\frac{1}{4}$, $\frac{1}{2}$, and $\frac{3}{4}$ full levels. Mark these on the tank centerline.

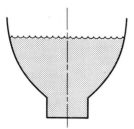

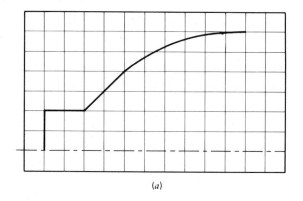

(a)

Solution

The cross section of the tank is irregular, but it is also symmetrical. Only the left half cross section of the tank is shown (a). The right half cross section is the mirror image of the left. Our procedure will be to integrate this half section and double its area to find the total cross-sectional area.

As a first step, the outline of the irregular tank is drawn carefully, to scale (b). The centerline is taken as the x-axis, with the origin at the left bottom of the tank. Integrating from left to right is equivalent to filling the tank from the bottom to the top. A pole distance $p = 5$ is chosen simply to allow the integral

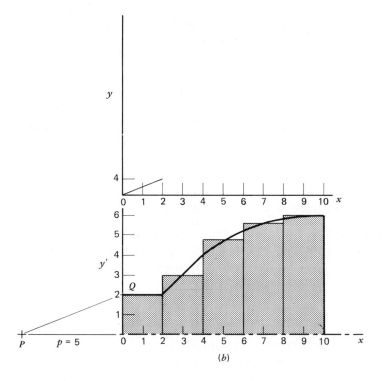

(b)

$\int y'\,dx$ to fit within our allowable space. The area is now blocked into five convenient increments. Note that in carrying a line parallel to line PQ to the upper drawing, the line will intercept the y-axis at four units, or four squares. Figure (c) shows the completed integration. The enclosed area is

$$A = \int_{0}^{10} y'\,dx = 43 \text{ squares}$$

You can easily check this by counting the squares in a rectangular grid, but other integrals are not found as easily.

The tank is calibrated by marking the $\frac{1}{8}$, $\frac{1}{4}$, $\frac{1}{2}$, and $\frac{3}{4}$ levels, or 5.4, 10.7, 21.5, and 32.3 units and transferring these marks to the tank centerline.

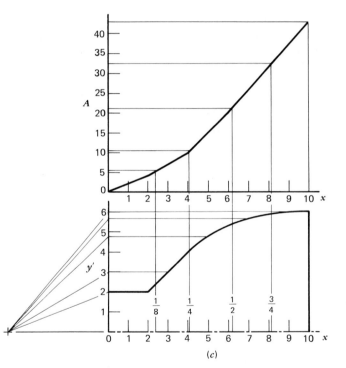

(c)

SAMPLE PROBLEM 9.41

A solid circular steel shaft 80 mm in diameter is simply supported between two bearings 4 m apart. The shaft supports a concentrated load of 8 kN, 1 m from the left bearing. The equation for bending in a beam is

$$\frac{d}{dx}\left(\frac{dy}{dx}\right) = \frac{d^2y}{dx^2} = -\frac{M}{EI}$$

where the deflection y is positive downward. Using graphic means, integrate the equation twice and determine the maximum deflection of the beam.

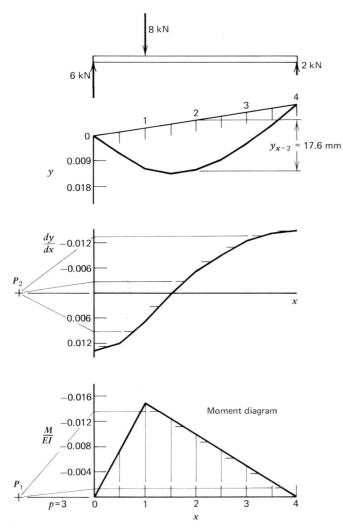

Solution

The modulus of elasticity E for steel is 205×10^9 N/m². The area moment of inertia I for a solid circular shaft 80 mm in diameter is 2.01×10^{-6} m⁴. These numbers are needed for the solution of the problem.

If we integrate this equation once, we will determine dy/dx or the slope of the beam. Integrating a second time will determine the deflection y.

The bending moment on the beam can be calculated from simple statics. For an 8-kN load applied at 1 m from the left end, the load at that end will be 6 kN, and the load at the other end will be 2 kN. The maximum bending moment will occur at the point where the load is applied; it will be 6 kN·m.

For the bending-moment diagram, which is the beginning of our graphic integration, the ordinate is M/EI, not M because EI, the flexural rigidity, is a constant.

The axes for the dy/dx curve are placed directly above the M/EI curve. A unit is selected as 0.5 m horizontally on the beam. A pole P_1 is set three units to the left of the origin. For the midpoint of each element on the abscissa, the bending moment can now be read graphically. At the point of maximum bending moment, $x = 1$ m and $M/EI = -0.01456$ m^{-1}. A scale of 0.004 m^{-1}/unit is used for the ordinate. Note that M/EI is a negative number (positive is downward).

The value of M/EI at the midpoint of each element is now carried horizontally to intercept the ordinate. Lines are drawn from pole P_1 to the various intercepts. Graphically, these represent the slope, or the derivative dy/dx, at each midpoint. Now, beginning at a convenient point where $x = 0$, directly above the M/EI curve, draw a line parallel to slope dy/dx of the point at the midsection of the first element. Here is the beginning of the dy/dx curve. At $x = 0.5$ m continue with another line parallel to slope dy/dx of the point at the midsection of the second element. Continue until the entire curve of dy/dx is drawn.

The problem now is what is a boundary condition? The boundary conditions are that $y = 0$, at $x = 0$ and $x = 4$ m, and these conditions are for deflection, not the slope.

Judgment is now introduced; let us pick the point of maximum deflection. Arbitrarily, let us take $x = 1.5$ m, halfway between the middle of the beam at $x = 2$ m, and $x = 1$ m, where the 8-kN load is applied. The actual point of maximum deflection will be near the center. A horizontal line is now drawn through the selected point. This will be the abscissa of the dy/dx versus x curve.

The scale for the dy/dx curve is now set by drawing a line from the origin where $x = 0$ and $dy/dx = 0$, parallel to a line from P_1 to the intercept $M/EI = 0.004$ m^{-1}. This line represents a slope $dy/dx = -0.006$ m/m, which is equivalent to -0.004 m^{-1}, multiplied by the pole distance $p = 3$ units, multiplied by the graphic scale of the abscissa, 0.5 m/unit. On the dy/dx ordinate, scale each graphic unit as 0.006 m/m.

The axes for the $y(x)$ curve are directly above the dy/dx curve. For simplicity, pole P_2 is also selected three units to the left of the origin, although this pole distance need not be identical to the pole distance of the previous curve. The value of dy/dx at the midpoint of each element is now carried horizontally to intercept the dy/dx ordinate. Lines are drawn from P_2 to these intercepts, and parallel lines are drawn from a point at $x = 0$ to develop the deflection $y(x)$.

The scale of the deflection curve is found as the scale for dy/dx was found. Each unit distance on the dy/dx scale is 0.006 m/m. Multiply this number by $p = 3$ units and by the graphic scale of the abscissa, 0.5 m/unit, to give a scale of 0.009 m/unit for the deflection. Note that the deflection is positive and downward, as expected. At $x = 2$ m, y scales to be 17.6 mm.

One further remark is necessary. The $y(x)$ curve is skewed because we did not locate the dy/dx baseline correctly. This is unimportant, although aesthetically we would like the abscissa to be horizontal. As long as we measure $y(x)$ vertically, there is no problem.

PROBLEMS

9.48 Show by graphic integration that

$$\int_1^5 \frac{dx}{x} = \ln x \Big]_1^5$$

9.49 Show by graphic integration that

$$\int_{0.2}^2 e^x \, dx = e^x \Big]_{0.2}^2$$

9.50 Determine the value of the following definite integral.

$$\int_0^\pi e^{-x^2} \cdot dx$$

Answer: 0.886

9.51 Determine the value of the following definite integral.

$$\int_0^{2\pi} \frac{\sin x}{x} \cdot dx$$

Answer: 1.417

9.52 Plot the given function and integrate graphically.

$$y = \frac{x^2}{2} + \sin x, \qquad 0 \le x \le \pi$$

Answer: 7.181   7.16 77132

9.53 Plot the given function and integrate graphically.

$$y = \frac{1}{x} - \log x, \qquad 1 \le x \le 4$$

9.54 Determine the average value for a sine wave over one half cycle.

Answer: 0.637 maximum

9.55 The charge on a capacitor is expressed as $q = \int_0^t i\, dt$, where q is the charge in coulombs and i is the current in amperes. Determine the charge after 10 μsec, either graphically or numerically.

i, amp	t, μsec
0	0
0.12	1
0.48	2
1.03	3
1.73	4
2.50	5
3.28	6
3.97	7
4.52	8
4.88	9
5.00	10

Answer: 25 coulombs

9.56 The following data are taken from a cross-sectional survey for a highway. Determine the elevation where the area cut will equal area filled, either graphically or numerically.

Ground Elevation, m	Horizontal Distance, m
30.1	0
28.1	5
25.9	10
23.7	15
17.0	20
13.1	25
7.2	30
5.6	35
8.1	40
18.3	45
21.4	50

Answer: 17.44 m

9.57 The graph shows a power demand in northern California during the month of April. Using graphic integration, determine the total power demand in megawatt-hours.

Answer: 21850 MWh

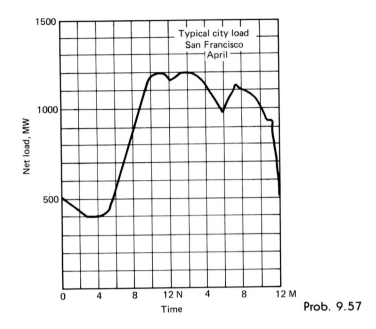

Prob. 9.57

9.58 The graph shows the same demand as Problem 9.57, here as a function of the duration of any given power demand instead of chronological time. Graphically integrate this curve, using power demand as the independent variable. The result will be the cumulative energy needed at any power demand. (*Hint.* Integrate with respect to the ordinate instead of the abscissa.)

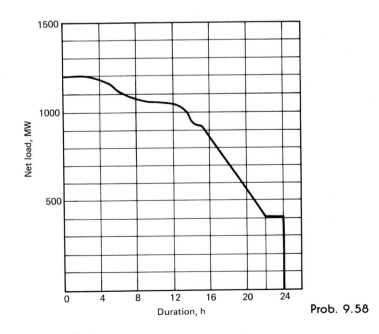

Prob. 9.58

9.59 The projected use of thermal energy from petroleum is shown. Graphically integrate this curve, which is an estimate of the total world supply of oil.

Answer: 1.1×10^{19} Btu

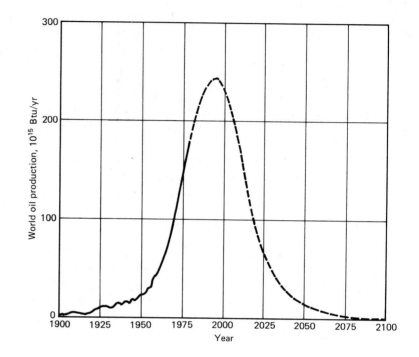

9.60 Determine graphically the deflection of the shaft in Sample Problem 9.47 if the load were placed in the center of the shaft.

Answer: 26 mm

9.61 Two steel I-beams with a total flexural rigidity of 18×10^6 N · m² rest on two bearing plates 8 m apart. They support a motor and a compressor; the motor has a 1000-kg mass, and the compressor has a 2000-kg mass. They can be considered as two concentrated loads, the motor at $x = 2$ m and the compressor at $x = 6$ m. Using the techniques of Sample Problem 9.47, determine the maximum deflection of the beams.

9.12 MOTION PROBLEMS

Kinematics is the study of motion without reference to force or mass. It is primarily concerned with the interrelation of displacement, velocity, acceleration, and time. It is one subject that is ideally studied through differential and integral calculus and is well suited to geometric interpretation. *Kinetics* is the study of the forces causing motion. Newton's laws of motion formulate the relation between force, mass, and absolute acceleration, but a thorough understanding of kinematics is necessary before any study of kinetics can be undertaken because the ability to describe motion must be gained before the forces causing motion can be included.

The position of a point P at any time t is described by the vector displacement **s** from any convenient reference point. If the particle moves a distance Δ**s** in an increment of time Δt, it is described as moving with a certain velocity. The instantaneous velocity **v** is the time rate of change of the displacement **s**.

$$\mathbf{v} = \lim_{\Delta t \to 0} \frac{\Delta \mathbf{s}}{\Delta t} = \frac{d\mathbf{s}}{dt} = \dot{\mathbf{s}} \qquad (9.26)$$

The dot over the variable is an accepted convention for indicating the first derivative with respect to time. Two dots over the variable would indicate the second derivative with respect to time.

Conversely,

$$d\mathbf{s} = \mathbf{v} \cdot dt \qquad (9.27)$$

The instantaneous acceleration is the time rate of change of velocity.

$$\mathbf{a} = \lim_{\Delta t \to 0} \frac{\Delta \mathbf{v}}{\Delta t} = \frac{d\mathbf{v}}{dt} = \dot{\mathbf{v}} = \ddot{\mathbf{s}} \qquad (9.28)$$

and

$$d\mathbf{v} = \mathbf{a} \cdot dt \qquad (9.29)$$

The definitions of velocity and acceleration involve the four variables displacement, velocity, acceleration, and time. The definitions of velocity and acceleration all involve differential equations of the type outlined in Sections 9.5 and 9.8. In these cases, however, the independent variable is time t, but the graphic methods used to determine the derivative and the integral are valid, whether the independent variable is t or x. Scalar displacement, velocity, and acceleration can be used if these are a function of a single coordinate. Scalar notations

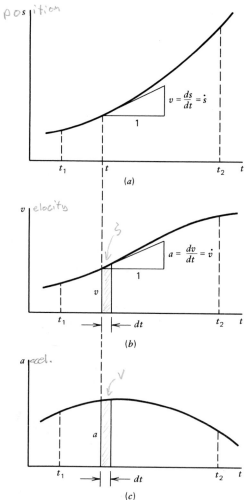

position

$$v = \frac{ds}{dt} = \dot{s}$$

t_1 t t_2 t

(a)

velocity

$$a = \frac{dv}{dt} = \dot{v}$$

t_1 dt t_2 t

(b)

accel.

t_1 dt t_2 t

(c)

Figure 9.14 Kinematic relationships.

$$\frac{ds}{v} = dt = \frac{dv}{a}$$

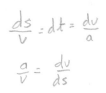

$$\frac{a}{v} = \frac{dv}{ds}$$

$$a \cdot ds = v \cdot dv$$

will be used from this point on because we are dealing with only one independent variable, but it is important not to lose sight of the fact that displacement, velocity, and acceleration are vectors.

An understanding of the graphic expression of kinematics is both useful and convenient. If the velocity is expressed as a function of time, the slope of the curve is the instantaneous acceleration, and the area under the velocity-time curve during an interval t_1 to t_2 is the change in displacement. These simple relations are universally understood. Less familiar are the expressions of acceleration. The area under the acceleration-time curve during the interval t_1 to t_2 is the velocity change between these limits, while the area under the acceleration-displacement curve occurring during a displacement from s_1 to s_2 represents half the difference in the squares of the velocity at t_1 and t_2.

In Figure 9.14a the variation of displacement s as a function of time is shown over the time interval from t_1 to t_2. The slope of the displacement-time curve at any time t is the velocity $v = ds/dt$. Figure 9.14b is the variation of velocity as a function of time for the same period. The slope of the velocity-time curve at any time t is the acceleration $a = dv/dt$. Figure 9.14c is the variation of acceleration as a function of time.

Reversing our thoughts, the area under the velocity-time curve during the increment of time dt is the incremental change in displacement ds. Over a longer interval from t_1 to t_2, the displacement is the total area under the velocity-time curve from t_1 to t_2, or the definite integral.

$$\int_{s_1}^{s_2} ds = s_2 - s_1 = \int_{t_1}^{t_2} v \cdot dt = \text{area under } v\text{–}t \text{ curve} \quad (9.30)$$

Similarly, the area under the acceleration-time curve during an increment of time dt is the incremental change in velocity dv. Over a time interval from t_1 to t_2, the velocity change would be total area under the acceleration-time curve from t_1 to t_2, or the definite integral.

$$\int_{v_1}^{v_2} dv = v_2 - v_1 = \int_{t_1}^{t_2} a \cdot dt = \text{area under } a\text{–}t \text{ curve} \quad (9.31)$$

Two other graphic relations are to be noted. From the definitions

$$v = \frac{ds}{dt}, \qquad a = \frac{dv}{dt}$$

combining and elimination of dt gives

$$v \, dv = a \, ds \quad (9.32)$$

In Figure 9.15 acceleration and velocity are plotted as functions of displacement. If velocity and acceleration were to be expressed as functions of displacement, the velocity change over the displacement interval from s_1 to s_2 would be the definite integral.

$$\int_{v_1}^{v_2} v \cdot dv = \frac{v_2^2}{2} - \frac{v_1^2}{2} = \int_{s_1}^{s_2} a \cdot ds = \text{area under } a\text{--}s \text{ curve} \quad (9.33)$$

It is not often that displacement is the independent variable instead of time, but in expressions involving changes in kinetic and potential energy, this kinematic relation is useful.

In Figure 9.15b the acceleration can be found constructing the normal AM and the tangent to the curve at point A. The tangent is dv/ds, and the acceleration can be found by noting similar triangles

$$\frac{a}{v} = \frac{dv/ds}{1}$$

or

$$a = v \cdot \frac{dv}{ds} \quad (9.34)$$

If the velocity and displacement had the same numerical scale, using similar triangles, the acceleration would be the subnormal. For example, if the velocity scale were 1 unit = 1 m/sec and the displacement scale were 1 unit = 1 m, the acceleration could be read as the length NM, directly in m/sec².

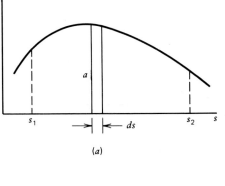

(a)

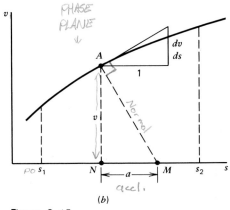

(b)

Figure 9.15

9.13 STATE SPACE AND THE PHASE PLANE

The graph of the integral is placed above the graph of the derivative for a very good reason. If the independent variable x, the dependent variable y, and the derivative y' are considered on three orthogonal axes, as in Figure 9.16, the integral will appear in the horizontal plane and the derivative will appear in the frontal plane. Continuing this analogy, logic leads us to conclude that a third view exists, the profile plane, which has coordinates y and y'. These coordinates have the independent variable x as a parameter.

Using three orthogonal planes to express two functions of a third is a very useful device, but y, y', and x is only one such set. Of course, there is an analogous set $\int y \cdot dx$, y, and x. In this case, $\int y \cdot dx$ would appear in the horizontal plane, and y versus $\int y \cdot dx$ in the profile plane. These are essentially the same, but it is worth the effort to consider that these two variations do exist.

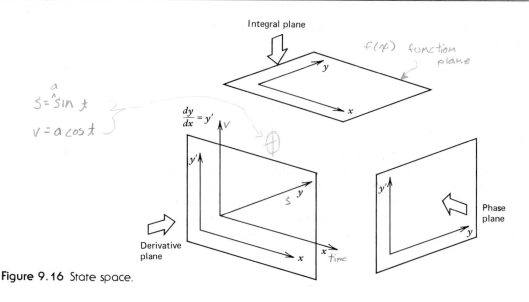

Figure 9.16 State space.

They are shown in Figure 9.17*a* and 9.17*b*. It is also appropriate to show that other behavioral combinations can be found. For example, in Figure 9.17*c* the second derivative y'', the first derivative y', and x are shown. There are others.

When a given function and its integral are known, or when a given function and its derivative are known, the three-variable combination can be visualized as a single three-dimensional space curve. Using y', y, and x, or the analogous set $\int y \cdot dx$, y, and x, results in a three-dimensional space interpretation of the three orthographic planes by the ordinary rules of projection geometry. The result is called *state space* or *phase space*. Much can be learned about the solution of differential equations by investigating the slopes and behavior of a state curve, particularly for first-order equations.

The key to the importance of a study of state space or phase space planes lies in the fundamental principle of orthographic projection: that only two orthographic views are needed to describe three-dimensional space completely. It is common to use the horizontal plane y versus x and the profile plane y' versus y.

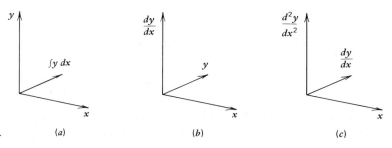

Figure 9.17 State space axes. (a) (b) (c)

In the profile plane, $y'(x)$ is expressed as a function of $y(x)$. Further,

$$\frac{dy'}{dy} = \frac{dy'}{dx} \cdot \frac{dx}{dy} = \frac{y''}{y'}$$ (9.35)

This states that in the profile plane, slope dy'/dy is a measure of the second derivative y''. Lines with the same slope are isoclines or isoclinics.

State or phase space takes its name from motion problems involving displacement y, velocity $\dot{y}$, and time t as the independent coordinates. In the profile plane, the velocity and displacement would be the coordinates, and they would be phases of each other. Each point in three-dimensional space represents one and only one point in time, one velocity, and one displacement. In phase space, the profile plane is known by a special name—the phase plane.

SAMPLE PROBLEM 9.62

Experimental observations of an earth-penetrating projectile yielded the following graphic data of deceleration measured as a function of time (a). The projectile entered the earth's surface at a velocity of Mach 1 and was recovered at a depth of 51 ft. Are the data valid?

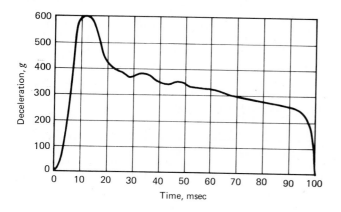

Solution

The initial condition is an initial velocity of Mach 1, at $s = 0$ and $t = 0$. Using the kinematic equations of motion

$$\int_{v_0}^{v} dv = \int_{0}^{0.1} a\, dt = v - v_0$$

The area under the deceleration-time curve in (b) represents the velocity change of the projectile. Integrating numerically, with 10-msec intervals, the velocity-time curve can be estimated. Each elemental area A_n will be the deceleration at time t, multiplied by 10 msec $A_n = a_n \, 0.01$.

Element, n	Time, msec	Deceleration, g	A_n, g-sec
1	0	0	0
2	10	600	6
3	20	420	4.2
4	30	380	3.8
5	40	360	3.6
6	50	340	3.4
7	60	320	3.2
8	70	300	3.0
9	80	280	2.8
10	90	260	2.6
			32.6

$$0 - v_0 = g \sum_0^{10} a \, \Delta t = 32.6(-32.2) = -1050 \text{ ft/sec}$$

$$v_0 = 1050 \text{ ft/sec}$$

This agrees with the statement of initial condition that v_0 is Mach 1. The total velocity change is 1050 ft/sec. The curve of v, where $v = v_0 - \int_0^t a \, dt$ can now be plotted. Note that this begins at $v = 1050$ and decreases to zero, which is consistent with integrating the velocity-time curve. The area under the velocity-time curve is the change in displacement of the projectile over the 100-msec interval.

In the following table, each elemental area A_n will be the velocity at time t multiplied by 10 msec.

Element	Time, msec	Velocity, ft/sec	A_n, ft
1	0	1050	10.50
2	10	1010	10.10
3	20	857	8.57
4	30	721	7.21
5	40	599	5.99
6	50	483	4.83
7	60	374	3.74
8	70	271	2.71
9	80	174	1.74
10	90	84	.84
			56.23

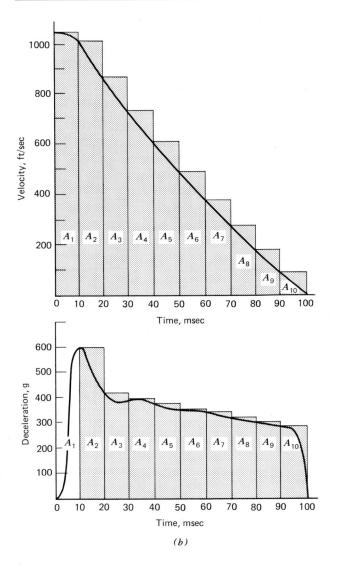

(b)

$$s - 0 = \sum_{0}^{10} v \, \Delta t = 56.2 \text{ ft}$$

The depth of 51 ft, at which the projectile was recovered, agrees quite well with observed data, and it would be reasonable to believe the data to be valid. Note that the calculated figure is 10 percent high, but we used the method of rectangles, and our approximation of the area of the $v(t)$ curve is about that much larger than the real curve. A more accurate method, or more intervals, would have yielded a more accurate result.

SAMPLE PROBLEM 9.63

Plot the function $y = a \sin x$ in all three orthographic planes described by the coordinates x, y, y'. The initial condition is at $x = 0$, $y = 0$, and $y' = a$.

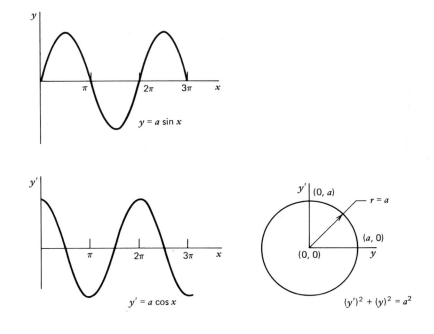

$y = a \sin x$

$y' = a \cos x$

$(y')^2 + (y)^2 = a^2$

Solution

First, plot $y = a \sin x$ in the horizontal plane. Second, obtain the derivative, $y' = a \cos x$, either analytically or graphically. Plot $y'(x)$ in the frontal plane.

In the profile plane, the abscissa $y = a \sin x$ and the ordinate $y' = a \cos x$. The following trigonometric equations will hold.

$$\cos^2 x + \sin^2 x = 1$$
$$\left(\frac{y'}{a}\right)^2 + \left(\frac{y}{a}\right)^2 = 1$$
$$(y')^2 + (y)^2 = a^2$$

The projection of each point of the state space curve will be a circle. The circle will have a radius a. The initial condition is at $x = 0$, $y = 0$, and $y' = a$. The state space curve is generated clockwise from that point.

Problems

9.64 A relation between velocity v in miles per hour and time t in minutes is shown in the following table. Plot curve $v(t)$ and integrate the curve to determine $s(t)$. Include a scale for $s(t)$. What is the acceleration at $t = 4$, $t = 12$, $t = 18$?

v, mi/hr	t, min
0	0
13	1
20	2
25	4
27	6
32	8
40	10
43	12
40	14
35	16
30	18
25	20

Answer: 1.1, 0, -2.5, mph/min

9.65 Differentiate the velocity-time relation of Problem 9.64 to determine acceleration as a function of time.

9.66 In the following table, displacement-time data are presented for a particle P moving in a straight line. Determine the velocity-time curve and the acceleration-time curve.

s, m	t, sec
4	0
2	1
2	2
5	3
10	4
17	5
23	6

9.67 A record of rectilinear acceleration of an experimental rocket sled, measured as a function of time by an accelerometer, is shown. Using this curve, draw the v–t curve for the 20-sec interval and determine the total distance traveled.

Answer: 16,550 ft

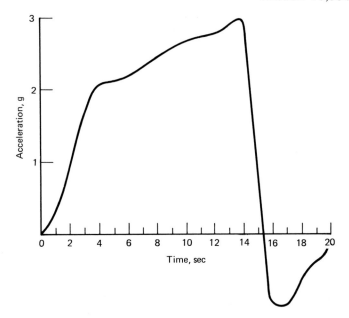

9.68 Experimental data for a particle moving in a straight line reveal the relation between velocity and displacement as shown. Determine the acceleration of the particle at a displacement of 2 m.

Answer: 3.9 m/s

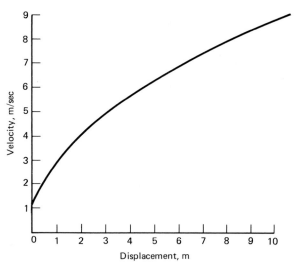

9.69 Measurements of the angular acceleration α of a flywheel are recorded for various angular displacements θ of the wheel and are plotted as shown. If the angular velocity was 6 rad/sec clockwise at $\theta = 5$ rad, estimate with the aid of the graph the angular velocity when $\theta = 10$ rad. The graph refers to clockwise motion.

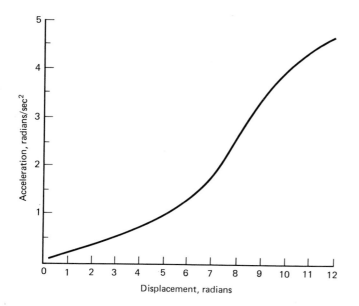

9.70 A particle moves along a straight line with constant acceleration. The displacement measured from a convenient position is $+4$ ft at $t = 0$ and $+4$ ft at $t = 4$ sec. Further, the velocity and displacement are both zero at $t = 2$ sec. What is the constant acceleration?

Answer: $a = 2$ ft/sec²

9.71 The motion of a body can be defined in terms of a single coordinate x and moves with a constant acceleration in the direction of that coordinate. At $t = 0$, the displacement is -6 ft from a convenient origin. At $t = 2$ sec, the displacement is zero. At $t = 4$ sec, the velocity is zero. What is the velocity at $t = 6$ sec?

9.72 A glass thermometer is immersed in a bath of boiling water. Beginning immediately after immersion and continuing for 3 min, the temperature is recorded every 5 sec. After 20 sec it is noted that the temperature on the thermometer has risen one half the total difference between the known ambient temperature and the temperature of the boiling water. The first-order equation

$$\theta = \theta_0 \left(1 - e^{-(t/b)}\right)$$

represents the temperature rise θ if the original difference in temperature was θ_0. Time t is in seconds; b is a time constant, which depends on thermal conductivity and other factors, also expressed in seconds. Plot the function $\theta(t)$ in all three orthographic planes using the coordinates θ, $\dot{\theta}$, and t.

Final im Etch

10
GRAPHIC
FUNCTIONS

A *scale* is the common means of measuring the values of a functional relationship. Scales are very quick and easy to use. When used alongside a mercury-in-glass thermometer, a scale gives us a visual measure of temperature. On the dashboard of a car, a scale is used to give a visual measure of the speed of the car, the engine oil temperature, or the level of fuel in the fuel tank. Other scales are used to measure our height, weight, and grade-point average.

A scale is defined as a one-dimensional representation of a function $f(x)$, where the positions of the graduations on the scale correspond to values of the function $f(x)$, but the graduations are identified by values of the variable x.

Specific examples of scales are shown in Figure 10.1. The logarithmic scale is an example of a functional scale that is particularly valuable to engineers.

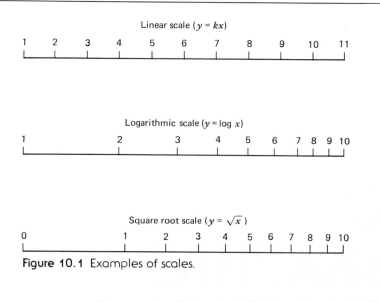

Figure 10.1 Examples of scales.

$f(x) = x^2$; $x = 1$ to 15

```
 1 24  16        DON'T
 +++++  +  ┄┄┄┄┄┄┄┄┄┄┄┄┄┄
 0   10  20  30  40
```

10.2 A FUNCTIONAL SCALE

Many scales are linear, but it is not necessary for a scale to be linear. Consider the scale used to measure the remaining fuel in a fuel tank. The fuel gage indicates the level of a float in the tank. But is this a direct measure of the volume? If the tank has rectangular sides, as in Figure 10.2a, it is, but if the tank has anything but rectangular sides, as in Figure 10.2b, equal incremental changes in the fuel level in the tank do not mean equal incremental changes in volume. To indicate the correct volume, the scale is calibrated to the tank and will be nonlinear. This is a *functional scale*. The function quantity, in this case the gage position, is visually identified by the direct indication of the variable, which is fuel volume. The advantage of using a functional scale is obvious. We could have used gadgetry or circuitry to convert the nonlinear reading to a linear reading, but it is far simpler to adjust the scale to the function than to adjust the function to the scale.

Consider a typical scale shown in Figure 10.3. For every graduation of the variable x there is a corresponding value of the function $f(x)$. The scale distance L, which places the graduation corresponding to x on the scale line, is proportional to $f(x)$, not x. If the values of the variable range from x_1 to x_2, then the values of the function will vary from $f(x_1)$ to $f(x_2)$. This range is represented graphically by the total length of the scale, which we will define as L_t. To define the correspondence between function value and scale length, a *scale factor M* can be calculated as

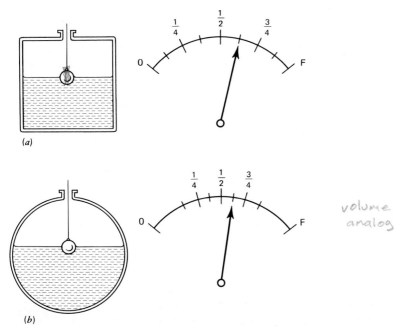

Figure 10.2 Use of a functional scale. *(a)* Rectangular fuel tank. *(b)* Spherical fuel tank.

$x = 6y \; : \; [2\emptyset, 6\emptyset]$

$$\frac{36\emptyset - 12\emptyset}{1\emptyset\emptyset \; mm} = 2.4$$

x	$f(y)$	L_t
$2\emptyset$	$12\emptyset$	$\emptyset$
$\downarrow$	$\downarrow$	$\downarrow$
$6\emptyset$	$36\emptyset$	$1\emptyset\emptyset$

Values of the variable x

Graduations on scale line

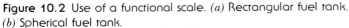

Scale distance, L

Total scale length, L_t

Figure 10.3 Scale for the function $f(x)$.

scale factor

$$M = \frac{f(x_2) - f(x_1)}{L_t} = \frac{\text{function units}}{\text{length units}} \qquad (10.1)$$

Once the scale factor M is determined, the scale distance L to any function value can be determined as

$$L = \frac{f(x) - f(x_1)}{M} \qquad (10.2)$$

To reverse this, the value of the function $f(x)$ at x can be determined, knowing the length L and the value of the function $f(x_1)$ at the end point x_1.

$$ML = f(x) - f(x_1) \qquad (10.3a)$$

If $f(x_1) = 0$ we have a special case and

$$f(x) = ML \qquad (10.3b)$$

For a *linear scale*, the graduations between the values of the variable are equal. Consider the linear function $6x$ where x ranges from 20 to 60. The function $f(x)$ then ranges from 120 to 360. Figure 10.4 presents a scale for this function that is 100 mm long. The scale factor is, from Eq. 10.1,

$$M = \frac{6(60) - 6(20)}{100} = 2.4 \frac{\text{function units}}{\text{mm}}$$

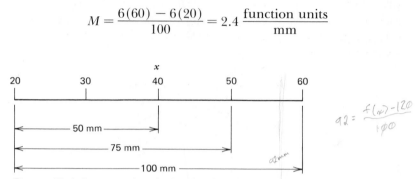

Figure 10.4 Functional scale for the function 6x.

We can now use M to construct the scale. Referring to Figure 10.4 and using Eq. 10.2,

$$L_{20 \text{ to } 40} = \frac{6(40) - 6(20)}{2.4} = 50 \text{ mm}$$

This locates the graduation for $x = 40$.

$$L_{20 \text{ to } 50} = \frac{6(50) - 6(20)}{2.4} = 75 \text{ mm}$$

This locates the graduation for $x = 50$.

Now let us determine the value of $f(x) = 6x$ at the graduation $x = 30$. The scale length between $x = 20$ and $x = 30$ is 25 mm. Therefore,

$$f(x) = 2.4(25) + 6(20) = 180$$

Of course, this is ridiculously simple, but it does illustrate the principle.

Probably the second most familiar scale after the linear scale is the *logarithmic scale*. If you can locate a slide rule (a precalculator device), inspect the C and D scales, which are logarithmic scales. Let us construct a scale for the function $f(x) = \log x$. We will choose a length of 3 in. and a variable range of $1 \leq x \leq 10$. Therefore, the scale factor is, from Eq. 10.1,

$$M = \frac{\log 10 - \log 1}{3} = \frac{1}{3} \frac{\text{function units}}{\text{in.}}$$

We are constructing the scale by making all measurements from the left end. Using Eq. 10.2,

$$L = \frac{\log x - \log 1}{1/3} = 3 \log x$$

To construct a scale, it is useful first to make a table of variables and function values for the range of interest. After the table is prepared, the positions of the variable x on the logarithmic scale can be located. The first column lists values of x, the second column lists values of $f(x)$, and the third column shows the values of the scale distance L.

x	$f(x)$ $\log x$	$L(in.)$ $3 \log x$
1	0.000	0.000
2	0.301	0.903
4	0.602	1.806
6	0.778	2.334
8	0.903	2.709
10	1.000	3.000

The resulting scale and some example measurements are shown in Figure 10.5.

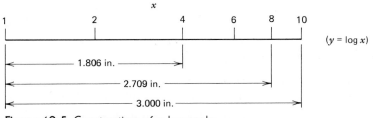

Figure 10.5 Construction of a log scale.

A third function, with which you already have some familiarity, is the squared function x^2, but can you immediately visualize what its functional scale looks like? Two decisions must be made before you can proceed. What is the variable range, and how long will your scale be? Let us assume that x varies from 1 to 5 and the length of the scale is to be 100 mm. The scale factor is therefore, from Eq. 10.1,

$$M = \frac{5^2 - 1^2}{100} = 0.24 \ \frac{\text{function units}}{\text{mm}}$$

and

$$L = \frac{x^2 - 1^2}{0.24}$$

To establish a variable location, $x = 3$, for example, from the left end, the difference between the function values is divided by M. Using Eq. 10.2,

$$L = \frac{3^2 - 1^2}{0.24} = 33.3 \text{ mm}$$

Again, a table can be established.

x	x²	L(mm)
1	1	0
2	4	12.5
3	9	33.3
4	16	62.5
5	25	100.0

Figure 10.6 presents the completed scale and some example measurements.

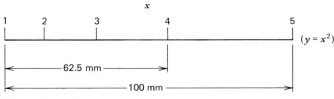

Figure 10.6 Squared scale.

Refer to Figures 10.5 and 10.6, and note how the scales give a visualization of the function. The logarithmic scale crowds the variable to the right, and the squared scale crowds the variables to the left.

As a further example, consider a cubic function $f(x) = x^3$ and let x vary from -5 to 5. The range of the function value is then $5^3 - (-5)^3$ or 250. Usually a specific scale length is not required, but available space and the proposed use will definitely influence how long the scale should be. If a scale length of 4 to 8 in. is convenient, it might be logical to use a 5-in. length to simplify the scale factor. In this case, the scale factor M is, using Eq. 10.1,

$$M = \frac{5^3 - (-5)^3}{5} = 50 \frac{\text{function units}}{\text{in.}}$$

and using Eq. 10.2,

$$L = \frac{f(x) - f(x_1)}{M} = \frac{x^3 - (-5)^3}{50}$$

which leads to the table for the functional scale.

x	x^3	L(in.)
−5	−125	0.00
−4	−64	1.22
−3	−27	1.95
−2	−8	2.34
−1	−1	2.48
0	0	2.50
1	1	2.52
2	8	2.66
3	27	3.04
4	64	3.78
5	125	5.00

Figure 10.7 shows the finished cubic scale. Note how the cubed scale crowds the variables toward zero even more than the squared scale.

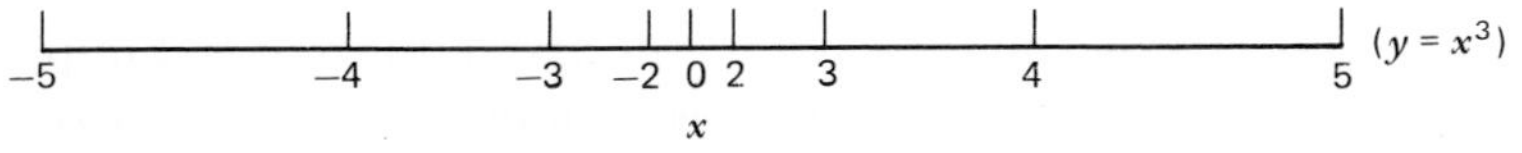

Figure 10.7 Cubed scale.

One statement about scales ought to be obvious, but it is so important that we cannot accept that it is obvious. A scale cannot represent a multivalued function. For example, in a squared function, where x varies from −2 to 2, $(-2)^2$ and $(2)^2$ both equal 4, you cannot account for both +2 and −2 in the construction of a scale.

10.3 ADJACENT SCALES

The functional scales that we have developed all serve the purpose for which they were constructed. They represent the function $f(x)$ on a line. If we have $f(x)$, we can read the corresponding value of x. For instrument panels, digitizing units, and gages, such as the gasoline gage cited in the previous section, these

$$\pi r^2 = A = f(r) \quad [0, \varphi]$$

$$g(A) = f(r)$$

functional scales will read x from $f(x)$. But we cannot read the value of $f(x)$. For that we need a means to measure scale distance. We need to know the scale factor M, and at least one point on the scale. We need to exercise Eq. 10.3a

$$f(x) = ML + f(x_1)$$

Now, let us go back to our original functional scale construction. If we had the function $y = f(x)$, we could have plotted $f(x)$ versus x using the x- and y-coordinates as orthogonal linear scales. In Figure 10.8a the function $y = x^2$, $0 \leq x \leq 5$ is plotted. You are familiar with the function and its graph. You recognize the graph as a parabola.

As an alternative, we could have constructed the graph of $f(x)$ as a straight line. Why not? Think of it as straightening out the graph of the function. If the graph is to be a straight line, one or the other of the axes would need to be distorted, but not both. Since we may want to read x from $f(x)$, let us select $f(x)$ to have the linear scale and distort the scale for x. This is what we have done in Figure 10.8b, where the abscissa is the same functional scale for $y = x^2$ as in Figure 10.6. As we will see in Chapter 11, distorting one coordinate is one more way of presenting information, and it is used extensively in engineering.

The y- and x-axes in Figure 10.8b have direct correspondence. If the function is a straight line, why is there need for a graph? The y-axis could be placed alongside the x-axis. Simply rotate the y-axis 90 deg. When this is done, we create *adjacent scales* (see Figure 10.8c). Of course, to place the y-axis alongside the x-axis, both must be exactly the same length. The straight-line graph of $f(x)$ would be inclined at 45 deg, as it is in Figure 10.8b.

Adjacent functional scales can be constructed by laying out two scales on opposite sides of a line in such a way that there is a direct correspondence between points. That is, $y = f(x)$ everywhere. This principle can be further extended to solve equations of the form $f(x) = g(y)$.

Mathematically, this is quite simple. Referring again to Eq. 10.3a,

$$f(x) = ML + f(x_1)$$

If $f(x) = g(y)$, and the scale lengths are equal, the scale factors must also be equal. This means that

$$g(y) = ML + g(y_1) = f(x) = ML + f(x_1)$$
$$f(x_1) = g(y_1) \qquad (10.5)$$

Adjacent scales are very useful for conversions. Figure 10.9 shows two adjacent scales used for unit conversions.

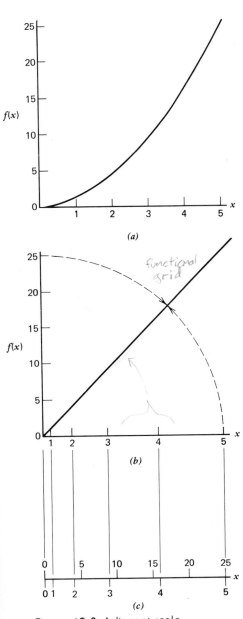

Figure 10.8 Adjacent scale.

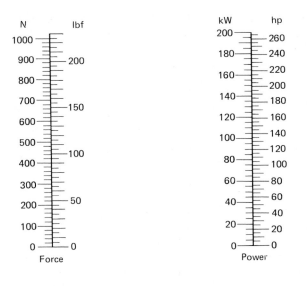

Figure 10.9 Adjacent scales used for unit conversions.

10.4 ALIGNMENT CHARTS

Adjacent scales can also be placed at right angles, or they can be placed parallel to one another. Neither of these configurations has an advantage over the other, but if one of the scales is inverted, as in Figure 10.10, the parallel scales need not be the same length. The length of one scale as compared to the other is controlled by the location of a turning point P. If it is closer to the x scale, the y scale is larger than the x scale. Geometrically, this would mean that

$$\frac{M_y}{M_x} = \frac{d_1}{d_2} \qquad M_x\,d_1 = M_y\,d_2 \qquad (10.6)$$

This set of parallel scales is called an *alignment chart*. That is, to read $y = f(x)$, three points must be in a straight line—the variable x, $f(x)$ or y, and the turning point P. This is also called a *nomograph*, taken from *nomos* ("law") and *graphein* ("write"). Entire books and courses of study are available on nomographs. This section only introduces what a nomograph is and how it is constructed.

Let us now consider a problem that is one step increased in difficulty: the addition of three functions such that

$$f(x) + g(y) = h(z) \qquad (10.7)$$

These three functions can be related by a nomograph consisting of three parallel scales.

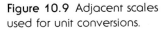

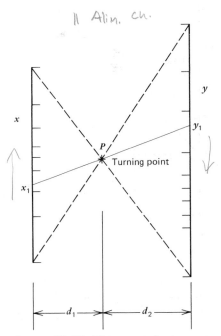

Figure 10.10 Alignment chart.

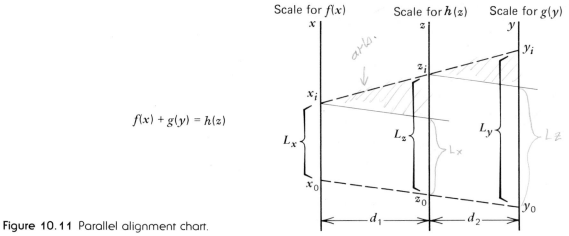

$$f(x) + g(y) = h(z)$$

Figure 10.11 Parallel alignment chart.

It will not be necessary to use the whole scale, but it is necessary to arrange these scales so that there will be direct correspondence for all values of x, y, and z. We must first select the origin of each scale as zero.

$$f(x_0) = g(y_0) = h(z_0) = 0$$

Referring to Eq. 10.2, the scale equations become

$$L_x = \frac{f(x) - f(x_0)}{M_x} = \frac{f(x)}{M_x}$$

$$L_y = \frac{g(y) - g(y_0)}{M_y} = \frac{g(y)}{M_y}$$

$$L_z = \frac{h(z) - h(z_0)}{M_z} = \frac{h(z)}{M_z}$$

Referring to Figure 10.11, to continue the derivations, lay out three convenient parallel lines and align x_0, y_0, and z_0. This is an important and necessary construction. Now, construct two lines, one through point x_i and one through z_i, both parallel to the baseline x_0-y_0. From similar triangles.

$$\frac{(L_y - L_z)}{d_2} = \frac{(L_z - L_x)}{d_1}$$

Cross-multiplying and dividing by $d_1 d_2$ gives the algebraic equation

$$\frac{L_y}{d_2} + \frac{L_x}{d_1} = L_z\left(\frac{d_1 + d_2}{d_1 d_2}\right) \tag{10.9}$$

Using the previously defined scale factors, M_x, M_y, and M_z, and substituting for $f(x)$, $g(y)$, and $h(z)$ in Eq. 10.7,

$$L_x M_x + L_y M_y = L_z M_z \tag{10.10}$$

We now have two equations. If we include Eq. 10.6, Eqs. 10.6, 10.9, and 10.10 permit us to solve for M_z in terms of M_x and M_y.

not arb.

$$M_z = M_x + M_y \qquad (10.11)$$

To construct a parallel alignment chart, you are free to choose the location of the x and y scales and the scale factors M_x and M_y. If we establish d_1 and d_2, this fixes the z scale and its scale factor M_z.

SAMPLE PROBLEM 10.1

Construct two adjacent functional scales that would allow you to read directly the areas of circles with radii ranging from 0 to 10 m. A convenient scale length would be approximately 150 mm.

Solution
This problem can be expressed analytically as

$$f(r) = g(A)$$

or

$$\pi r^2 = A$$

The limits for r and A are $0 \le r \le 10$ and $0 \le A \le 314.2$.
  From Eq. 10.1, the scale factor M_A is

$$M_A = \frac{314.2 - 0}{\sim 150} = 2.09 \frac{m^2}{mm}$$

For the sake of convenience, let us select a scale factor of $M_A = 2$ because 2.09 is inconvenient. Thus the scale length will be a little more than 150 mm.
  Since these are to be adjacent scales, $M_A = M_r$, and we have two functional relationships. From Eq. 10.2,

$$L_r = \frac{f(r) - 0}{M_r} = \frac{\pi r^2}{2}$$

$$L_A = \frac{g(A) - 0}{M_A} = \frac{A}{2}$$

We can tabulate each independently.

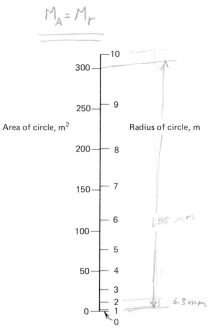

$M_A = M_r$

Area of circle, m^2 Radius of circle, m

r	f(r)	L	A	g(A)	L
0	0	0	0	0	0
2	12.6	6.3 mm	50	50	25
4	50.3	25.1	100	100	50
6	113.1	56.6	150	150	75
8	201.1	100.5	200	200	100
10	314.2	157.2	250	250	125
			300	300	150

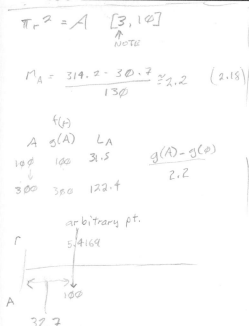

$\pi_r{}^2 = A \quad [3, 10]$

$M_A = \dfrac{314.2 - 30.7}{130} \cong 2.2 \quad (2.18)$

$f(A)$

A	$g(A)$	L_A	$\dfrac{g(A) - g(\emptyset)}{2.2}$
100	100	31.5	
300	300	122.4	

arbitrary pt.

r 5/4169

A

32.7 100

The completed adjacent scales are shown. They are as precise as our graphic techniques can make them. Note that although there is direct correspondence, that is, $f(r) = g(A)$, the individual graduations do not line up because each functional scale was determined independently.

One alternative would be to locate radii for a series of areas and *interpolate* the location of the graduations for the function r. This is not precise, and it is not a use of functional scales or a good use of graphics.

SAMPLE PROBLEM 10.2

Construct a parallel scale alignment chart for a Francis weir formula

$$\log \left[Q = (3.33\, L H^{3/2}) \right] \log$$

where Q is the quantity of flow in ft^3/sec, L is the width of the weir in feet, and H is the height of the water surface at the weir in feet.

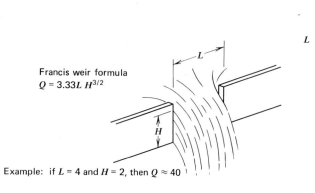

Francis weir formula
$Q = 3.33L\, H^{3/2}$

Example: if $L = 4$ and $H = 2$, then $Q \approx 40$

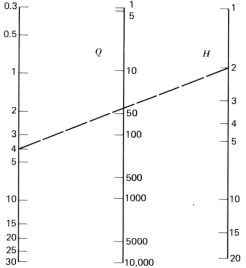

Solution
The Francis weir flow equation may be rewritten as

$$\log (3.33L) + \frac{3}{2} \log H = \log Q \quad \text{or} \quad f(L) + g(H) = h(Q)$$

This puts it in the mathematical form $f(x) + g(y) = h(z)$, which can be represented with a parallel alignment chart.

If we assume the range of L to be $0.3 \le L \le 30$, then the range of log $(3.33L)$ is approximately 0 to 2. If we choose a scale length of 100 mm, the scale factor then becomes

$$M_L = \frac{2-0}{100} = 0.02$$

Assuming the range of H to be $1 \le H \le 20$ gives the approximate range of $g(H)$ to be $0 \le (3/2) \log H \le 2$. Again, taking the scale length as 100 mm gives a scale factor of

$$M_H = \frac{2-0}{100} = 0.02$$

Therefore,

$$M_Q = M_L + M_H = 0.04$$

and the approximate range of Q is $1 \le Q \le 9000$.
This table of values was used to construct the completed alignment chart.

L	log(3.33L)	L_L(mm)		H	$\frac{3}{2}$ log H	L_H(mm)		Q	log Q	L_Q(mm)
0.3	0	0		1	0	0		1	0	0
0.5	0.22	11		2	0.45	23		5	0.70	1.8 17.5
1	0.52	26		3	0.72	36		10	1.00	25
2	0.82	41		4	0.90	45		50	1.70	42
3	1.00	50		5	1.05	52		100	2.00	50
4	1.12	56		10	1.50	75		500	2.70	67
5	1.22	61		15	1.76	88		1,000	3.00	75
10	1.52	76		20	1.95	98		5,000	3.70	92
15	1.70	85						10,000	4.00	100
20	1.82	91								
25	1.92	96								
30	2.00	100								

PROBLEMS

Construct scales for the following functions. Use the variable ranges given.

10.3 $\sqrt{x}$　　　　$(0.5 \le x \le 5)$

10.4 $x^4 + 2x^2$　　　$(0 \le x \le 1.5)$

10.5 e^x　　　　$(0.5 \le x \le 3)$

10.6 $\ln x$　　　　$(2 \le x \le 20)$

10.7 $\cos x$ $\quad\quad\quad (0 \le x \le \pi)$

10.8 $\dfrac{1}{x}$ $\quad\quad\quad (0.5 \le x \le 2)$

10.9 Match the scales with the functions shown.

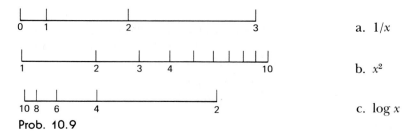

a. $1/x$

b. x^2

c. $\log x$

Prob. 10.9

10.10 The radiant energy coming from a surface is proportional to the absolute temperature of the surface T, ($^0R = {}^0F + 460$), to the fourth power (T^4). Construct a scale for T^4 ($0 \le T \le 600$).

10.11 Sketch scales directly from the following graphs. First establish length scales for the $f(x)$ and x axes.

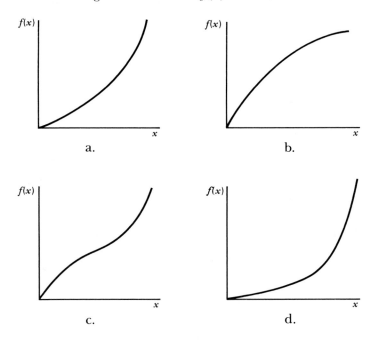

a.

b.

c.

d.

10.12 Construct adjacent scales for the volume of a sphere as a function of the sphere diameter $(0 \leq d \leq 2 \text{ m})$.

10.13 Draw adjacent functional scales for the function $\ln R = \sqrt{s}$. The scale should be approximately 10 cm long and should cover the range from $s = 2$ to $s = 20$. Your functional scale for R should show the five values of R, for $R = 5, 10, 20, 50,$ and 100. Your functional scale for s should show the values of s, for $s = 2, 5, 10, 15, 20$.

10.14 The normal acceleration for circular motion is $a_n = \omega^2 r$ where,

$a_n = $ acceleration toward center, m/sec²

$\omega = $ angular velocity, radians/sec

$r = $ radius of circular path, m

Construct adjacent functional scales to give the values of a_n (m/sec²) when ω varies from 20 to 100 rev/min, and $r = 0.5$ m is a constant. Make the scale length approximately 60 mm. Label the axes properly and include units. Locate the speeds 20, 40, 60, 80, and 100 rev/min and the accelerations 2, 5, 10, 20, and 50 m/sec².

10.15 Construct a parallel alignment chart for the function $a_n = r\omega^2$ in Problem 10.14. Use scale lengths of your own choice but make the lengths of the r and ω scales about equal. Use the ranges

$$0 \leq r(\text{m}) \leq 10$$
$$0 \leq \omega \text{ (rad/sec)} \leq 10$$

Remember to first take the logarithm of the equation.

10.16 Construct a parallel alignment chart to determine the area of a circle, with its center at the origin, from the coordinates of any point P on its circumference. In other words,

$$\pi x^2 + \pi y^2 = A$$

x and y vary from 1 to 9, scale length = 16 cm.

10.17 The lengths of the sides of a right triangle are related by

$$x^2 + y^2 = L^2$$

Construct an alignment chart for this relationship for the ranges

$$0 \leq x \leq 10$$
$$0 \leq y \leq 10$$

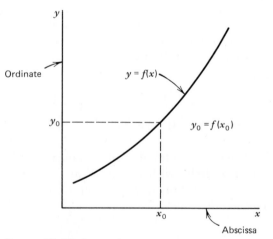

10.5 GRAPHING A FUNCTION

A *graph* is a two-dimensional plot that uses functional scales to determine the position of a data point in two-dimensional space. The values of the *independent variable x* are placed on the horizontal axis, or the *abscissa* (Figure 10.12). The values of the function $y = f(x)$, the *dependent variable,* are placed on the vertical axis, or the *ordinate*. These positions can be interchanged, but the normal arrangement is for the x-axis to be the horizontal axis and the y-axis to be the vertical axis. The correspondence between x and y is then governed by a two-dimensional curve.

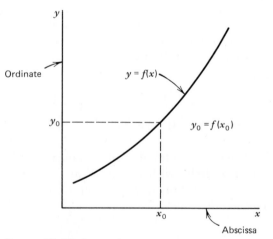

Figure 10.12 A graph.

The locus of points that satisfy the equation of the function is called the *graph* of the function. Graphs are usually shown as lines, but the lines need not be straight.

Now, let us compare the graphic representation of functions by functional scales and graphs. Both functional scales and graphs have specific uses. The graph is more familiar, because we can recognize the locus of points that satisfy the function. You are less familiar with a functional scale and less likely to recognize it. When we are interested in an analysis of the function $f(x)$, that is, when we want to know its derivative or its maximum value, minimum value, or roots, a graph of the function is necessary. When we are only interested in determining y or $f(x)$ for various values of x, the functional scale is easier to use and less prone to errors in reading.

One further use of graphs is to develop mathematical expressions for $f(x)$ from observed data when no expression is known. This is our next subject.

10.6 EMPIRICAL EQUATIONS

The previous sections in this chapter have presented methods to represent mathematical functions graphically in the form of functional scales. The discussion that now follows presents methods by which mathematical functions are determined from graphs. The actual graphing of real data is discussed at length in Chapter 11.

Data come directly from experiments and observations. If the data follow a regular curve, the functional relation between the dependent and independent variables can be stated *empirically*, provided the scales, usually linear or logarithmic, can be identified. Remember, a logarithmic scale, by our definition, is a functional scale. The equation that expresses this functional relation is called an *empirical equation*.

The three common combinations of linear and logarithmic scales are linear-linear, logarithmic-logarithmic (log-log), and linear-logarithmic (semilog). Consider the experimental observations shown in Figure 10.13. These data represent some function $f(x)$ that varies in what looks to be a linear function of the dependent variable x. For this linear-linear graph both axes are linear scales. Formal curve-fitting techniques could be used to determine a mathematical expression for the best straight line through these data. They are usually a part of the study of statistics. We will limit our methods of curve representa-

[handwritten margin notes:]
scale = linear
  logarithmic
      ↑
    fxn scale

lin - lin
log - log
log - lin

plot w. open circle of error
don't have to put line on
  graph

1) eyeball
2) know eqt'n
3) remainder

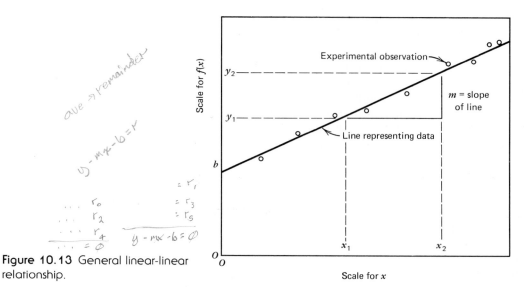

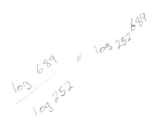

Figure 10.13 General linear-linear relationship.

tion to the simpler ones because they are actually the methods most used.

One simple method is to draw a line that looks to be the best straight line to fit the data. Data points on either side of the line should be balanced. There should be about as many on one side as on the other.

The ordinate or vertical scale in Figure 10.13 is a linear scale for the general function $f(x)$. Recall that the length along the scale corresponds to values of the function $y = f(x)$. Lengths along the abscissa or horizontal scale correspond to values of the independent variable x. The equation of the line is

$$y = mx + b \qquad (10.12)$$

Equation 10.12 is the familiar expression for a straight line where m is the *slope* of the line and b is the ordinate *intercept* (i.e., the value of y when $x = 0$). The slope of the line in Figure 10.13 is

$$\frac{f(x_2) - f(x_1)}{x_2 - x_1} = \frac{y_2 - y_1}{x_2 - x_1} = \frac{\Delta y}{\Delta x} = m \qquad (10.13)$$

where Δ indicates a finite length along the corresponding axis.

Two approaches can be used to find the values of m and b. The slope and intercept can be taken directly from the plot, as shown in Figure 10.13. If the range of the x-axis does not include $x = 0$, so that intercept b cannot be obtained directly, the coordinates of any two points on the line can be substituted into the equation and the values of m and b can be calculated. In

Figure 10.13, the two points (x_1, y_1) and (x_2, y_2) are selected points on the accepted line. They therefore satisfy Eq. 10.12; that is,

$$y_1 = m x_1 + b$$

and

$$y_2 = m x_2 + b$$

Since x_1, x_2, y_1, and y_2 are all known, two simultaneous equations can be solved for the two unknowns m and b. This method of finding m and b from two points (x_1, y_1) and (x_2, y_2) is called the selected ordinate method.

Data that can be represented by a straight line when both the ordinate and the abscissa are logarithmic scales fit the exponential equation

$$y = bx^m \qquad (10.14a)$$

Expressed in logarithms, this is

$$\log y = m \log x + \log b \qquad (10.14b)$$

and the graph that it represents is called a logarithmic-logarithmic or log-log graph. Figure 10.14 presents an example of a logarithmic-logarithmic plot. Again, intercept b and slope m can be obtained directly from the plot. The intercept is $\log b$ and it is the value of y when $\log x = 0$ or $x = 1$. The intercept is the value of y when $f(x) = 0$ and $\log x = f(x)$ in this case. The slope is the same general relationship as before.

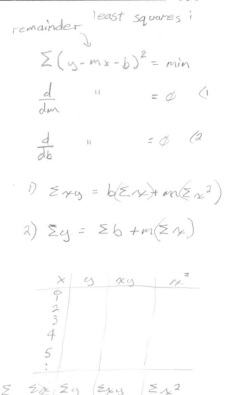

remainder · least squares

$$\Sigma (y - mx - b)^2 = \min$$

$$\frac{d}{dm} \quad '' \quad = \emptyset \quad (1)$$

$$\frac{d}{db} \quad '' \quad = \emptyset \quad (2)$$

1) $\Sigma xy = b(\Sigma x) + m(\Sigma x^2)$

2) $\Sigma y = \Sigma b + m(\Sigma x)$

x	y	xy	x^2
0			
1			
2			
3			
4			
5			
⋮			

$\Sigma \quad \Sigma x \quad \Sigma y \quad \Sigma xy \quad \Sigma x^2$

selected ordinate

0.434 $\log_{10}$ to $\log_e$

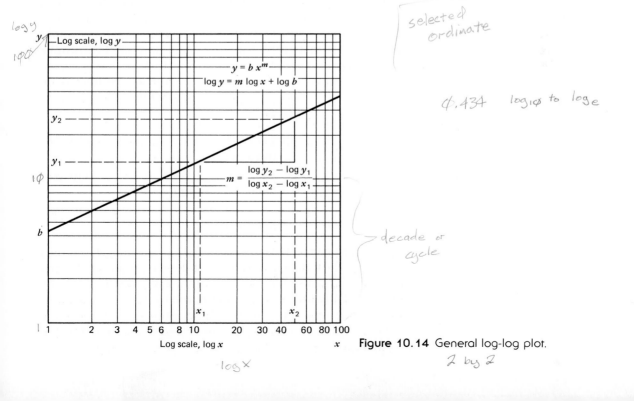

decade or cycle

Figure **10.14** General log-log plot.

2 by 2

$\log y$

$\log x$

$$m = \frac{\Delta y}{\Delta x} = \frac{\log y_2 - \log y_1}{\log x_2 - \log x_1}$$

Note that the slope is related to the functions (log y, log x), not the variables (y, x). It may not be convenient to determine the intercept, log b, from the value corresponding to $f(x) = 0$, where $\log x = 0$ or $x = 1$ is off scale. Instead, the method of selected ordinates can be used. Two points are arbitrarily chosen, and their coordinates are used to solve for b and m. Taking points (x_1, y_1) and (x_2, y_2) from Figure 10.14, Eq. 10.14b gives

$$\log y_1 = m \log x_1 + \log b$$
$$\log y_2 = m \log x_2 + \log b$$

Again, these are two simultaneous equations where b, or log b, and slope m are the unknowns.

Natural logarithms can be used with equal results. Equation 10.14a can be expressed as

$$\ln y = m \ln x + \ln b \qquad (10.14c)$$

The relation between natural logarithms and $\log_{10}$ is a constant, $\ln b = 2.3026 \log_{10}b$. The intercept would, of course, be $\ln b$ and the slope

$$m = \frac{\ln y_2 - \ln y_1}{\ln x_2 - \ln x_2}$$

If the ordinate is a logarithmic scale and the abscissa is a linear scale, the graph is called a linear-logarithmic graph or a semilogarithmic graph. Four possible analytic equations would be straight lines on a graph with one linear and one logarithmic scale. Although they express the same line, they are not the same equations. Expressing each in exponential and logarithmic form,

1. $y = be^{mx}$ $\qquad\qquad (10.15a)$
   $\ln y = mx + \ln b$

2. $y = e^{(mx+b)}$ $\qquad\qquad (10.15b)$
   $\ln y = mx + b$

3. $y = b\,10^{mx}$ $\qquad\qquad (10.15c)$
   $\log y = mx + \log b$

4. $y = 10^{(mx+b)}$ $\qquad\qquad (10.15d)$
   $\log y = mx + b$

If the data appear to fall along a straight line when placed on a semilog plot, the data will fit all four of these equations. The first is preferred. The differences between these equations are in the expression of the intercept and in the use of natural logarithms or $\log_{10}$. They will not yield the same intercepts or the same values of slopes. Figure 10.15 shows a semilog plot for Eq.

FINAL

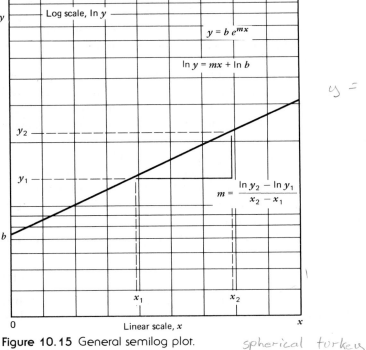

Figure **10.15** General semilog plot.

spherical turkey °C

$$y = b e^{mx}$$

$$poss. \quad y = e^{(mx + b)}$$

$$y = b(10)^{mx}$$

10.15*a* and suggests how the slope and intercept can be obtained.

Linear-linear, semilog, and log-log plots are the three most common ways of presenting empirical data. The relationships

$$y = mx + b$$
$$y = be^{mx}$$
$$y = b x^m$$

are relatively simple relationships leading to easy understanding and interpretation. Two-dimensional plots are by no means limited to these three relationships, but these three commonly used plots are usually chosen because they are simple. Referring to Figure 10.16 and remembering that the length of a scale corresponds with the value of the function, we can write the general relationship

$$g(y) = mf(x) + g(b) \qquad (10.16)$$

where the functions $g(y)$ and $f(x)$ cause the data to fall on a straight line. Experience and experimentation will help you choose the appropriate scales for a particular plot of empirical data.

To choose appropriate scales for a given situation, you need some feel for the relationship between functional scales and the

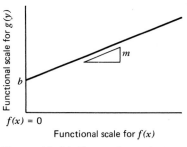

Figure **10.16** General graph.

graphic distortion they represent. For example, the distances between graduations on a logarithmic scale become smaller as the variable values become larger. A squared scale has the opposite effect. The smaller values of the variable are crowded closer together.

We are not going to offer rules of thumb regarding choice of scales for an empirical plot. There are too many possibilities and combinations to consider. A start in this direction is the ability to read and recognize scales and to plot scales—that is, having a feel for scale distortion and the graphic relationship of various functions. A useful step in the choice of scales is to plot the data first using two linear scales. The general shape of the resulting curve can help you decide what scales to try next.

SAMPLE PROBLEM 10.18

Determine the empirical equation for the linear-linear plot.

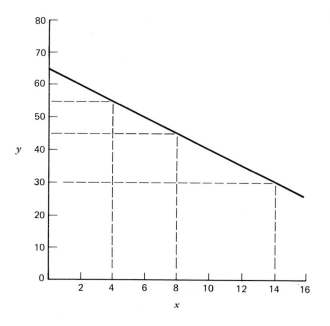

Solution

The general relationship for this graph is $y = mx + b$, the y-intercept b is 65, and the slope is

$$m = \frac{\Delta y}{\Delta x} = \frac{y_2 - y_1}{x_2 - x_1} = \frac{45 - 55}{8 - 4} = -2.5$$

Note that the slope is negative. The specific equation of this line is then $y = -2.5x + 65$. To check the results, coordinates of points on the line can be substituted into the equation

$$x = \ 4, y = 55 \qquad 55 \stackrel{?}{=} -2.5(4) + 65$$
$$55 = \ 55$$

$$x = 14, y = 30 \qquad 30 \stackrel{?}{=} -2.5(14) + 65$$
$$30 = 30$$

SAMPLE PROBLEM 10.19

Determine the empirical equation for the line in the linear-linear plot.

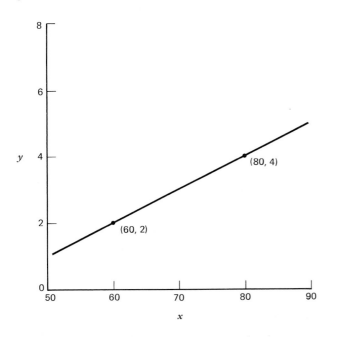

Solution

The x-axis does not include the value $x = 0$. The coordinates of two points on the line can be substituted into the general equation to find the values of m and b. Using the points $(60, 2)$ and $(80, 4)$ and the general equation $y = mx + b$ gives

$$\begin{aligned} 2 &= m\ 60 + b \\ 4 &= m\ 80 + b \\ \hline 2 &= m\ 20 \end{aligned}$$

$$m = 0.1$$

Then solving for b,

$$2 = 0.1(60) + b \qquad b = -4$$

So the equation of the line is

$$y = 0.1x - 4$$

SAMPLE PROBLEM 10.20

Determine the empirical equation for the log-log plot.

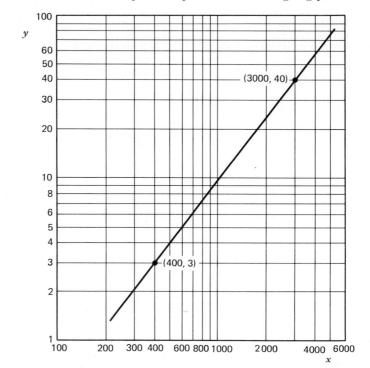

Solution

The intercept cannot be determined directly from the plot because the range of x does not include $x = 1$ ($\log x = 0$). The coordinates of two points on the line can be used to determine the equation. The general expression for this graph is

$$y = b\,x^m$$

Using the two points $(400, 3)$ and $(3000, 40)$ gives

$$3 = b(400)^m$$

$$40 = b(3000)^m$$

Then dividing one equation by the other results in

$$\frac{40}{3} = 13.33 = \left(\frac{3000}{400}\right)^m = 7.5^m$$

$$m = 1.29$$

So

$$b = \frac{3}{400^{1.29}} = 0.00132$$

and the equation is

$$y = 0.00132x^{1.29}$$

SAMPLE PROBLEM 10.21

Determine the empirical equation in two forms for the semilog plot. Use the two general forms

$$y = be^{mx} \qquad \text{and} \qquad y = b\,10^{mx}$$

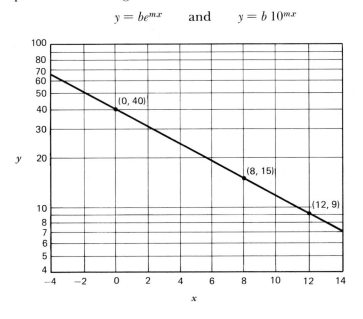

Solution

Using the coordinates of two points on the line $(0, 40)$ and $(12, 9)$ and the natural log form of the equation gives

$$40 = be^{m(0)} \qquad \text{and} \qquad 9 = be^{m(12)}$$

From the first equation $b = 40$, the second is

$$9 = 40e^{m(12)}$$

and $m = -0.124$; therefore,

$$y = 40e^{-0.124x}$$

Using the direct determination of b and m from the graph for the $\log_{10}$ form gives

$$x = 0, \; y = b = 40$$

and

$$m = \frac{\log 9 - \log 15}{12 - 8} = -0.0555$$

so

$$y = (40) \; 10^{-0.0555\,x}$$

To compare these results the two equations are written in log form

$$\ln y = \ln 40 - 0.124x$$
$$\log y = \log 40 - 0.0555x$$

because

$$\ln A = 2.3026 \log_{10} A$$
$$2.3026 \; (0.0555) \stackrel{?}{=} 0.124$$
$$0.123 \quad\;\; = 0.124$$

Remember that values used in this problem were picked off the graph, resulting in some inaccuracy.

SAMPLE PROBLEM 10.22

Determine the empirical equation for the graph shown.

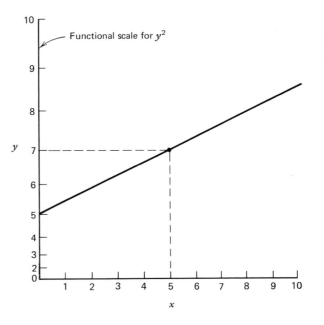

Solution

The abscissa is a linear scale and the ordinate is a squared scale.
Therefore, the general empirical equation

$$g(y) = m f(x) + g(b)$$

for these two scales is

$$y^2 = mx + b^2$$

When $x = 0$, $y = 5$; therefore, $b = 5$. The slope becomes

$$m = \frac{\Delta y}{\Delta x} = \frac{7^2 - 5^2}{5 - 0} = 4.8$$

The equation of the line is

$$y^2 = 4.8x + 25$$

To check this result, choose $x = 10$

$$y^2 = 4.8(10) + 25 = 73$$
$$y = 8.54$$

which checks with the graph.

SAMPLE PROBLEM 10.23

Determine the empirical equation for the graph shown.

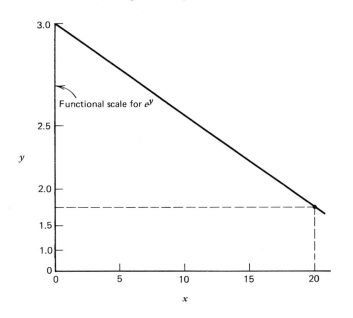

Solution

The abscissa is a linear scale, and the ordinate is a scale for the function e^y. From the general empirical equation for a straight line

$$g(y) = m f(x) + g(b)$$

The specific form for this graph becomes

$$e^y = mx + e^b$$

The values of m and b can be determined directly from the plot when $x = 0$, $y = 3$, so $b = 3$ and $e^3 = 20.1$. Also,

$$m = \frac{e^{1.8} - e^3}{20 - 0} = \frac{6.05 - 20.09}{20} = -0.702$$

Therefore, the empirical equation is

$$e^y = -0.70x + 20$$

PROBLEMS

10.24 Determine an empirical equation for the degree days per year as a function of latitude for the United States and Canada. (Values from America Society of Heating, Refrigeration and Air Conditioning *Engineers Handbook.*)

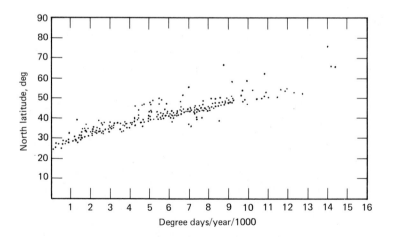

North latitude, deg

Degree days/year/1000

10.25 Determine an empirical equation for the following.

    a. Primates and mammals. ●

    b. Birds. ▲

    c. Bony fish. ○

    d. Reptiles. △

In the figure, brain size is plotted against body size for some 200 species of living vertebrates. (Figure based on graph from *Scientific American* "Paleoncurology and the Evolution of Mind" by Harry J. Jerison, January 1976, p. 94)

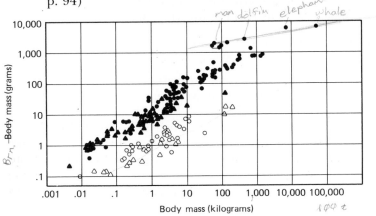

10.26 For the graph shown, plot maximum amplitude on a logarithmic vertical (ordinate) scale and the number of cycles on a linear horizontal (abscissa) scale. Then determine the functional relationship between amplitude and number of cycles.

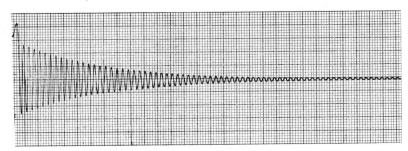

10.27 The energy radiating from a surface has been measured for several surface temperatures. The results are as follows.

E (Btu/h)	°F
200	0
300	32
450	72
600	110

The radiant energy coming from a surface is proportional to the absolute temperature of the surface (T, °R = °F + 460) to the fourth power. Therefore,

$$E = \frac{C}{10^{10}} T^4$$

Using the data given, estimate the value of the constant C for this particular experiment.

10.28 Given the following data, determine an empirical equation of the algebraic form $y = be^{mx}$ that will best plot the data as a straight line. Graph it. Check the graph for the point $x = 9$.

y	x
165	2
136	4
110	6
70	10
60	12
47	14

10.29 Determine which type of graph yields a straight line for the data shown and then determine the empirical equation for this straight line.

y	x
135	3
320	4
625	5
1080	6
1715	7
2560	8
3645	9

10.30 Determine the empirical equation for the relation between V, the volume in cubic meters of 1 kilogram of steam, and p, the pressure in newtons per square meter.

p, N/m^2	V, m^3
102730	1.690
121350	1.400
144100	1.192
172370	1.024
199260	0.880
233040	0.762

10.31 Data for the total load that a vehicle can carry are obtained for various speeds. An empirical equation for the data is needed for computer simulation. Determine the numerical values of a, b, and m that will best fit the given data $(L = a + b\,v^m)$.

Load L, tonnes	Speed v, km/h
5.00	0
8.10	10
12.6	20
17.8	30
23.4	40
29.5	50
35.9	60
50.0	80
65.0	100

10.32 Both scales for the following curves are linear. Replot each curve using scales that result in a straight-line plot.

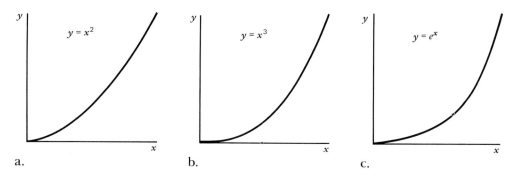

a. b. c.

10.33 Both scales for the following curve are linear. Replot the curve using scales that result in a straight-line plot.

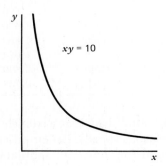

11
GRAPHIC PRESENTATION OF DATA AND RELATIONSHIPS

11.1 WHAT ARE DATA?

In the past few chapters, we have studied how functions are created, how to classify them, how they behave, and how, as one part of that study of behavior, to use calculus to integrate and differentiate them. These are all ways in which the information that you are given, or that you have gathered, can be better understood. Let us go back one step and look at our information as it was received.

Numerical information in its raw form is known by the collective noun *data*. This is a plural noun. We used the word *datum* earlier to designate a reference plane. Here we define a datum as one numerical piece of information, but we rarely have only one piece of information. Data are organized in various ways, but the most usual form is the table or matrix. Each piece of information, such as time, termperature, pressure, number, voltage, magnetic flux, or numerical rank, is listed in a column with all temperatures being in one column, all pressures in another, etc. Pieces of related information are

Charts: bar, pie, spag.

Graphs: lin-lim, lin-log
log-log

nomograph

TABLE 11.1
Rankings of U.S. Standard Metropolitan Statistical Areas

Rank Area	Population
1. New York	9,081,000
2. Los Angeles–Long Beach	7,445,000
3. Chicago	7,058,000
4. Philadelphia	4,701,000
5. Detroit	4,340,000
6. San Francisco–Oakland	3,226,000
7. Washington, D.C.	3,042,000
8. Dallas–Fort Worth	2,964,000
9. Houston	2,891,000
10. Boston	2,760,000
11. Nassau–Suffolk, N. Y.	2,604,000
12. St. Louis	2,341,000
13. Pittsburgh	2,260,000
14. Baltimore	2,165,000
15. Minneapolis–St. Paul	2,109,000
16. Atlanta	2,010,000
17. Newark	1,964,000
18. Orange County, Calif.	1,926,000
19. Cleveland	1,895,000
20. San Diego	1,857,000
21. Denver–Boulder	1,614,000
22. Seattle–Everett	1,601,000
23. Miami	1,573,000
24. Tampa–St. Petersburg	1,550,000
25. Riverside–San Bernardino–Ontario	1,538,000
26. Phoenix	1,512,000
27. Milwaukee	1,393,000
28. Cincinnati	1,390,000
29. Kansas City	1,322,000
30. San Jose, Calif.	1,290,000

Source. Bureau of the Census.

located in the same row. For example, when a particular reading of data is taken, all information is recorded in one row. To locate a particular piece of numerical information, simply find it by its corresponding row. Several examples of tabularized data are given.

Table 11.1 lists 30 major metropolitan areas in the United States and ranks them by population. The name of the city identifies the area. The ranks and population are the pieces of numerical information corresponding to the identified city. For example, Boston, Massachusetts, ranks 10th in population with a population of 2,760,000. Table 11.2 is a physical test of a load cell. The load applied to the cell is in the identifying column. The measured strain is the corresponding information. Table

TABLE 11.2
Dynamometer Calibration

Load, lbf	Strain, ϵ, 10^{-6} in./in.	
	Load Increasing	Load Decreasing
500	28	33
1000	65	71
1500	102	103
2000	138	138
2500	170	174
3000	202	207
3500	237	241
4000	273	278
4500	304	310
5000	338	343
5500	376	380
6000	404	412
6500	443	449
7000	480	—

11.3 is the pig-iron and steel output in the United States tabulated by year. The year is the identifying column, and three pieces of information are given for each year.

Tabularized data transmit facts, but it is difficult, if not impossible, to transmit a perspective of what data imply in tables. Perspectives are given in charts and graphs, which are used to convey information. A chart or graph that does not do this, or that confuses as much or more than it informs, is of little use. Charts and graphs are an engineering art form, and as an art form they can be pleasing and utilitarian or displeasing and useless.

TABLE 11.3
United States Pig-Iron and Steel Output (in net tons)

Year	Total Pig Iron	Pig Iron and Ferro-Alloys	Raw Steel
1940	46,071,666	47,398,529	66,982,686
1945	53,223,169	54,919,029	79,701,648
1950	64,586,907	66,400,311	96,836,075
1955	76,857,417	79,263,865	117,036,085
1960	66,480,648	68,566,384	99,281,601
1965	88,184,901	90,918,040	131,461,601
1970	91,435,000	93,851,000	131,514,000
1975	101,208,000	103,345,000	116,642,000
1980	68,721,000	70,329,000	111,835,000

Source. American Iron and Steel Institute.

11.2 CHARTS

Charts are one way to present numerical information. They give a perspective of the numerical information presented on a comparative basis.

The simplest chart is the *pie chart*. It shows how the whole is divided into separate parts. The usual form of the pie chart is a circle, a pie, with the various parts being sectors of the pie. The bigger the sector, the larger the proportion of the whole the sector represents. The most common example is the use of a pie chart to show the collection and distribution of money—a familiar chart that accompanies tax announcements. In Figure 11.1 the collection and distribution of taxes for the fiscal year 1983 are shown. Note that in each case, the pie chart permits easy and quick comparisons.

Figure 11.1 Budget dollar for 1982–1983. (*a*) Where the dollar comes from. (*b*) Where the dollar goes.

Excise taxes

Other

Corporation income taxes

6¢   4¢

9¢

Borrowing
12¢

29¢
Social
insurance
receipts

40¢
Individual
income
taxes

(*a*)

Grants to states
and cities

Federal
operation

11¢

4¢

Interest
13¢

Entitlements
43¢

National
defense
29¢

(*b*)

Figure 11.2 The 1982–1983 federal budget— where it goes.

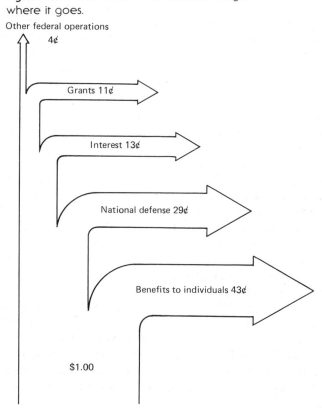

Other federal operations
4¢

Grants 11¢

Interest 13¢

National defense 29¢

Benefits to individuals 43¢

$1.00

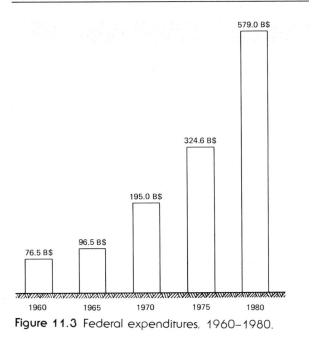

Figure 11.3 Federal expenditures, 1960–1980.

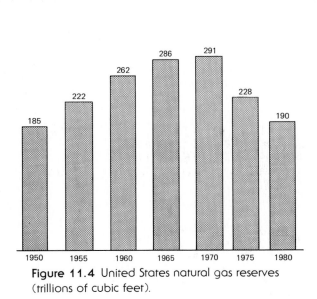

Figure 11.4 United States natural gas reserves (trillions of cubic feet).

The *bar chart* is the next step in sophistication. Bar charts permit comparisons, but they also permit a third dimension, such as time, rank, or location. The segmented bar chart is a form of the pie chart. It is used to show how parts of the whole are taken away or added. In the segmented chart, the width of the bar segment gives an added dimension for comparison. Figure 11.2 shows the same data as Figure 11.1*b*. Which figure is better? The bar chart of Figure 11.3 shows the dramatic increase in federal expenditures from 1960 to 1980; a bar chart shows this very well. The bar chart of Figure 11.4 shows how America's natural gas reserves are declining. This is a visual display of the gradual increase in reserves until the year 1970 and a sharp decline since then. Each bar depicts one year, and the height of the bar is proportional to the trillions of cubic feet of reserves known to exist in that year. These data could be adequately shown in a table, but a bar chart is more effective. The bar chart is more sophisticated than the pie chart, because it can show the changes in distribution visually as well as show distribution, as the pie chart can.

In Figure 11.5, the forest production in the United States is shown, as achieved in 1952, 1962, and 1970, and as projected by the U.S. Forest Service for 1985 and 2000. This bar chart shows that the total production is expected to increase, but the production of hardwoods will increase more than the production of softwoods. Two thoughts are contained in this one bar chart!

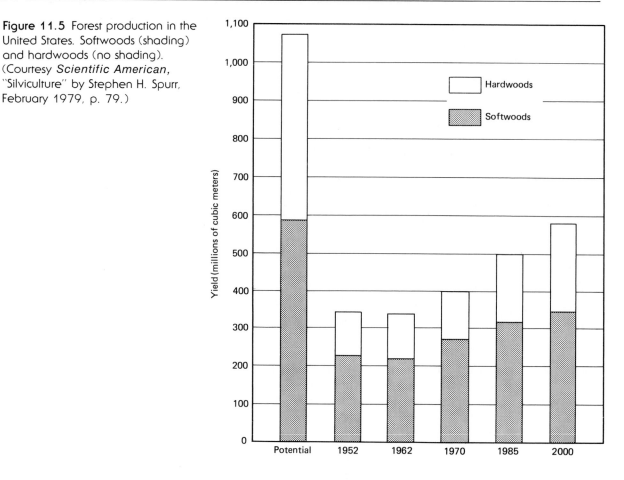

Figure 11.5 Forest production in the United States. Softwoods (shading) and hardwoods (no shading). (Courtesy *Scientific American,* "Silviculture" by Stephen H. Spurr, February 1979, p. 79.)

Sometimes the bar chart is combined with the segmented chart. Figure 11.6 shows the sources of energy and its consumption. At the left, the chart shows the sources of energy as a segmented chart. This segmented chart changes into a series of bars, which show the consumption of energy in 1979. At the right, the chart segments show where the energy goes. This chart has been called a *spaghetti chart,* a name that is both descriptive and appropriate.

Pictorial charts are a chart form where data are imposed on a map or picture to give a visual image of the information contained in the data. Maps containing geographic or economic data are pictorial charts. In Figure 11.7 the world's major crude-oil movements by tanker in 1980 are shown as projected by petroleum economists. The figure is a combination of a pictorial chart and a segmented bar chart. Without referring to the complete data, and without words, an image of dependency on Middle Eastern oil is quickly drawn.

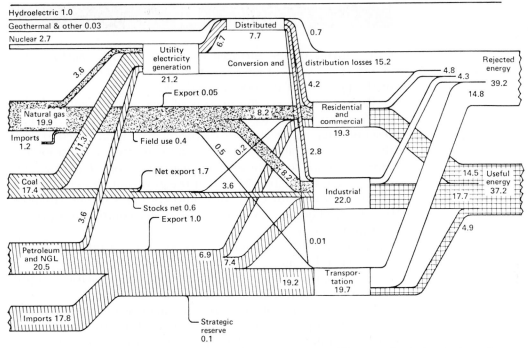

Figure 11.6 Sources and consumption of energy in 1979. (Courtesy Dr. Arthur L. Austin)

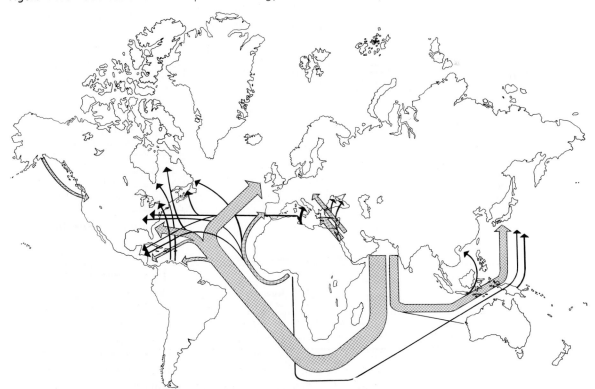

Figure 11.7 The map depicts the world's major crude movements by tanker in 1980.

11.3 RELATIONSHIP CHARTS

Other forms of the pictorial chart are the *flow chart* and the *organization chart*. These show the flow of material, or events, or the position of people, respectively.

Figure 11.8 shows the process of how a bill becomes law in the New York State legislature. It is a flow chart. Each box represents a step in the legislative process. Arrows are used to show the flow of the bill from conception to enactment. Occasionally, more than one arrow is used to mean that more than one thing can happen to the bill. Note that the governor can send it back to the legislature or can sign it into law. This simple visualization is an effective way to describe a very complicated process and to do it quickly.

Figure 11.9 shows another process: making pipeline-quality gas from oil shale. Note how the blocks are shaped like reactors, fractionators, and separators to add a pictorial representation to the chart.

Figure 11.10 is a block diagram of an experimental apparatus for photographing the response of a tensile specimen to explosive loading. It shows the position of each piece of experimental equipment in the experimental layout and is thus an organization chart for an experiment, with parts of the apparatus located instead of people.

Figure 11.8 New York legislature— process of bill to law. (Courtesy *Mechanical Engineering* magazine)

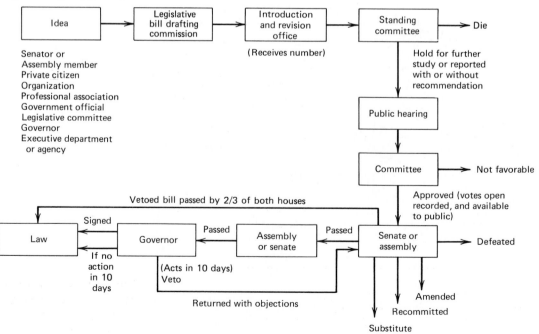

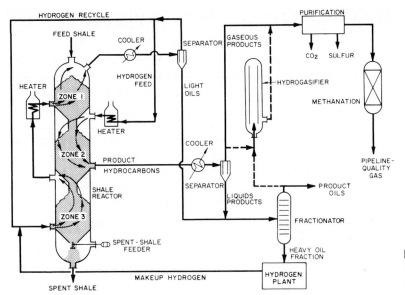
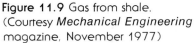

Figure 11.9 Gas from shale. (Courtesy *Mechanical Engineering* magazine, November 1977)

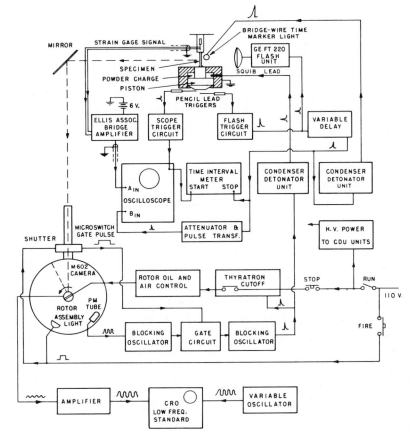

Figure 11.10 Experimental apparatus to obtain response to explosive loading. (Courtesy Dr. Arthur L. Austin)

PROBLEMS

11.1 Draw a bar chart showing the energy consumed in 10^6 GJ/year for each year for the following residential uses.

End Use	Energy Consumption (10^6 GJ/year)				
	1970	1980	1985	1990	2000
Space heating	8,890	10,740	11,400	12,130	13,590
Space cooling	530	1,710	2,220	2,640	3,010
Hot-water heating	2,530	3,330	3,630	3,760	4,230
Cooking	700	810	1,230	1,070	1,190
Appliances and lighting	2,680	3,700	4,300	5,240	5,880
Total	15,330	20,290	22,780	24,840	27,900

Source. Oak Ridge National Laboratory ONRL 5124.

11.2 Draw a bar chart showing the fuel consumed, measured in energy units of 10^6 GJ/year, for each year.

	Energy Consumption (10^6 GJ/year)				
	1970	1980	1985	1990	2000
Gas	9,060	11,940	11,950	11,730	12,190
Oil	5,910	8,090	8,930	9,740	11,290
Coal	4,380	6,810	7,080	7,160	6,610
Nuclear	390	2,720	6,200	10,320	17,180
Hydro	440	610	660	680	680
Total	21,180	30,170	34,820	39,630	47,950

Source. Oak Ridge National Laboratory ONRL 5124.

11.3 Draw a bar chart to show the number of visitors to the National Park system each year.

Year	Millions of Visitors
1950	33.3
1955	56.6
1960	79.2
1965	121.3
1970	172.0
1974	217.4

11.4 Draw a bar graph that shows the following data.

Production of Electric Energy in the United States

Calendar Year	Total, 1000 kwh	Hydro, 1000 kwh	Steam, 1000 kwh	Gas Turbine, 1000 kwh	Internal Combustion, 1000 kwh
1965	1,055,251,929	193,850,603	856,312,128	—	5,089,198
1968	1,329,443,027	222,490,584	1,101,767,366	—	5,185,000
1969	1,442,182,474	250,192,655	1,178,182,761	8,227,148	5,579,910
1970	1,531,608,921	247,456,119	1,262,358,866	15,732,082	6,061,854
1971	1,613,935,744	266,320,232	1,319,291,654	22,072,221	6,251,637
1972	1,747,322,933	272,733,504	1,438,420,059	29,493,248	—
1973	1,856,216,160	271,633,700	1,548,701,518	29,533,899	6,347,043
1974 (prelim.)	1,864,961,337	300,447,428	1,526,442,169	32,082,824	5,988,916

Source: Federal Power Commission.

11.5 According to a national survey, the following proportions represented the educational backgrounds of the engineers practicing in 1976. Draw a pie chart to show these data.

	Percent
No degree	9
Associate degree	2
Bachelor's degree	67
Master's degree	17
Doctoral degree	4
Other	1

11.6 Draw a segmented chart to show the distribution of funds in the United Fund crusade.

	Percent
Group services	30
Individual and family services	26
Multiservices agencies	25
Health services	14
Community development and planning	5

11.7 Show the growth of the public debt in the United States over the last 100 years with a bar chart.

Fiscal Year	Gross Debt
1880	$ 2,090,908,872
1890	1,132,396,584
1900	1,263,416,913
1910	1,146,939,969
1920	24,299,321,467
1930	16,185,309,831
1940	42,967,531,038
1950	256,087,352,351
1960	284,092,760,848
1970	370,093,706,950
1980	907,701,290,900

11.8 Show the growth of federal appropriations in the United States over the last 100 years with a bar chart.

Year	Appropriations
1890	$ 395,430,284.26
1895	492,477,759.97
1900	698,912,982.83
1905	781,288,215.95
1910	1,044,433,622.64
1915	1,122,471,919.12
1920	6,454,596,649.56
1925	3,748,651,750.35
1930	4,665,236,678.04
1935	7,527,559,327.66
1940	13,349,202,681.73
1944	118,411,173,965.24
1945	73,067,712,071.39
1950	52,867,672,466.21
1952	127,788,153,262.97
1955	54,761,172,461.58
1960	80,169,728,902.87
1965	107,555,087,622.62
1970	222,200,021,901.52
1975	374,124,469,875.62
1980	690,391,124,920.77

What are you going to do about 1944 and 1952?

11.9 Graphically show the data for employment and unemployment in the United States. The data are in thousands.

Year	Civilian Labor Force	Employed	Unemployed
1940	52,705	45,070	7,635
1950	62,208	58,918	3,288
1960	69,628	65,778	3,852
1970	82,715	78,627	4,088
1980	104,500	97,000	7,500

11.4 GRAPHS

Graphs of real data are usually developed in one of two ways. In the first, the coordinate scales are linear, and the behavior of the function is interpreted in terms of these linear scales. Normally, the origin is located at the lower left corner, and the independent variable is plotted as the abscissa and the dependent variable as the ordinate. This conforms to the usual functional relation $y = f(x)$.

Figure 11.11 is a standard, conventional graph with linear coordinates. It is a graphic expression of the data contained in Table 11.2, a dynamometer calibration. Since the strain measure on the load cell will be used to determine load, strain is the independent variable and load is the dependent variable. An actual field strain reading of 424 μin./in. translates to a load of 6250 lbf on a normal rectangular coordinate graph with linear scales. Variations from this norm are used when the perspective to be conveyed requires that the norm be set aside.

Figure 11.12 is a graph of the temperature distribution in a geothermal well. The origin for both temperature and depth in the well is the upper left corner because wells are dug down, not up. It would seem odd to show that temperature increases with depth in a well by plotting depth in the well upward.

Another variation is used when two or more graphs are plotted on the same set of coordinates. In Figure 11.13 the primary set of coordinates are the left ordinate and the abscissa. A secondary set of coordinates are the right ordinate and the abscissa. This variation avoids plotting two graphs. The set of curves shows the maximum pressures and rates of pressure rise produced by explosions of clouds of dry wheat starch. The intent is to show the effect of grain-dust concentration on grain-dust explosions. Maximum pressure, maximum rate, and aver-

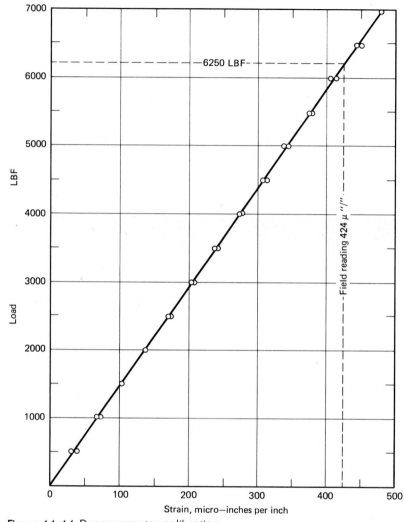

Figure 11.11 Dynamometer calibration.

age rate are all important, and all are shown on the graph. It would be disconcerting to shuffle three separate sheets with one curve on each sheet. In this case, plotting three curves on one graph is considered good practice.

One other aspect of this particular graph should be noticed. The grid lines have been removed. Tick marks have been placed at the major divisions. The effect is to give the graph a less cluttered appearance. To be sure, it is more difficult for the reader to read a precise value from the curve, but it is probably the intent of the originator of the graph that precise values were *not* to be read. This graph, without grid lines, is designed to show trends and not precise values.

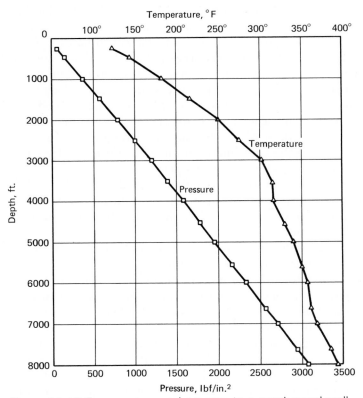

Figure 11.12 Temperature and pressure in a geothermal well. Imperial Valley, California.

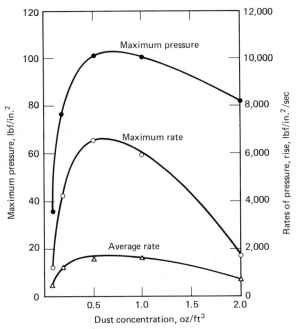

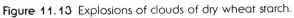
Figure 11.13 Explosions of clouds of dry wheat starch.

The second way of developing data is to plot the data, as nearly as possible, as a straight line. If the data do not follow a straight line, perhaps the scales of the coordinates can be changed to make the function a straight line. The most common nonlinear scale is the logarithmic scale. Scales other than linear or logarithmic do exist, but they are seldom used. They were the subject of Chapter 10. Figure 11.14 shows one such oddity: the viscosity of lubricating oil expressed as a function of temperature. The graph of oil viscosity is a straight line on these coordinates, but the coordinates are neither linear nor logarithmic. In fact, the grid is not even square!

There are three possible combinations of linear and logarithmic scales: linear-linear, logarithmic-logarithmic, and linear-logarithmic. All three have been discussed as the source of empirical equations in Chapter 10.

The first should be very familiar. Figure 11.11 is an excellent example of a linear-linear graph.

The second form is the logarithmic-logarithmic graph. It is often used in engineering, but it is not so often used outside of engineering and science. The concept of a logarithm must be understood by both the graph originator and the reader. Logarithmic scales should be used to express logarithmic functions, such as multiplication, division, and where one function is raised to some power. Logarithmic scales should not be used to express linear functions, such as addition and subtraction. Figure 11.15 shows a logarithmic-logarithmic plot of the cancer death rate as a function of age. Note how linear this curve is with logarithmic scales. Can you develop an empirical equation for this graph?

The third form is the linear-logarithmic or semilogarithmic graph, which is less used than either the linear-linear or the logarithmic-logarithmic graph. As it implies, one coordinate is linear, while the other is logarithmic. As an example of a semilogarithmic, consider Figure 11.16, which is a plot of the number of earthquakes per year at or above a specific Richter magnitude. The Richter scale is a measure of recorded ground motion. Each increase of number—from 5 to 6, for example—means the ground motion is 10 times greater. The actual amount of energy released may be 30 times greater. Earthquakes of Richter magnitude 4 number in the hundreds per year, but several years may go by before a violent earthquake of Richter magnitude 8 is felt. Earthquakes above Richter magnitude 9 have not been recorded. Obviously, you cannot use the same scales for both coordinates. A semilogarithmic scale collapses a linear scale into something more acceptable by decreasing the scale divisions for larger numbers. Note that for

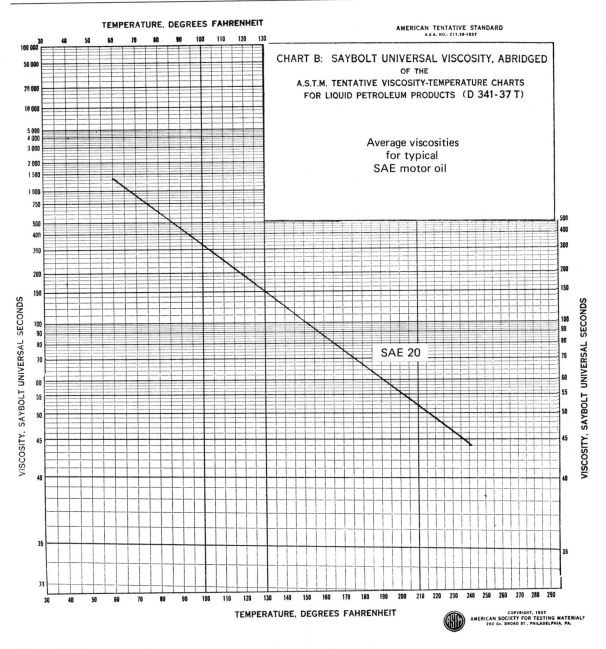

Figure 11.14

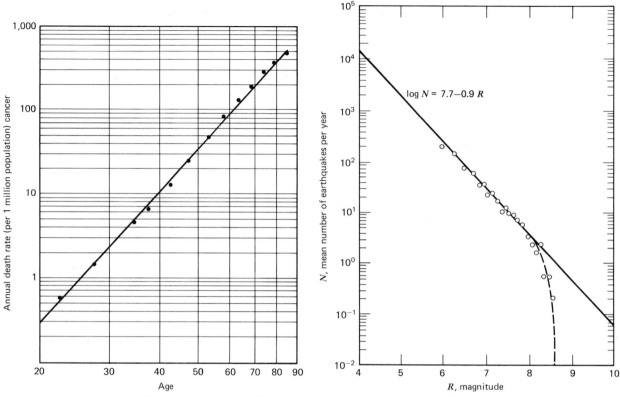

Figure 11.15 Annual death rate from cancer. (Courtesy *Scientific American*, "The Cancer Problem" by John Cains, November 1975, p 67)

Figure 11.16 Number of earthquakes per year versus magnitude for the entire earth. (Courtesy Professor George W. Housner)

these data, the originator has developed an empirical equation that fits the following.

$$\log N = 7.7 - 0.9R$$

Above the Richter magnitude 8, however, the semilogarithmic linearity no longer exists. Again, tick marks are used, not a grid.

In some cases, the use of coordinates other than rectangular coordinates enhance understanding of the presented data. It would be silly to use orthogonal coordinates to plot data given in polar coordinates. Figure 11.17 is a polar chart. The coordinates are r and θ.

In Figure 11.18 the thermal stress pattern in the hull of a ship is shown. One coordinate is the position on the hull, and a pictorial cross section of the ship's hull is used to locate this position. The other coordinate is stress, measured as a linear distance from the hull outline. It is an odd plot, but note that the single significant feature of the information presented is the marked change in stress at the hull waterline. At all other points, the measured stress is quite small. As a visualization of the phenomenon, this graph is effective.

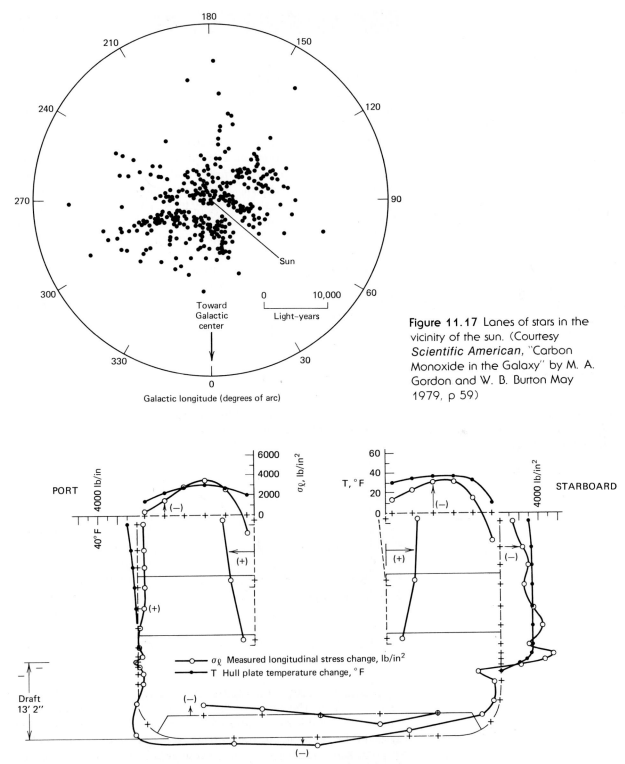

Figure 11.17 Lanes of stars in the vicinity of the sun. (Courtesy *Scientific American*, "Carbon Monoxide in the Galaxy" by M. A. Gordon and W. B. Burton May 1979, p 59)

○———○ σ_ℓ Measured longitudinal stress change, lb/in²
●———● T Hull plate temperature change, °F

Figure 11.18 Thermal stresses on a transverse section — S. S. Boulder Victory.

11.5 HOW TO DRAW A GRAPH

The first step in drawing any graph is to decide what information is to be presented and how. Take some time to sketch what you have in mind. Which of the pieces of given information are dependent on others? What are the maximum and minimum ranges of the data?

When these questions are answered, a grid can be selected. Remember, the purpose of a graph is to present data or information in an organized, simple, and logical manner. Choose scales that are convenient. Again, this means that the smallest division on the scale should be 1, 2, 5, 10, or some decade of these numbers. At this time you should also decide whether to use linear or logarithmic scales.

The data should now be plotted by placing each data point on the grid and identifying it by a symbol. Where only one set of data is to be placed on the graph, a small open circle is the identifying symbol. Originally, this circle was meant to show the circle of error, but this interpretation has been lost. Now, identification is the sole purpose of the symbol, and circles, squares, and triangles—open or filled—are all used.

If the data are discrete, which means that each data point represents an independent piece of factual information separate and distinct from every other, the data points are connected to one another. Figure 11.19 is a discrete plot of the hourly temperature variation in Chicago, Illinois, over one

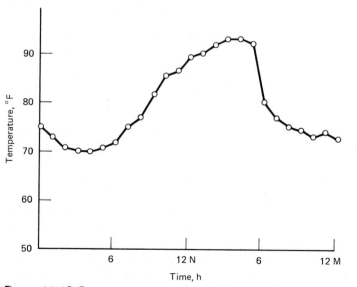

Figure 11.19 Temperature variation in Chicago, Illinois, on a typical summer day.

summer day. In a discrete curve, the graph passes through each and every point. Figures 11.12 and 11.18 are also discrete graphs. If the data consist of many points, but each has an error or uncertainty and no point is more definite than any other, a curve is "fitted" to the plotted data points, balancing the data so that the distribution of plotted points is the same on both sides of the curve. This is called a continuous curve, and there are various schemes for it. Figures 11.11, 11.15, and 11.16 are continuous curves. Fitting a functional curve to a set of data yields an empirical function, or probably an empirical equation.

Last, the axes should be independently labeled, including units with major divisions marked. For example, if 5 had been chosen as the smallest division, perhaps labeling the grid as 10, 20, 30, 40, or 50, 100, 150, etc., would be appropriate. The completed graph should be properly identified. In doing this, some judgment must be used to avoid cluttering the graph with unnecessary or superficial notes that distract the eye from its first purpose, which is to absorb information. There must be enough information to describe the graph clearly, particularly since it may be viewed with no supporting text long after the reason for drawing the graph is forgotten, but not so much information as to make the graph distracting, ineffective, or unappealing.

11.6 WHAT NOT TO DO!

It is as important to recognize poor graphic technique as it is to understand what is good technique. Examples of bad charts, poor graphs, and improper practices abound, mostly through the careless treatment of graphics as a medium of expression. If you know what is good practice, you should be able to recognize bad practice.

The most abused practice is the broken scale. In plotting a graph, there is often the question of whether one or the other of the axes could be shortened. There are times when one or the other should be shortened, but a scale must be broken only for a reason. For example, in Figure 11.20 if the purpose of a graph is to show the variation in speed from a nominal speed of 3600 rev/min, why not plot the curve from 3595 to 3605 as in (a) instead of from 0 to 4000 in (b)? A vertical ordinate varying from 3595 to 3605 is sensible. Certainly, on an absolute scale you would not be able to see a speed variation of 10 rev/min in 3600. On the other hand, if your purpose is to show constant or non-constant speed, perhaps the absolute scale would be better.

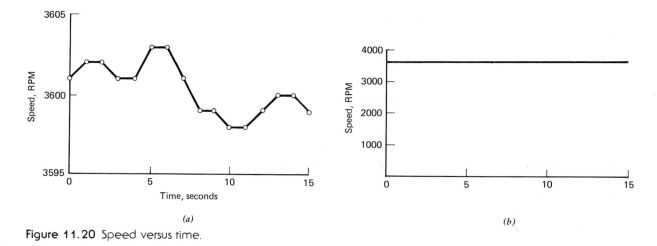

Figure 11.20 Speed versus time.

As another example, breaking any temperature scale is usually quite acceptable. An absolute scale is only necessary if absolute temperature is being treated. The Fahrenheit and the Celsius scales are broken scales, so shortening them cannot be misleading. In Figure 11.19 the temperature scale is broken. Note that it begins at 50°F. In the Chicago area, where the temperatures were measured, 50°F is about the lowest temperature observed during the summer. This graph is a correct use of a broken scale.

Figure 11.21*a* shows an incorrectly used broken scale. From 0 to $20 000, the graph is cut out, with the lower section being removed. The impression from the graph is that there is more of a change of salary with years than there really is, and it is misleading because most readers will not recognize what has been done to the graph. Figure 11.21*b* shows the data of Figure 11.21*a* correctly redrawn without the broken scale. It gives a much different impression.

Figure 11.22 is similar. It is a curve showing the number of juveniles paroled to a state youth authority between 1955 to 1975. Beginning the graph in 1955 at 3000 gives the mistaken impression that 1955 was some sort of threshold year, which in fact it was not. You will also notice that the first three divisions have 1000 per division, the fourth 1500 per division, and the last three 2500 per division. The graph, as it exists, is incorrect and is misleading, defeating its purpose.

Another practice is to place too much information on one graph. The mind and the eye cannot absorb more than three or four separate pieces of graphic information at any one time, particularly if there are separate scales for each curve.

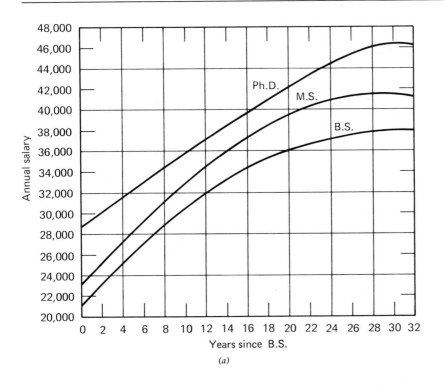

Years since B.S.

(a)

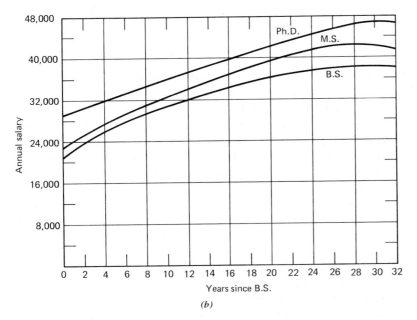

Years since B.S.

(b)

Figure 11.21 Comparisons of salaries by degree level, without regard to supervisory status. (Part (a) courtesy *Mechanical Engineering* magazine, February 1982)

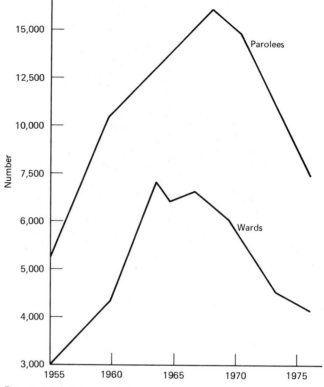

Figure 11.22 A graph from a newspaper.

In Figure 11.23, twenty separate representations of medical data are placed on one graph—far too much information for one graph and very bad practice.

In Figure 11.24 three separate scales are used on the same graph. As a further affront, they are placed end to end. The ordinate for 100,000 lb/in.² on the first scale is the origin for the second, and so on. The graph is busy as well as improper. It contains too much information.

In Figure 11.25 the bar is broken as well as the scale. Neither is warranted. Figure 11.26 is a bar graph showing the total number of new members admitted to an engineering society, but it visually misleads the reader because the scale starts at 100 members. To be accurate, the bar chart should have been less than 10 mm larger! One could argue that breaking a scale saves space, but what advantage is there in saving 10 mm out of 150 mm?

Figure 11.27 is a graph on the correction of carbon dating through a study of tree rings. Whimsically, the lower right cor-

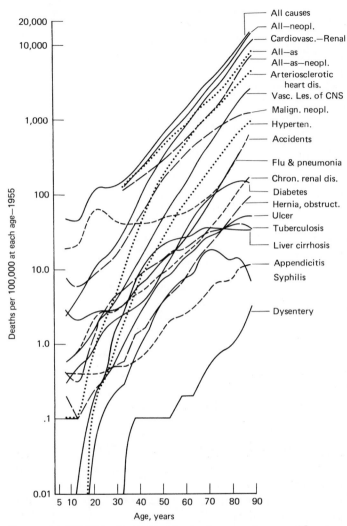

20,000
10,000

1,000

100

10.0

1.0

.1

0.01

Deaths per 100,000 at each age—1955

5 10   20   30   40   50   60   70   80   90

Age, years

All causes
All—neopl.
Cardiovasc.—Renal
All—as
All—as—neopl.
Arteriosclerotic
  heart dis.
Vasc. Les. of CNS
Malign. neopl.
Hyperten.
Accidents

Flu & pneumonia
Chron. renal dis.
Diabetes
Hernia, obstruct.
Ulcer
Tuberculosis
Liver cirrhosis

Appendicitis
Syphilis

Dysentery

Figure 11.23 Mortality from selected causes by age (Courtesy *Creative Science and Technology*, Vol. 1, No. 1, p 14)

ner is taken as the origin. Corrected radiocarbon dates go negatively, to the left, backward in time. This is reasonable, but uncorrected radiocarbon dates go positively, upward in time. One line confuses the other. This graph would have been both eye-catching and easier to read if the upper right corner had been taken as the origin. If you become aware of what the author is trying to do, you can judge whether the graphs help or hinder. You may also become aware of some very bad graphs in places where you would not expect them to be.

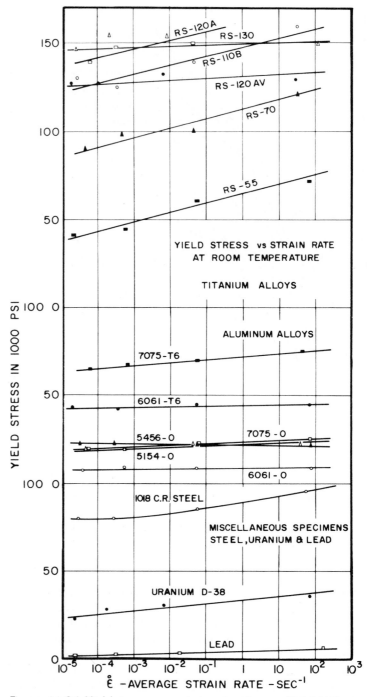

Figure 11.24 Yield stress versus strain rate at room temperature.

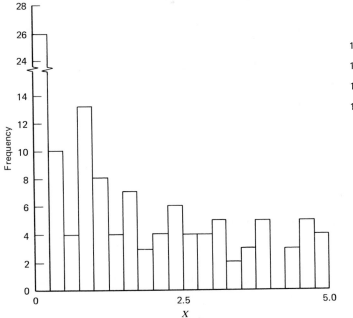

Figure 11.25 Histogram of fracture location. X = distance in millimeters from left-hand load point.

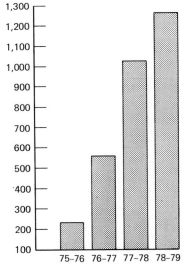

Figure 11.26 Total new members by year. (Courtesy *Engineering News*, October 1979)

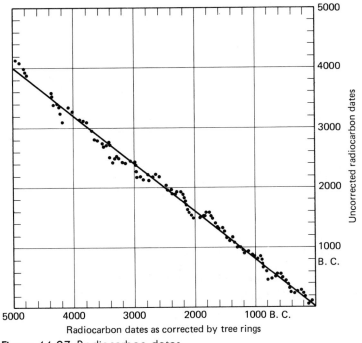

Figure 11.27 Radiocarbon dates.

PROBLEMS

11.10 Graph the years of life expected at birth from 1900 to 1980.

Year	Total Population	Male	Female
1900	47.3	46.3	48.3
1910	50.0	48.4	51.8
1920	54.1	53.6	54.6
1930	59.7	58.1	61.6
1940	62.9	60.8	65.2
1950	68.2	65.6	71.1
1960	69.7	66.6	73.1
1970	70.8	67.1	74.7
1980	74.1	70.2	78.1

11.11 Graph the combustion efficiency as a function of air-to-fuel ratio.

Air / Fuel	Combustion Efficiency, percent
11	66.7
12	73.8
13	81.5
14	89.6
15	93.8
16	94.8
17	95.5
18	96.2
19	96.5
20	96.9

11.12 Make a linear-linear graph that shows the production of automobile factory sales by year.

Year	Passenger Cars, Number	Motor Trucks and Buses, Number	Total
1900	4,192	—	4,192
1905	24,250	350	24,600
1910	181,000	6,000	187,000
1915	895,930	74,000	969,930
1920	1,905,560	321,789	2,227,349
1925	3,735,171	530,659	4,265,830
1930	2,787,456	575,364	3,362,820
1935	3,273,874	697,367	3,971,241
1940	3,717,385	754,901	4,472,286
1945	69,532	655,683	725,215
1950	6,665,863	1,337,193	8,003,056
1955	7,920,186	1,249,106	9,169,292
1960	6,674,796	1,194,475	7,869,271
1965	9,305,561	1,751,805	11,057,366
1970	6,546,817	1,692,440	8,239,257
1975	7,331,256	2,727,313	10,058,569
1980	6,400,026	1,667,283	8,067,309

11.13 The flow in a mountain stream is read on a remote indicator on the first day of every month. The following data are for a typical year. Draw a discrete graph of the annual flow in cubic feet per second. This is called a hydrograph, and it is used to determine the potential power available.

Month	Annual flow, ft^3/sec
January	200
February	195
March	200
April	475
May	700
June	600
July	510
August	340
September	200
October	200
November	320
December	250

11.14 Graph the United States employment statistics over the 25-year period 1950–1975.

	Employed	Unemployed	Armed Services	Total
1950	58,920,000	3,288,000	1,650,000	63,858,000
1955	62,171,000	2,852,000	3,049,000	68,072,000
1960	65,778,000	3,852,000	2,512,000	72,142,000
1965	71,088,000	3,366,000	2,724,000	77,178,000
1970	78,627,000	4,088,000	3,188,000	85,903,000
1975	83,549,000	7,820,000	2,195,000	93,564,000

Source. U.S. Bureau of Labor Statistics.

11.15 Prepare a graph of the data in Problem 11.8.

11.16 Plot the following data on linear-logarithmic axes for 25 United States companies with the largest annual sales or revenues.

Rank	Company	Sales or Revenues
1	Exxon Corp.	$45,020,800,000
2	General Motors Corp.	31,549,500,000
3	American Tel. & Tel.	26,174,400,000
4	Texaco Inc.	25,417,200,000
5	Ford Motor Co.	23,620,600,000
6	Mobil Oil Corp.	20,284,000,000
7	Gulf Oil Corp.	17,952,000,000
8	Standard Oil Co. of Cal.	17,924,400,000
9	General Electric Co.	13,413,100,000
10	Sears, Roebuck & Co.	13,101,200,000
11	Int'l. Business Machines	12,675,300,000
12	Int'l. Tel. & Tel. Corp.	11,154,400,000
13	Chrysler Corp.	10,971,400,000
14	Standard Oil Co. (Ind.)	10,024,600,000
15	U.S. Steel Corp.	9,186,400,000
16	Shell Oil Co.	8,418,300,000
17	Safeway Stores, Inc.	8,185,200,000
18	Continental Oil Co.	7,279,600,000
19	Atlantic Richfield Co.	7,166,900,000
20	Penney (J.C.) Co.	6,935,700,000
21	Du Pont de Nemours (E.I.)	6,910,100,000
22	Great A & P Tea Co.	6,874,600,000
23	Westinghouse Electric	5,798,500,000
24	General Tel. & Electronics	5,661,500,000
25	Occidental Petroleum Corp.	5,537,500,000

11.17 Sample automobile speed records are listed in tabular form. Make a graph that clearly shows how the speed record changed with time.

Year	Driver	Speed, mi /h
1898	Chasseloup-Laubat	39.24
1904	Henry Ford	91.37
1910	Barney Oldfield	131.72
1914	L. G. Hornsted	124.10
1922	K. L. Guinness	129.17
1926	J. G. Parry-Thomas	168.08
1932	Sir Malcolm Campbell	253.96
1938	John Cobb	350.20
1947	John Cobb	394.20
1963	Craig Breedlove	407.45
1970	Gary Gabelich	622.41
1979	Stan Barrett	638.64

11.18 Graphically present the changes in the first-class mail rate between 1917 and the present. Include the effect of time in your presentation.

Nov 3, 1917	3¢
July 1, 1919	2¢
July 6, 1932	3¢
Aug. 1, 1958	4¢
Jan. 7, 1963	5¢
Jan. 7, 1968	6¢
May 16, 1971	8¢
Mar. 2, 1974	10¢
Dec. 11, 1975	13¢
May 29, 1978	15¢
Mar. 22, 1981	18¢
Nov. 1, 1981	20¢

11.19 Graphically present the data for United States corporate profits (in billions of dollars) in an appropriate way.

Year	Total Pretax Profits	Tax Liability	After-Tax Profits	Dividends	Undistributed Profits
1929	10.0	1.4	8.6	5.8	2.8
1930	3.7	0.8	2.9	5.5	−2.6
1931	−0.4	0.5	−0.9	4.1	−4.9
1932	−2.3	0.4	−2.7	2.5	−5.2
1933	1.0	0.5	0.4	2.0	−1.6
1935	3.6	1.0	2.6	2.8	−0.2
1940	10.0	2.8	7.2	4.0	3.2
1945	19.7	10.7	9.0	4.6	4.4
1950	42.6	17.9	24.7	8.8	15.9
1955	48.4	22.0	26.4	10.3	16.1
1960	48.5	22.7	25.8	12.9	13.0
1965	75.2	30.9	44.3	19.1	25.2
1967	77.3	32.5	44.9	20.1	24.7
1968	85.6	39.4	46.2	21.9	24.2
1969	83.4	39.7	43.8	22.6	21.2
1970	71.5	34.5	37.0	22.9	14.1
1971	82.0	37.7	44.3	23.0	21.3
1972	96.2	41.5	54.6	24.6	30.0
1973	115.8	48.7	67.1	27.8	39.3
1974	126.9	52.4	74.5	31.0	43.6
1975	120.4	49.8	70.6	31.9	38.7
1976	155.9	64.3	91.7	37.9	53.8
1977	173.9	71.8	102.1	43.7	58.4
1978	202.6	84.1	118.5	49.3	59.2

Source: U.S. Department of Commerce, Bureau of Economic Analysis.

11.20 Graphically present the data for degrees conferred to college and university graduates in an appropriate way.

School Year	All Degrees	Bachelor's and First Professional	Master's	Doctoral
1869–1870	9,372	9,371	0	1
1889–1890	16,703	15,539	1,015	149
1899–1900	29,375	27,410	1,583	382
1909–1910	39,755	37,199	2,113	443
1919–1920	53,516	48,622	4,279	615
1929–1930	139,752	122,484	14,969	2,299
1939–1940	216,521	186,500	26,731	3,290
1949–1950	496,661	432,058	58,183	6,420
1959–1960	476,704	392,440	74,435	9,829
1969–1970	1,065,391	827,234	208,291	29,866
1979–1980	1,305,000	923,000	290,000	33,000

Source: National Center for Education Statistics.

11.21 The shaded-pole electric motor is the simplest alternating current (ac) motor. The speed of this low-efficiency motor is governed by (synchronous with) the frequency of the ac current supplied to the motor. These low-horsepower ($< 1/200$) motors are used for such devices as fans where low starting torque can be tolerated. Data from a test of a shaded-pole motor are tabulated. Prepare a performance curve for this motor; a performance curve is a single graph that presents all the relationships in the table. Plot speed, power, and efficiency on the ordinate and torque on the abscissa.

Torque, oz-in.	Speed, rev/min	Power, hp	Efficiency, percent
1	1400	0.0015	5
2	1360	0.0028	9
3	1320	0.0041	13
4	1290	0.0056	16
5	1220	0.0066	17
6	1090	0.0070	17

11.22 The force F on a sphere of diameter D and frontal area (projected area) A immersed in a fluid of density ρ moving past the sphere with velocity V is given by

$$F = C_D \, \rho \, \frac{V^2}{2} \, A$$

where C_D is the so-called drag coefficient. A common dimensionless parameter used in fluid mechanics is Reynolds Re, which represents the ratio of inertia forces to friction forces.

$$Re = \frac{DV}{\nu}$$

where ν is the viscosity of the fluid. For low Reynolds numbers, the following data were obtained experimentally.

Re	C_D
0.4	60
0.8	35
1.0	30
2.0	15
4.0	9

Plot these data on logarithmic-logarithmic paper along with the corresponding theoretical relationship

$$C_D = \frac{24}{Re} \quad \text{(Stokes law)}$$

Prob. 11.23

11.23 When a bending moment M is applied to a solid circular member of diameter D, the resulting maximum stress τ is

$$\tau = \frac{32}{\pi} \frac{M}{D^3}$$

Prepare a logarithmic-logarithmic plot that shows the relationship between τ and D for four discrete values of M (1000, 2000, 3000, 4000 lb-in.). Let D vary from 1 in. to 3 in. for each of the four lines you plot.

11.24 The electrical resistance R of a homogeneous conductor of constant cross section A and length L is given by

$$R = \rho \frac{L}{A}$$

where ρ is the resistivity of the material. The resistivity of a particular copper at 20°C is 1.75×10^{-8} ohm·m and it increases 0.4 percent for every °C increase. Prepare a graph to show ρ versus T (temperature) for this copper ($0 \leq T \leq 80°C$).

11.25 Noise level β is expressed in decibels or db and is defined as

$$\beta = 10 \log \frac{I}{I_0}$$

where I is sound intensity and I_0 is an arbitrary reference intensity that corresponds to the lowest intensity that can be heard. Some typical levels are as follows.

	Noise Level, db
Pain threshold	120
Elevated train	90
Conversation	60
Low whisper	30
Hearing threshold	0

Plot the relationship between β and I two ways using: (1) two linear scales, and (2) semilogarithmic paper. Which method is best?

11.26 One form of the ideal gas law is

$$pv = RT$$

where

p = pressure, N/m²

v = molar volume, m³/mole

R = universal gas constant, 8.317 J/mole · °K

T = temperature, °K

Prepare a pictorial of the three-dimensional surface that represents the ideal gas law. Let T vary from 200 °K to 350 °K.

11.27 Read an engineering, scientific, or technical article that contains data in the form of charts, curves, graphs, or some other graphic method. Then make comments on the graphic presentation of the data. Are they clearly presented? What is the message of the author? Why did the author present a point of view graphically in that way? Can you make a suggestion that could improve the presentation? Is there too much material? Too little?

11.28 Locate an article in the popular press (magazine, newspaper) that uses some form of graphic presentation of data or relationships. Comment on the presentation. What is good about it? How can it be improved? How did this graphic presentation help the reader understand the article?

12
THE ENGINEERING REPORT

More than any other communication vehicle, your *engineering reports* will be your major means of transmitting your engineering information and results to others. In your career in engineering, you will write hundreds of reports of all shapes and sizes, possibly thousands, depending on your job. Many reports will be short, a few will be long, some will be major achievements and used as reference works for a long time, some will be swept under a rug and forgotten, and a few may be published. The composite lot of them will be you on paper, a written history of your career. Surprising though it may be, over most of your professional life, paper documents will be your principal product. Others will judge you by the quality of your reports. You will give yourself and your career a considerable lift if you learn to write a good engineering report, and if you learn to write a good report early in your career.

It is important to develop a good writing style, but it is difficult to be taught a good writing style. So much depends on the career you choose. You can learn some obvious things that should or should not be done. In writing a report, you should always know and consider your reader and always consider that your reader will not read detail. We can suggest a proper format for your engineering reports, and we can suggest some outside reading. We can correct your grammar and punctuation. Beyond that, the communication skills that you develop are up to you. The statement is redundant, but any time that you can devote to improving the clarity, accuracy, completeness, and neatness of your engineering reports is a good investment.

Good writing should be coupled with good graphics. It is not possible to transmit complete engineering information through prose alone. Charts, graphs, figures, photographs, drawings, sketches, layouts, prints, and diagrams may all be included in an engineering report. Good graphics can convey much information, but poor graphics can turn off your reader. You must know what to include as graphics, when to include them, how to use them effectively, what constitutes good graphics, and what constitutes correct graphics.

You will now put all the things you learned in the previous chapters together using the best graphics to communicate your ideas in an engineering report.

12.2 TYPES OF ENGINEERING REPORTS

There are as many types of engineering reports as there are engineering topics and engineering organizations, but some of the common categories are listed here. Each has a different purpose.

Letters.
Memorandums.
Engineering notes.
Reviews.
Proposals.
Progress reports.
Technical reports.
Technical articles.

Letters and *memorandums* are the simplest forms of engineering reports. They are directed to a single person or to a very few persons, usually with a single, simple objective. Letters and memorandums are much the same, and they are used for the same purposes; the only difference between them is that letters are used externally and memorandums are used internally. Memorandums are apt to be a little less formal, since the writer and reader may know each other, and they are part of the same organization.

Figure 12.1 is an example of a letter written to transmit engineering information. Figure 12.2 is an internal memorandum to

William H. Kimball, D.Eng.
Consulting Engineer
813 MALAGA AVENUE
DAVIS, CALIFORNIA 95616
(916) 756-3585

August 16, 1982
Job. No.82/103.3

Capitol Engineering Laboratories
1828 Tribute Road
Sacramento, CA 95815

Attention Mr. Howard Anderson

Re: Lab No.559-995
    Job No.8-201

LABORATORY TEST REPORT

Observations of the fracture surfaces of the part involved in your project referenced above show that fatigue fracture as a mechanism of failure was evident on the fracture surfaces. Fatigue patterns were observed on both the long fracture surface and on the shorter surface. The fracture surface morphology indicates multiple origins with no obvious defects at the point of initiation of the fatigue cracks. Further observations indicate that the cracks were propagating on the "push" cycle rather than on the "pull" cycle.

With regard to the other part on the crane that matches the part which failed, any indications of cracking should be considered as fatigue cracking until determined to be otherwise. Further, grinding or arcing and rewelding of any cracks found is not a recommended repair method since this will in most cases leave either a portion of the crack at the root of the repair weld or fatigue damaged material at the root of the weld. The weld can then increase residual stress near the crack or damaged area, resulting in a greatly reduced fatigue life.

These results are based on observations at this point; further testing can be performed to verify these observations if needed. If you need additional analysis, please let me know.

Respectfully submitted,

W. H. Kimball
W.H. Kimball, D. Eng.

Figure 12.1 Letter report

October 2, 1975

Steam Generator J Tube Test at Trojan
File No. 140.020
Units 1 and 2- Diablo Canyon Site

MEMORANDUM TO THE FILE:

On September 28, 1975, I attended tests at Trojan Power Plant demonstrating the "J tube" modification of the steam generator feedwater distribution ring as a solution to the steam generator water hammer problem. The tests were completely successful in that no water hammer occurred at any time.

The tests consisted of initiating auxiliary feedwater flow to a steam generator under a number of simulated accident conditions. Flows tested were nominally 120 gpm, 220 gpm, and 440 gpm. 220 gpm is the minimum they need for their safety analysis criteria to be met.

Tests were run at approximately 1000 psig, 900 psig, and 400 psig at saturated temperature. These represented auxiliary feedwater flow initiated by a safety injection signal caused by a LOCA at low or no load, a LOCA at full load, and a steam line or feedwater line blowdown accident. The tests were run with the steam generator water level no higher than the bottom of the feedwater ring.

A chart recording was taken for steam generator pressure, pump discharge pressure, and flow to the steam generator. Another chart recorded temperature at five points around the circumference of the feedwater line close to the steam generator feedwater nozzle. Westinghouse had instrumented the feedwater line with strain gauges and accelerometers which would have recorded the forcing function of the water hammer if there had been one.

Most of the tests were run after a thirty minute waiting period to allow the feedwater ring to drain. This was agreed upon with the NRC rather than proving that the ring had drained. It appeared that the header did not drain completely, as indicated by the temperature around the feedwater line. One test was run after a two hour wait for draining to occur. In this test there was no question that the header was drained completely.

The feedwater lines at Trojan are similar to those at Diablo Canyon. I believe that there is no question that the tests are representative of and valid for our plant.

The tests at Trojan demonstated that the J tube modification will prevent steam generator water hammer over the range of auxiliary feedwater flow and steam generator pressure which might be anticipated in service. It is the only known modification which will do so. Therefore, it seems obvious that we should incorporate this modification at Diablo Canyon.

Ross M. Laverty

RML/jm

510

file, which means that it was written to record information but it was not directed to a specific individual. It is a common practice to file information for the record.

Engineering notes are more formal than letters and memorandums but less formal than a fully developed engineering report. They are usually longer than letters or memoranda and are also more complete and more technical. A note may be directed to one person, but a note is usually written for wider distribution than a letter, and persons other than the addressee receive copies.

Engineering notes range from handwritten notes to formally printed reports. In Figure 12.3 two pages of a four-page engineering note are reproduced. Note that it is handwritten but neatly printed. It is meant to be used as a reference document, but not by very many people.

p. 512

Proposals, progress reports, and *reviews* all have specific purposes. Each is a short report. A proposal is a specific plan presented for acceptance or rejection. It carries a brief statement or offer, arguments, reasons for acceptance, and, usually, costs in time and money. Figure 12.4 is a short letter proposal. A progress report states the progress of work over a brief period. Monthly progress reports are common to many projects and are a primary way for project engineers to inform others. Figure 12.5 is a short progress report. A review covers more time and is more comprehensive than a progress report. A review occurs at some definitive milestone or at the end of a project, but not at weekly or monthly intervals.

e.g. Theatre review, or on a test, article

The formal engineering report is usually a published document that is either internally published as a *technical report* or externally published as a *technical article,* which is usually published in a technical journal or magazine. Technical reports and technical articles have the same structure. The only real difference between the two forms is that the technical article is usually reviewed by others before publication. Most journals have an independent panel consisting of at least three members to which technical articles are submitted for an independent, objective review before publication. Internal reports may also be reviewed, but they are reviewed more for grammar and policy than for technical content.

← annual or monthly

9) CONTRACT

W.O. STYLE

10) SPECIFICATION

Figure 12.2 (opposite) Internal memorandum to file

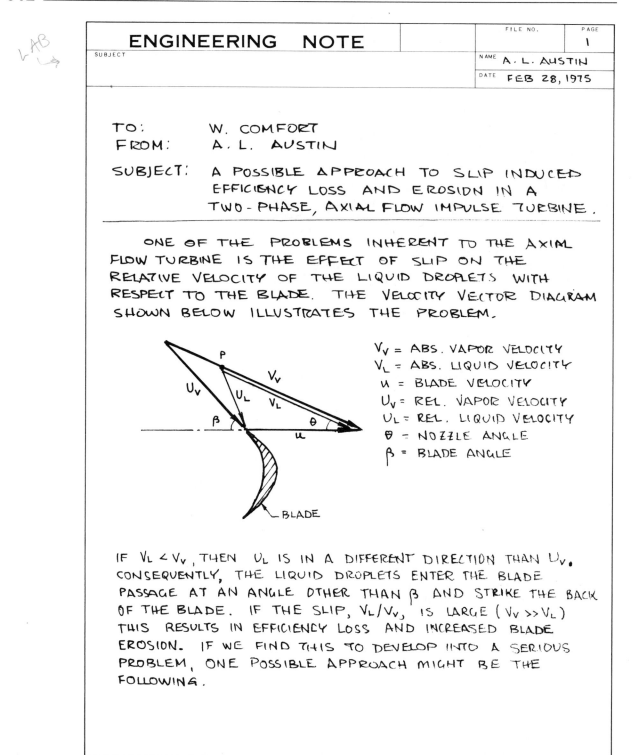

Figure 12.3 Engineering note (Courtesy Dr. Arthur L. Austin)

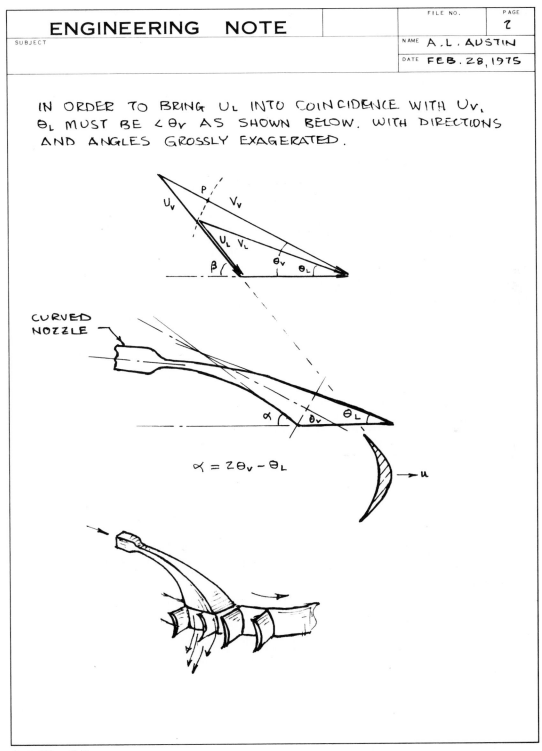

ENGINEERING NOTE

SUBJECT

FILE NO.

PAGE
2

NAME A.L. AUSTIN

DATE FEB. 28, 1975

IN ORDER TO BRING U_L INTO COINCIDENCE WITH U_V, θ_L MUST BE $< \theta_V$ AS SHOWN BELOW, WITH DIRECTIONS AND ANGLES GROSSLY EXAGERATED.

CURVED NOZZLE

$$\alpha = 2\theta_V - \theta_L$$

July 19, 1982

Mr. Donald G. Harris
Energy Conversion Coordinator
Process Engineering Department
Chevron Research Company
576 Standard Avenue
Richmond, California 94802

Dear Don:

This is a proposal to assess the loss of efficiency in a Curtis impulse bladed turbine through blade erosion. I am joined in this proposal by Dr. David Pankow.

Our objective is simple. We propose to install the turbine in a conventional manner and measure the mechanical efficiency of the turbine rotors with different amounts of blade erosion. The inlet pressure that we can provide in Hesse Hall, at Berkeley, is slightly below the turbine design condition, but we feel that by lowering the exhaust pressure, the velocity of steam across the blading will be essentially the same. Using readily available instrumentation will provide uncertainties in the order of 2-3%. A new rotor and an eroded rotor would be the minimum needed for our proposed tests, although it would be more satisfactory to measure the mechanical efficiency of several rotors with varying degrees of erosion. Once the turbine and brake are installed, changing the turbine rotor would be a minimum effort. If you accept our proposal, we will have to make some firm arrangements and come to an agreement on which and what rotors are to be tested.

We plan to install the Elliott 2-AYR Turbine, Serial No. C2585, in Room 120 Hesse Hall, and direct couple it to the Telma model CA-135 Eddy current dynamometer. Saturated steam at 125 psia (.18 MPa) is readily available at the selected test site. Exhaust pressure will be actively controlled, discharging to an existing condenser, and condensate flow will be calibrated using existing weight tanks. We have a Lebow model 1605-2K torque cell for measurement of the power output of the turbine. Other instrumentation will have to be purchased. Once installed, it will be permanently a part of the fabricated test stand.

The total cost for the project is projected to be $36,785. This includes the cost for the fabrication of the test stand, installation of the turbine and brake, and the purchase and installation of instrumentation and controls.

This cost is projected to be $7,625, and is included in the total. The projected time for completion of the testing and the preparation of a report is twelve months. A budget and other details, such as an instrument and equipment list, are attached.

Sincerely yours,

Robert F. Steidel, Jr.
Professor of Mechanical Engineering

RFS:crm
Attachments

Figure 12.4 Letter proposal

February 24, 1982

Mr. John Gonzales
Geothermal Energy Division
Department of Energy
San Francisco Operations Office
1333 Broadway
Oakland, CA 94612

Dear John:

   This our progress report for the months of October, November and
December, 1981. You will recall that we agreed to report our progress
quarterly during the academic year.

   Our efforts during these three months were directed to hybrid staging
of the Lysholm engine with conventional Curtis turbines. The available
thermal energy is dramatically increased by expanding steam to sub-atmospheric
pressures, but our test results indicate the Lysholm is not particularly
suited for low exhaust pressures, thus the Lysholm-turbine combination. To
shorten the data collection process, we are presently developing a computer
model of the Lysholm engine, and a computer model of the staged system.

   The models will be constructed by curve-fitting our observed data and
developing interpolation programs to fit between our observations. This will
eliminate any need for further testing within the regimes of 4500 to 9000
rpm, 0:1 to 5:1 water to steam mass flow ratios, and pressure ratios from 3:1
to 13:1. In order to pin something down, we are fixing our inlet pressure at
120 psia. The exhaust pressure will vary. In order to proceed with the
construction of the computer model, we did have to make additional Lysholm
runs at high exhaust pressures, e.g. 20-30 psia. We expect optimum performance to be within this range of exhaust pressures. These runs have been completed.

                              Sincerely yours,

                              Robert F. Steidel, Jr.
                              Professor of Mechanical Engineering

RFS/cdc

Figure 12.5 Progress report

12.3 KNOW YOUR READER

The first and most important guideline to writing a good report is to know your reader. To whom is the report to be written? What is the reader's background? How much should you explain? Can you use commonly accepted engineering terms or do you need to introduce them? Who is going to use the information you have assembled? Where will it reside a year after it is written?

Your reader may be a specific person—your supervisor, the project engineer, the chief engineer, or a group of people, perhaps the president of the company and an advisory staff. Few, if any, engineering reports are written to be read by the general public. Some engineering reports are disseminated widely, but even these have some specificity. For example, the report may go to a distribution or mailing list of tens or even hundreds, but there is some reason for each and every name to be on that mailing list. Obtain that distribution list before you start writing! Find out who these people are and why they will read your report!

Second, you cannot underestimate what your readers will know about your subject. After all, it is your subject, not theirs. You are the expert, giving out information that you have. If the information were otherwise available, you would probably not be writing your report. This is a delicate matter and one that is more often mishandled than not. In preparing for an uninformed reader, most report writers treat the reader as unintelligent instead of uninformed. Your reader may be, and probably is, better prepared than you. Most engineering reports are read by people above the engineer in a corporate structure rather than below. It is a good bet that your reader will have knowledge that you do not have about economics, personnel, company plans or policies, other disciplines, and other experiences. Your reader is interested in your information and what it will add to the subject. Your reader does not wish to read abbreviated textbook material, nor does your reader expect the abrupt introduction of unfamiliar terms and technology. Introduce only what is necessary and no more.

Finally, as a last comment, do not write for peer review of your work. You can be assured that you will be judged by your writing, but you should not write an engineering report solely to impress others. It will not. Nothing could be more contrary to the purpose of an engineering report, which is to transmit information to other engineers who want and need that information. "Snowing" your reader is a mistake that few engineers have an opportunity to make twice.

12.4 THE FORM OF A REPORT

The form of an engineering report varies with its purpose, the intended reader, and the circumstances under which it is written. The philosophy of the parent organization must be considered. Permanence and distribution could be factors. Time or the size of the project may be important. A small project may deserve only a short note. A large project may require a formal report that takes many months of preparation. These are all possible circumstances affecting the construction of the report. Most reports, however, do have most or all of the following six features.

1. Abstract or summary.
2. Introduction.
3. Report body.
4. Conclusions and (recommendations.) NOT ALWAYS ⊕ , DON'T WEASEL
5. Bibliography or references.
6. Appendixes. NOT READ

This list is a general outline; all reports should be outlined before writing is started. The particular outline for your reports may vary, but all reports have a beginning, a middle, and an end. The abstract or summary and introduction are the beginning. The report body is the middle and the main section of the report. The conclusions, with or without any recommendations, are the end. This construction is normal for engineering reports, and some readers develop a pattern of reading only beginnings and conclusions. Although this a bad pattern to set yourself, you should be prepared for the fact that it does exist in others. Your beginning should be a clear and accurate statement of the problem and your objective. Your conclusions should be a precise and brief statement of your findings. Each should be written to entice the reader to read the main body of the report.

12.5 AN ABSTRACT

ENTICE

An abstract states the results and conclusion of the report concisely and gives any important recommendations. The purpose of an abstract is to describe the contents of the report in a few sentences so that someone searching for information would be able to determine whether or not to read the report in detail.

An abstract that is a part of a report is sometimes called a summary.

An abstract is written after the rest of the report. It describes not only the contents of the report but also some details and numerical values from the results and conclusions.

The importance of a good abstract cannot be overstated. In most cases, the abstract will be all that a reader will read. It should be written in a manner that assumes that a reader may not proceed beyond this point. An abstract should have no explanations or undefined symbols. It is often published separately, so it should stand by itself. Figure 12.6 is a published abstract that appeared in *Solar Energy Update* (1981), a compendium of 10,406 abstracts on solar energy. Do you see the need for abstracts?

Today, an abstract is a vital part of an engineering report. Unfortunately, many engineers are piqued by such scant use of their effort and time. They do not appreciate the need or the use of the abstract. They write incomplete abstracts, expecting the reader to read the main report, or they write long abstracts when they do not expect the reader to read on. Neither is a good practice.

6207   SOLAR-FOSSIL COMBINED CYCLE POWER PLANT. Darnell, J. R.; Lam, E. Y.; Westsik, J. H. (Bechtel National, Inc., San Francisco, CA). Proceedings of the Annual Meeting—American Section of the International Solar Energy Society ; 3.1: 563-567 (1980). (CONF-800604--P2).

From American section of the International Solar Energy Society conference; Phoenix, AZ, USA (2 Jun 1980).

The results of a DOE-funded study of solar-fossil hybrid power systems are summarized. The concept uses an air-cooled central receiver, with the collected solar energy displacing a portion of the fossil fuel requirements of a combined cycle power plant. The conceptual design of a first of its kind commercial size (112.6 MWe) power plant is described. Operational and performance characteristics and cost estimates of this plant are presented. An assessment of this concept shows significant technical advantages and a potential for utility application by the year 1990.

Figure 12.6 An abstract

12.6 INTRODUCTION

An introduction is a general statement, paragraph, or section concerning the importance or the objective of the report. The introduction supplies the background material to help the reader understand the problem covered by the report and the need for a study. It should include a brief review of any previous work or results that may be available from a literature survey. There should be a brief explanation of any theory that may be available for predicting results, but there should be no lengthy derivations of theory. Derivations should be deferred to an appendix, or given in a reference, using only the final relationships in the introduction.

The introduction should contain a statement of the purpose or the objective of the report. What are you going to accomplish by writing this report, and why are you writing it? This statement must be specific. You cannot proceed to the main body of the report with any ambiguity lingering in your reader's mind. Many times specific words are suggested, such as, "the objective of this report is to evaluate the performance of . . . under the conditions of. . . ." Such rigid wording is not attractive, but it does ensure a concise statement of the purpose of the report.

12.7 THE REPORT BODY

The body of a report is also called the discussion, results, or some other descriptive title, depending on the purpose of the report. This section communicates the basic information for which the report is written and it does so in detail.

The body of a report should be written in simple statements, using simple words. Resist any temptation to use big sentences or big words. Tell your reader what you did, what you propose to do, your plans, your structure, or your point of view. Document your facts, data, and technical information. List your sources of information.

In this section, data and calculations should be presented in concise and final form. Leave out irrelevant information. Data and calculations are not shown, unless they are essential for general understanding. Complete data should be placed in an appendix. The results are best presented in tabular or graphic form, with both tables and graphs properly identified as to what is being presented.

When an equation is included in the text, it should appear on a single line. All symbols used in the equation should be defined, either immediately before or immediately after the equation statement. One alternative would be to include a general section on nomenclature to define all symbols in one place—a good practice if there are many symbols. The report body is then not interrupted, giving the report more continuity. The reader will have to look up unfamiliar symbols, flipping pages in order to read the report, which is inconvenient, however.

The body of the report should also emphasize the interpretation of the results regarding accuracy. It should emphasize the agreement or disagreement with expected theoretical estimates and give possible causes for large deviations from expected or theoretical results, including results from data that are suspected of being in error. The existence of questionable results should be brought to the reader's attention, and the possible reasons for such observations should be discussed. The discussion should comment on the success or failure in attaining results.

Finally, for successful report writing, you should cultivate an impression of intellectual honesty. You must convey the thought that you have presented the data completely, exactly as they were observed, without bias and without embellishment.

12.8 CONCLUSIONS AND RECOMMENDATIONS

Conclusions are the end product of an engineering project. All engineering reports should have them. Conclusions should be phrased in clearly stated, definite statements. If at all possible, conclusions should be listed by number and supported by specific reference to tabulations, curves, and prior discussion. Conclusions are the results of your work. If you have them, state your conclusions boldly. These are your judgments, based on the facts you gathered and the data you obtained. Did you or did you not meet the stated objective of the report? This is not the place to hedge or to be coy, nor is it the place to be modest. If you did it, say so!

If you do not have conclusions, then make that statement. Yours would not be the first inconclusive piece of work in engineering. A negative statement can be very useful, and you may prevent further unproductive expenditures. Whether you have conclusions or not, you should decide one way or the other.

Recommendations follow your conclusions. Recommendations should pertain to any suggestions made as a consequence of this work. They can be either positive or negative, but they should be definite. Remember, you did not undertake this exercise to pass time. Someone will read your report and commit time, people, and resources on the basis of what you recommend. You can be wrong, but you cannot be uncertain!

12.9 BIBLIOGRAPHY OR REFERENCES

References are numbered and are listed alphabetically by the author's surname and initials or in order of use, followed by the title of the paper or text, the publisher or organization, the volume number or edition, the year published, and finally the page number, if the reference is being made inclusive page numbers for a journal article or book.

Listed in Figure 12.7 are the references for an engineering report on a comparison of turbomachines for geothermal energy utilization. Note the differing reference techniques for a book, a journal, and a report. In all cases, the title is in italics (underlined in the typewritten manuscript) to set it apart. Article titles are in quotation marks. In references 2, 3, and 6, specific page numbers are given.

REFERENCES

1. D. G. Shepherd, *Principles of Turbomachinery,* Macmillan Company, New York, N.Y., 1956.
2. O. E. Balje, "A Study on Design Criteria and Matching Turbomachines, Part A—Similarity and Design Criteria of Turbines," *Journal of Engineering for Power, Trans. ASME,* Vol. 84, January 1962, pp. 83–102.
3. O. Cordier, "Aenlichkeitsbedingun gen fur Stromungsmaschinen," *V.D. I Berichte,* Vol. 3, 1955, pp. 85–88.
4. O. E. Balje and D. H. Silvern, *A Study of High Energy Level, Low Power Output Turbines,* AMF/TD No. 1196, Department of the Navy, Office of Naval Research, Contract No. NONR-2292(00), Task No. NR 094-343.
5. O. E. Balje and R. L. Binsley, *Final Report, Turbine Performance Prediction: Optimization Using Fluid Dynamic Criteria,* R-6805 Rocketdyne, Office of Naval Research, Contract No. NONR-4507(00).
6. W. Rice, "An Analytical and Experimental Investigation of Multiple-Disk Turbines," *Journal of Engineering for Power, Trans. ASME,* Vol. 87, Series A, No. 1, January 1965, pp.29–36.
7. H. Weiss, R. Steidel, and A. Lundberg, *Performance Test of a Lysholm Engine,* UCRL-51861, Lawrence Livermore Laboratory, 1975.

Figure 12.7

12.10 APPENDIXES

There may be one or several appendixes, depending on the quantity of information to be placed in the section. The following information should be put in an appendix: details of the experimental apparatus, including the length of runs and frequency of readings; the consistency of observed data; any lengthy mathematical developments of special equations or relationships used in the reduction of data; any computer programs used and printouts of computer results; all original data sheets, sketches, or diagrams of the system and of system analysis. Sample calculations, unless very brief, belong in an appendix. If they are very brief, the sample calculations may be incorporated in the main body of the report.

Special descriptions, detailed drawings, and special tests or control techniques belong in an appendix. If the evaluation or test was carried out in accordance with some particular test code or contract, detailed information on the test code or contract statements belongs in an appendix.

Appendix material is included to make the report complete, but it is not necessary to include this information in the body of the report for clarity. The main story transmitted by the report is given by the introduction, body, and conclusions. Put as much material in the appendix as you possibly can. A major portion of a report is often appendixes.

12.11 GRAPHICS

Particular attention should be paid to graphics. Your graphics should be both interesting and informative. Drawings in an engineering report may be of various kinds. In most cases, they are only simple sketches to show the arrangement of the various parts of the apparatus and their relationship to one another. Detailed drawings and assembly drawings are not usually a part of an engineering report. They are separate documents. If they are included in a report, they are usually relegated to an appendix.

You should consider graphics to be an integral part of any engineering report. Sketches, photographs, and diagrams are just as important as text material, and coordinating graphics with the text deserves both your effort and your time.

Including photographs in a report is one quick and certain way to present information. Photographs of apparatus rarely

Figure 12.8 Wind-induced vibration fatigue failure.

are helpful. Photographs of cracks, failures, people, physical conditions, test arrangements, etc., are all important. One photograph can be the equal of many pages of description if it conveys the necessary information. Some photographs convey confusion; photographs should show detail, but not confusion. Photographs also date an engineering report, which may or may not be desirable. Figure 12.8 is a frontispiece from an engineering report on the failure of suspended cables as a result of wind-induced vibration. In this case, words could not create the same impact as a photograph. This particularly choice photograph shows the fatigue failure of all but one of the aluminum strands, the last one, which failed in tension. No verbal description is necessary. How old is this photograph?

12.12 STYLE

Style is the most personal characteristic of your report, but it is an elusive characteristic. It is tied to your personality, your writing skill, your background and knowledge, the material presented, and the objective of the report. Change any one of these factors and the style of a report changes. Yet style is distinguishable even if it is impersonal.

Consider the following two quotations. You know nothing of the material content, but look at the two styles. The first is a simple and crisp report of a special master on a court case. Note that the second paragraph has only one sentence—an unusual and dramatic device.

> Seeking an explanation of these phenomena, Inco's metallurgists subjected the metals which they had produced to microscopic analysis. Instead of the usual graphite flakes, they found varying percentages of what came to be called graphite spheroids.
>
> They understood the significance of what they saw.
>
> The inventors pursued their experiments. They quickly learned that the improvement in physical properties of the casting varied with the quantity of magnesium retained and the consequent alteration of the form of the graphite. Small quantities of retained magnesium resulted in the shortening, thickening and curving of the graphite flakes with commensurate improvement in properties. As the retained magnesium increased, graphite spheroids appeared and increased in number, while physical properties steadily improved. (*Report of Special Master*, Civil Action No. 78-76, United States District Court, Southern District of New York.)

The second is from a general article on technology in society.

> The familiar and seemingly straightforward phrase "the effects of technology on society" actually subsumes under the single word "technology" three disparate concepts called: "technics," "manufacturing technologies," and "sociotechnical systems" (see box on basic terms). Such pernicious semantics can, and has, led to much noncoplanar discussion. However, the full utility or disutility of more careful semantics is seldom seen ab initio; it emerges rather from where the concepts lead. We have to ask: "Do they encourage and facilitate clearer thinking, better understanding, or do they merely confuse us on some other level?" Hence we need to move on. Let us go first to the "basic pattern" between technics, manufacturing technologies, and human use of sociotechnical systems. (*Mechanical Engineering*, Vol. 99, No. 3, March 1977, p. 42.)

The differences in style are obvious.

There are a few basic rules for good style.

1. *Use Good Grammar and Good Spelling.* You should be acquainted with the rudiments of English grammar. Sentences have subjects and predicates, and the predicate must agree with the subject. There are nouns, pronouns, verbs, and prepositions. Prepositions and verbs have objects. Grammar may be dull and uninteresting to you, but nothing destroys continuity of thought more completely than a grammatical error or a boldly misspelled word. When a reader wants a reason to avoid reading your report, if there is bad grammar or poor spelling, you have provided sufficient reason. Some readers look for grammatical errors and misspelled words with more zeal than they look for the meaning of the report.

2. *Be Concise.* In engineering and science we tend to be interested in details. This is in the nature of engineers. But details can be boring if you do not want them. Don't use unnecessary words, and don't explain the obvious. Concise writing does not mean that your sentences are terse, such as telegraph statements. It simply means that unnecessary words, sentences, paragraphs, and pages are not used, if they do not contribute to the report. Here, judgment is needed, and judgment comes with experience. You must write, write, and write some more. You must obtain criticism from others, and you must be critical of yourself.

3. *Be Specific.* There is a great tendency to be nonspecific in engineering. Words such as *most, few,* and *in many cases* are used to modify specific statements, probably to take care of the one or two instances when a statement would be in some doubt. These are "weasel-words," used to avoid specificity. Lawyers are taught to be specific. Once a decision in law is made, it is made with emphasis. You need not emulate the law profession and its words, but you can be specific. If you want to be specific, you are going to have to make positive and specific statements. You can do this and still avoid criticism if you make your statements clearly and precisely, basing them on your data and facts.

4. *Be Yourself.* Use your own language. Write so that your personality comes through. Although you may learn from someone else, don't mimic anyone. Humor also has a place, if it is restrained, and it can make the reading of a report more pleasant. Your readers are human beings,

[handwritten margin notes: "STRUNK", "Coolidge said nothing", "700 WORDS / PAGE"]

with their own problems and foibles. Your writing objective is to get through to your readers and convey your information and thoughts to them. The objective is difficult, not unlike writing a book on engineering graphics for first- or second-year engineering students, who may not be too interested in the subject in the first place.

PROBLEMS

12.1 Obtain two yardsticks (or meter sticks), a C-clamp, and a mass of about 3.5 oz (100 gm), more if the yardstick is thick. Clamp the yardstick to a table about 6 in. from one end and place the mass at the other. Measure the deflection of the free end of the clamped yardstick under the weight of the mass with the second yardstick. Shorten the suspended length of the yardstick and repeat the measurement. Do this at least four times. Determine the relation between the suspended length of the yardstick and the deflection of its free end. Write a short report on this experiment, presenting your data numerically and graphically.

12.2 From advertisements, published commercial literature, or visits to sales offices, determine the selling price, the predicted gas mileage, and the weight of at least 10 new automobiles. Prepare a table that shows the data, and prepare a graph that shows how price, gas mileage, and weight are related. Write a short report on your findings.

12.3 On a trip by car of 50 mi, or at least 1 h, record the elapsed distance traveled and the speed of the car every 2 min. Prepare graphs that show your recorded data. Write a short report on your results. What differences would have been observed if you had made observations every 10 min instead of every 2 min?

12.4 Over a measured distance of one or more miles, maintain the speed of an automobile at 20, 30, 40, and 50 mi/h, according to its speedometer. Calculate its actual speed from the elapsed time for the measured distance. Write a short report on the calibration of the vehicle's speedometer. Present your data numerically and graphically.

12.5   Run a minimum of 1500 m (or 1 mi). Immediately afterward, record your pulse, measured as a function of time. Determine how your pulse rate declines with time. Write a short report on your experiment presenting your data numerically and graphically.

12.6   Place a glass thermometer in a small container of hot water. Read the thermometer at 15-sec intervals for a period of 2 min, every 30 sec for another 3 min, and at 1-min intervals for 5 min. Be careful to remove only enough of the thermometer stem to read it and replace it immediately. Be careful not to exceed the limit of the thermometer. Write a short report on this experiment presenting your data numerically and graphically.

12.7   Place a glass thermometer in a hot cup of coffee. Read the thermometer at 30-sec intervals for a period of 5 min and at 1-min intervals for 10 min. Be careful to remove only enough of the thermometer stem to read it and replace it immediately. The cup of coffee should be large so you can record the temperature when the coffee has cooled enough to drink, and when it is too cold to be drinkable, by direct sampling. Present your data numerically and graphically. Write a short report on your findings.

12.8   Record the number of cars parked in a specific parking lot on the hour for an 8-hr period. Present your data numerically and graphically. Write a short report and comment on your observations.

12.9   Record the daily and cumulative use of water for a private house or an apartment over a period of 1 week from a water meter. Present your data numerically and graphically. Write a short report and comment on your observations.

12.10 Support a known weight W with three cords in a Y configuration as shown. Tie one of the upper cords to a fixed support and the other to a spring scale. Hold the spring scale in a position that results in $\theta_1 = \theta_2$. Because of symmetry, the spring scale will measure the tension force or pull P in the two top cords. The principles of equilibrium indicate that

$$W = 2P \sin \theta \qquad \text{or} \qquad P = \frac{W}{2} \csc \theta$$

Measure the pull P for a range of values of θ. Compare your experimental data and theory in a short report including a graphic presentation of your results.

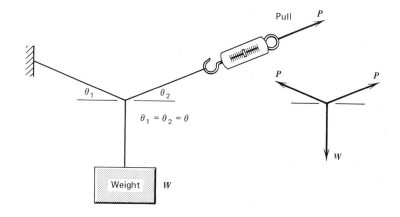
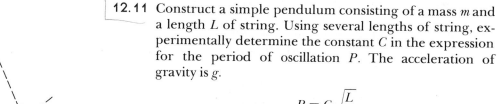

12.11 Construct a simple pendulum consisting of a mass m and a length L of string. Using several lengths of string, experimentally determine the constant C in the expression for the period of oscillation P. The acceleration of gravity is g.

$$P = C\sqrt{\frac{L}{g}}$$

Write a short report on this experiment, presenting your data numerically and graphically.

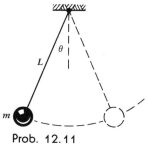

Prob. 12.11

12.12 In your house or apartment, record the temperature and time at which a thermostat turns the heating system on and off. Record time and temperature at regular intervals. Present your data numerically and graphically. Write a short report on your data.

12.13 The Filbonacci series is the sequence of numbers

$$1, 1, 2, 3, 5, 8, 13, 21, 34, 55, 89, \ldots, N_n, \ldots$$

where each number is the sum of the last two ($N_n = N_{n-2} + N_{n-1}$). With the exception of the first few numbers, the ratio of adjacent numers is constant.

$$\frac{21}{34} = \frac{34}{55} = \frac{55}{89} = 0.618$$

This unique ratio can be found numerous places in nature and is said to be pleasing to the eye. To check this theory, have 10 to 20 people sketch a rectangle that they feel has pleasing proportions. Calculate and tabulate the ratios of the lengths of the two sides of the rectangles sketched by each person. Present your results graphically and prepare a short report discussing your investigation and conclusions.

12.14 On a sunny day, record the hourly temperature outside your building in at least four locations. At least one should be on the south side and protected, one should be on the north side, and one should be where you think the maximum incidence of solar energy would occur. Tabulate your findings and write a short report on what you have learned.

12.15 Make a survey among your friends of the time required to complete their homework assignments per course, per week. Obtain data representing a minimum of 20 different courses from a minimum of 10 different people. Write a memorandum to one of your instructors presenting your data numerically and graphically.

12.16 Obtain a rubber band, a paperclip, a paper cup, and some string. Support the cup with the string from the rubber band, connecting the string and the band with the paperclip. Now place a measured amount of water in the cup and measure the length of the rubber band. Add another measured amount of water and record the length of the rubber band. Repeat this until the paper cup is filled. Now, reverse your procedure and pour a measured amount of water from the cup, recording the length of the rubber band each time until the cup is empty. Present your results in a single graph and write a short report. What are your conclusions?

12.17 Obtain a wood pencil, a scale, a length of string about 30 in. or 750 mm long, and a small solid mass such as a small rock.

a. Make a loop in one end of the string, with the diameter of the loop about twice the diameter of the pencil. Tie the rock to the other end. Insert the pencil in the loop of the string and suspend the rock as a pendulum. Oscillate the rock as a pendulum measuring the maximum linear displacement with the scale. On a linear-linear graph and a linear-logarithmic graph, show the decay of the amplitude of the pendulum with the number of cycles. Use the logarithmic scale for amplitude.

b. Tie the string to the pencil so that there is no slipping. Repeat your experiment. Present your results in a short report.

12.18 Obtain a yardstick or a meter stick, a scale, and six small cans, all the same size, but no larger than 8 oz. On a flat, horizontal surface, support the yardstick on two of the cans, one at each end. Place a third exactly at the un-supported center of the yardstick. Measure the vertical deflection of the yardstick and its unsupported length. Place a second can on top of the can in the middle. Bring in the cans at either end of the yardstick until the center deflection is the same as it was supporting one can. Again, record the unsupported length of the yardstick. Place a third can on top of the other two and again bring in the end cans. Record the unsupported length of the yardstick. If you can, repeat this experiment with the fourth can. Plot a logarithmic-logarithmic graph with the unsupported length of the yardstick as the ordinate and number of cans as the abscissa. Write a short report of your observations and findings.

12.19 Now that you have considered report writing and the graphics contained in engineering reports, locate a short technical report you or a student colleague have written, such as a lab write-up.

a. Criticize the written and graphic portion of the report.

b. Prepare an improved version of the report.

12.20 Locate a technical report or article on a subject familiar to you. Improve a few paragraphs and several of the graphic portions of the report.

INDEX